T0364139

Trade in Services in the Asia-Pacific Region

NBER–East Asia Seminar on Economics
Volume 11

National Bureau of Economic Research
Tokyo Center for Economic Research
Korea Development Institute
Chung-Hua Institution for Economic Research
Hong Kong University of Science and Technology
National University of Singapore

Trade in Services in the Asia-Pacific Region

Edited by **Takatoshi Ito and Anne O. Krueger**

The University of Chicago Press

Chicago and London

TAKATOSHI ITO is professor in the Institute of Economic Research at Hitotsubashi University, Tokyo, and a research associate of the National Bureau of Economic Research. ANNE O. KRUEGER is first deputy managing director of the International Monetary Fund, and a research associate of the National Bureau of Economic Research.

The University of Chicago Press, Chicago 60637
The University of Chicago Press, Ltd., London
© 2003 by the National Bureau of Economic Research
All rights reserved. Published 2003
Printed in the United States of America
12 11 10 09 08 07 06 05 04 03   1 2 3 4 5
ISBN: 0-226-38677-5 (cloth)

Library of Congress Cataloging-in-Publication Data

Trade in services to the Asia-Pacific region / edited by Takatoshi Ito
    and Anne O. Krueger.
        p.   cm. — (NBER East Asia seminar on economics ; 11)
    Includes bibliographical references and index.
    ISBN 0-226-38677-5 (cloth : alk. paper)
        1. Service industries—East Asia.   2. Service industries—Pacific
Area.   3. Banks and banking, International—East Asia.   4. Banks
and banking, International—Pacific Area.   5. Telecommunication—East Asia.   6. Telecommunication—Pacific Area.   I. Ito,
Takatoshi, 1950–   II. Krueger, Anne O.   III. NBER-East Asia
seminar on economics (Series) ; 11.

    HD9987 .E182 T7 2003
    380.I'45'00095—dc21

                                                                2003018125

Since this volume is a record of conference proceedings, it has been exempted from the rules governing critical review of manuscripts by the Board of Directors of the National Bureau (resolution adopted 8 June 1948, as revised 21 November 1949 and 20 April 1968).

Contents

Acknowledgments

This volume contains edited versions of papers presented at the NBER's East Asia Seminar on Economics eleventh annual conference, held in Seoul, Korea, 22–24 June, 2000.

We are indebted to members of the program committee who organized the conference, and to Chung-Hua Institution, Taipei; the Hong Kong University of Science and Technology; Korea Development Institute, Seoul; Korea Institute for International Economic Policy; and the Tokyo Center for Economic Research.

The KDI and KIEP were the local hosts. The conference arrangements were made by Kirsten Foss Davis and Brett Maranjian, and were superb. All the participants enjoyed the venue and the wonderful facilities in which the conference was held.

The National Bureau of Economic Research provided logistical support. We are greatly indebted to the NBER, and the Asian institutions which supported the research and the conference.

Introduction

Takatoshi Ito and Anne O. Krueger

Until recently, analysts of economic development thought of the process as one in which activity gradually shifted from agriculture to industry. Developed countries were regarded as industrial countries, that is, having large industrial sectors and relatively small agricultural sectors. Over the past several decades, however, it has become apparent that after a stage in development in which shifts of resources occur between agriculture and manufacturing, there comes a point when services growth accelerates and outweighs growth in manufacturing and industry as per capita incomes continue rising.

In the United States, for example, manufacturing accounted for 15.2 million jobs and 25.8 percent of total employment in 1950; by 1970, the corresponding figures were 19.4 million and 24.7 percent; by 1990, they were 19.0 million and 16.0 percent; and by 1998 they were 18.8 million jobs and 15 percent of total employment. Whereas less than half of all jobs were in services in 1950, by 1998, 76.4 percent of all jobs were in the service sector in the United States.[1] Indeed, even if one speaks of all goods-producing activities (including construction and mining), the picture is little different; and the same general pattern is being followed in other industrialized countries.

Although the weight of services industries has increased in the economy, trading of services was not historically considered to be important. Indeed, many products of the services industries have been regarded as "nontrad-

Takatoshi Ito is a professor in the Institute of Economic Research at Hitotsubashi University, Tokyo, and a research associate of the National Bureau of Economic Research. Anne O. Krueger is first deputy managing director of the International Monetary Fund and a research associate of the National Bureau of Economic Research.

1. Data are from Council of Economic Advisers, *Economic Report of the President*, February 2000, tables B-33 and B-44.

able." Economics textbooks are full of examples and anecdotes, from haircuts to city bus services. It used to be the case that in many countries electricity, telephone services, and water and sewerage networks were provided by the national monopoly and not considered to be tradable. Producers and exporters used to use their local banks, contract with local shippers, and buy their insurance from domestic firms, and they either sought local contractors or themselves provided other services in-house. Today, however, the greatly reduced costs of transport and communications have meant that many previously domestically produced services can now be obtained internationally. As technologies and competition policies have changed, these services have also become tradable. International airline services and international telephone services have been hotly contested trade disputes. By the late 1990s, penetration into domestic service sectors by foreign companies became commonplace for large construction projects and financial services as well.

There are four types of service trades according to the General Agreement on Trade in Services (GATS), which came into being as part of the Uruguay Round agreement. GATS was established to join the General Agreement on Tariffs and Trade (GATT) as part of the World Trade Organization (WTO). The four types are (a) cross-border supply in which services can be produced in one country and delivered to another; (b) consumption abroad, when domestic residents go abroad to consume the services (such as tourism); (c) activities provided to foreign nationals by foreign branches and subsidiaries of domestic firms; and (d) relocation of natural persons. The first category includes services flows from the territory of one country to another. For example, air transportation to carry foreigners is counted as an export of a service. Banking services supplied to foreigners transmitted via telephone or mail are also service exports. The second category includes such items as ship repairs and aircraft maintenance abroad. These first two types of cross-border trade are typically captured in the balance-of-payments statistics.

However, services trade is not limited to cross-border transactions. When firms and natural persons physically move abroad to provide services, that is also service trade. This would include foreign subsidiaries of insurance companies, hotel chains, department stores, and the like. These activities are not usually captured in the balance-of-payments statistics. The fourth category includes professionals (consultants, accountants, doctors, teachers, programmers, architects) who physically move abroad (and expect to return to their own countries) to provide services. When income is repatriated, this may or may not be captured in balance-of-payments statistics.

Reflecting the growing importance of services in industrial countries' gross domestic products (GDPs) as well as the reduced costs of trade in services, services trade (and especially business services) has grown more rapidly than merchandise trade in recent years. From 1990 to 1998, the value of world exports of merchandise trade rose at an average annual rate of 6.8 percent, whereas the value of trade in services rose at an average annual rate of 8.0 percent. The share of services in total international trade in goods and services has been steadily rising and reached 19.8 percent in 1998.[2]

As the importance of services has grown, economists have increasingly focused on policy issues raised by them and have sought to understand what, if any, differences there are between production and delivery of goods and of services.

A fundamental difficulty is the problem of identifying services output. If a manufacturer has a bookkeeping department that does the billing, the economic activity is regarded as part of the manufacturing. If, instead, the manufacturer hires an outside firm to do the same task, it is regarded as a service. The inability to split up activities into their goods component and their service component is one difficulty in coming to grips with services. A second problem arises because of the difficulty of measuring services output. Although matches, candles, and other goods may not be identical, estimating the quantity (or volume) of output (in terms of units such as weight) independent of its price is relatively straightforward, and there are means by which adjustments for quality can be made.

When, instead, the output of a firm is a service, there seems to be little way of measuring output other than by the value of the inputs (salaries of employees, rents, interest payments costs, etc.). Hence, it is difficult to disentangle the productivity of inputs into services industries because there is no separate measure of output. In turn, that makes many aspects of services activities difficult to measure and to understand.

As employment and output of services have increased as a proportion of GDP, it has become increasingly recognized that some services—such as accountancy, financial services, and so on—are generally provided to business users. Thus, the quality, timeliness of delivery, and other attributes of services are increasingly important as a determinant of a country's comparative advantage: exporters who must wait weeks for freight forwarders to ship their goods are at a strong competitive disadvantage, as are those who must rely on high-cost banking services relative to their foreign competitors.

Two interesting questions arise in this regard: on one hand, it can be asked how the availability, quality, and timeliness of services affect the comparative advantage of other traded goods industries; on the other, issues arise with regard to the degree to which there exists comparative advantage among countries with regard to the provision of services.

Although little is known about the first question, ignorance did not seem important until the cost of trading services internationally fell markedly.

2. Data are from World Trade Organization, *Annual Report 1998 International Trade Statistics.*

When it did fall, immediate questions arose as to the gains to be had from liberalizing trade in services. Because it is in the nature of services trade that the traditional instruments of trade policy do not constitute the major protective barriers, it is not surprising that little was known about the costs and benefits of services liberalization.

Internet trading, e-commerce, financial services, and a host of other services are now readily tradable internationally. However, the barriers to services trade—often initially erected in connection with regulation of domestic service providers when international trade in services was not economic—are different from those to trade in goods. Licensing regulations, regulatory environments (such as bank supervision), and a host of other impediments to services trade differ significantly in their nature from the tariffs and quantitative restrictions that the GATT and WTO have so successfully brought down.

To complicate the matter further, changing technologies have affected the nature of and the need for regulation. Just as the costs of overseas phone calls have dropped sharply, technological change has affected the feasibility of competition among firms in what was once regarded as a natural monopoly. Consequently, regulatory regimes are often themselves outdated by virtue of technological change at the same time that efforts are being made to liberalize trade in services.

Despite the paucity of our understanding of these issues, policy makers are moving ahead with deregulation of services domestically and liberalization of international trade in services. In this volume, we bring together thirteen papers that were presented in the eleventh annual East Asian Seminar on Economics, which took place on 22–24 June 2000.

The first set of papers gives overviews on the issue. The paper by Philippa Dee, Kevin Hanslow, and Tien Phamduc uses their newly developed comprehensive data set on trade barriers to services to estimate the costs of these barriers. They first consider the nature of these barriers (see their table 1.1) and the measures that can be used, depending on the type of restriction to trade. They then estimate the tariff equivalents of these barriers for nineteen countries (one of which is a group of four Latin American countries) for services imports and exports.

With these data, they then estimate the partial equilibrium effects that would occur with removal of these barriers to trade and foreign investment in services. They use these results to better understand the ways in which removal of barriers to trade and investment in services would affect primary, secondary, and tertiary activities in the various countries. Their estimates generally indicate that liberalization of services trade would contribute more to the world economy than is widely thought, although the authors give ample warning that the data are a "first pass" and that a great deal of additional research is needed both to improve estimates of barriers to trade in services and to improve the model from which the estimates are derived.

They also note that it seems—at least, based on their results to date—that the biggest gains to liberalizing trade in services would originate in the services industry itself, although in many countries there would be positive benefits for other sectors of the economy as well.

In his valuable paper, Aaditya Mattoo assesses the prospects for liberalization of trade in services through the GATS. He believes that multilateral cross-sectoral liberalization of services trade—through, for example, requiring national treatment for all foreign services providers—is not feasible. However, as he demonstrates, much could be achieved by more limited measures, such as the requirement that services barriers meet a "necessity": a test showing that the mechanism chosen is the least costly for achieving a desirable social objective (insuring that medical qualifications are adequate, for example). Mattoo's analysis illustrates both the problems inherent in achieving open multilateral trade in services and, despite the obstacles, the feasibility of achieving significant gains in further multilateral services liberalization.

Richard H. Snape analyzes the political economy aspects of the services trade deregulation. With the Snape paper, the reader is reminded of the importance of the political economy. The link between the objectives of policies and policy instruments is not so straightforward as economic theory suggests. The paper asks, "could stated policies justify what has been done in regulating service industries?" The paper consists of two sections: a discussion on whether negotiation should be generic or sector-specific and an examination of the correspondence between policy objectives and policy instruments. An ultimate objective and a stated objective can be differentiated, and optimal instruments to achieve an ultimate objective are examined.

The second set of papers investigates the economy-wide trade liberalization impact on the economies of Taiwan, Korea, and Hong Kong.

The paper by Ji Chou, Shiu-Tung Wang, Kun-Ming Chen, and Nai-Fong Kuo discusses the impact of Taiwan's WTO accession on the economy, using a general equilibrium analysis. Objectives of the paper are fourfold. First, Hoekman's method of estimating tariff equivalents in services trades in Taiwan is discussed. Second, a computable general equilibrium model is employed to estimate impacts of liberalization of services trade. Third, the paper analyzes technological spillovers from advanced countries to developing countries embodied in imports of intermediate inputs. Fourth, impacts from WTO accession by Taiwan and Mainland China are measured.

The Li-min Hsueh, An-loh Lin, and Su-wan Wang paper discusses the growth of Taiwan's services trades. The paper offers a good overview on the time-series changes of Taiwan's services trades. The paper shows that "increases in the imports of goods and per capita income, and decreases in the relative prices of import services, are major contributors to the growth in import services." It also finds that "the trade in goods, per capita income levels, and the relative price of services are three important determinants of the services trade found in our regression analysis."

The Jong-Il Kim and June-Dong Kim paper examines the impact of service liberalization on productivity and growth in Korea. The Kim and Kim paper investigates the impact of service-sector liberalization on productivity in the distribution sector itself and the manufacturing sector. Changes in labor productivity and total factor productivity (TFP) of the service and manufacturing sector are tabulated. Although some correspondence between liberalization and productivity increase in the service subsector (or the manufacturing sectors that use service-sector input) is suggested, it is weak, partly because liberalization took place only recently and full effects may not have materialized.

Although there has been much discussion of liberalization of services, and especially of financial services, there has been little empirical work at the microeconomic level that assesses the importance of barriers to trade in services and the potential benefits to their removal.

In their paper, Clement Yuk Pang Wong and Anming Zhang begin to fill this gap by analyzing the perceptions of the Hong Kong business community of the effects of restrictions on services trade on their economic prospects. To do this, the authors sent out a questionnaire to a number of Hong Kong businessmen. They then tabulated their results and, in addition, interviewed the respondents. They asked businessmen to indicate which types of liberalization would be most beneficial for them, and Wong and Zhang report the responses in their paper. Financial services was the area in which most businessmen thought that further liberalization would have the biggest benefit for them. However, business services and information technology services were not very far behind. Interestingly, at least for Hong Kong, transport services was a sector in which little benefit was expected from further liberalization of the sector.

The third set of papers deals with a specific service sector of a specific country. In their paper, Nae-Chan Lee and Han-Young Lie consider the liberalization of Korean telecommunications in the context of the WTO agreements. Until the end of the 1980s, Korean Telecommunications (KT) had a monopoly of telephone services, and the main objective of Korean telecoms policy was to provide universal access. Once that was achieved, there was an initial, relatively timid, market opening, but it met with considerable resistance from KT. The Uruguay Round agreement then provided the impetus for considerable reform, including permission for foreign ownership and competition in the domestic market.

Analysis of the role of the WTO in furthering liberalization is difficult because the Asian financial crisis also affected telephony at the same time that deregulation was taking place. Nonetheless, Lee and Lie point to the major benefits to consumers that have arisen due to liberalization and find the telecoms market today much more competitive and open than it was in 1997, when the second round started.

The area of financial services is one of the focuses of efforts to liberalize services trade. It has been prominently featured in Uruguay Round and subsequent negotiations, and many analysts believe that there are very sizable benefits to be gained by financial liberalization. In their paper, Sang In Hwang, Inseok Shin, and Jungho Yoo review the literature on the benefits and costs of services liberalization and then apply that analysis to the case of Korean financial services.

Korean financial liberalization came about only slowly until the 1990s. Then, after the Uruguay Round and, more importantly, accession to the Organization for Economic Cooperation and Development (OECD), liberalization accelerated. Before it had gone very far, however, the Asian financial crisis erupted. In response, the Korean authorities accelerated financial liberalization. The authors first present an account of financial liberalization in Korea at its various stages and document the changes made in the 1990s.

Hwang, Shin, and Yoo then attempt to assess the positive and negative impacts of financial services liberalization in Korea from the time of the Asian financial crisis until 2000, but find that three years is too short a time from which to reach very many firm conclusions. They cannot find evidence that the presence of foreign banks improved the efficiency of financial intermediation, although they point out that that does not prove that there were no such improvements. Likewise, any benefits that might have taken place because of improved financial intermediation have been overwhelmed by the effects of the financial crisis in Korea.

The area of tourist services is one of the most rapidly growing areas of international trade in services. In most regards it is also one of the areas to which there are fewer trade barriers, in part because foreigners come to the country that exports the services and in part because most countries want to attract tourist expenditures. In their paper, Kuo-Liang Wang and Chung-Shu Wu study the determinants of the relative tourist flows to eight Southeast and East Asian countries from the United States and Japan.

They show, first, that barriers do matter: When China began to encourage tourists, the attractiveness of other destinations in the region fell. Second, individual tourist decisions appear to be quite responsive to changes in the relative costs of different tourist destinations, although business and group tourism is less so.

Southeast and East Asian countries have been increasing their share in the world tourist trade. Although the Want and Wu paper investigates the relative competitiveness of tourism among regional destinations, much work remains to be done to understand the determinants of the overall share of the region in the world's rapidly growing tourist industry.

Among services, there has been considerable attention devoted to accounting. Calls for international standards have come from those who believe they are important for strengthening financial services and corporate governance in developing countries as well as from those who seek more transparency to encourage foreign direct investment (FDI) and to enable meaningful comparisons across countries. Accounting services were among those services pinpointed in the GATS, which set forth guidelines for services liberalization, and standards are in the process of being agreed upon.

In his paper, Fukunari Kimura provides an analysis of some of the difficult issues involved in liberalizing accounting standards and services in the case of Japan. Although he covers only accounting, the nature of the problems that are encountered in liberalizing services trade is similar for many other areas.

By the 1990s, the Japanese accounting system had become "increasingly incompatible" with modern corporate management, especially in using the book value of assets and treatment of affiliates. It was also incompatible with the emerging International Accounting Standards (including provisions preventing foreigners from qualifying to undertake accounting services in Japan), and an "accounting Big Bang" is planned, led by the Business Accounting Council of the Ministry of Finance. Three major reforms are contemplated: consolidating financial statements, calculating assets based on market values, and adding cash flow statements to income and balance sheet statements. Kimura reports on the many difficulties and resistances that are encountered and the very significant changes that are likely to follow in management practices as a consequence of the Big Bang.

The fourth set of papers extends a usual concept of service trades measured in the balance-of-payments (BOP) statistics. The paper by Shujiro Urata and Kozo Kiyota carries out an interesting analysis of services trades. First, services trades (BOP-based net exports, exports, imports) are regressed on factor inputs (capital, labor, and human capital) in order to investigate the validity of Heckscher-Ohlin predictions. The authors find that the explanatory power of the Heckscher-Ohlin theorem on the patterns of service trade is quite limited. The authors guess that the failure of theory is due to the presence of restrictions and barriers and intraindustry trade due to product differentiation. Second, the authors investigated services trades embodied in goods trade (indirect services trade). Using the input-output table, they estimate service input contents for traded goods. The authors compared embodied services trade and disembodied services trade. The ratio of embodied services trade to total services trade is low for the Philippines and high for China. For Japan, disembodied services imports exceed disembodied services exports; that is, services trade in the BOP statistics shows deficits. However, embodied services trade surpluses are large enough to more than offset disembodied services trade deficits. The situation is the opposite for Singapore. Then the authors calculated the service contents for 1 million dollars' worth of goods exports and imports for East Asian countries. They found that they are similar for Japan, Singapore, Taiwan, and Malaysia. However, when the indirect trade ratios for exports and imports are compared, the authors found that Singapore and Taiwan "export services through goods trade," whereas Japan, Malaysia, the Philippines, and China "import services through trade." Although Japan is a developed country, services trade shows deficits. The "anomaly" may be explained by the Japanese preference for the consumption of services (including imports). By sectors of total (embodied and disembodied) trades, Japan is a net importer of electricity, gas, water supply, and education and research, and a net exporter of wholesale and retail and transportation services.

Kyoji Fukao and Keiko Ito use the FDI data, based on the Toyo Keizai firm-level database, in order to estimate activities through foreign subsidiaries and affiliates. Fukao and Ito investigate sales and employment of Japanese affiliates of foreign firms (JAFFs) and foreign affiliates of Japanese firms (FAJFs) in the service sector at the three-digit industry level for the year 1995. They also compare Japan's purchase of services from foreigners with U.S. purchases from foreigners. Their findings indicate that foreign activities in Japan are much greater than those reported in the Ministry of International Trade and Industry's survey on Japanese subsidiaries of foreign firms. The study suggests that the frequently cited FDI statistics of Japan, which rely on cross-border capital flows on the reporting basis, grossly underestimate activities of foreign affiliates in Japan. However, when activities of foreign firms in Japan are compared to those in the United States, the latter are much larger than the former.

The last two papers are complementary in deepening understanding of services trades beyond just the measured balance of payments. The Urata and Kiyota paper calculates the embodied (indirect) services trade, using the input-output table, as well as the disembodied (measured in balance of payments) services trade. Although this type of embodied service is not a concept covered in the GATS, it provides us with new information to reconsider what Heckscher-Ohlin predictions mean with respect to services trade. The Fukao and Ito paper directly addresses the third aspect of the GATS services trade concept, namely, activities through subsidiaries and affiliates abroad. Taking advantage of the rich data set, the authors describe Japanese affiliates abroad and foreign affiliates in Japan in much more detail than before. These papers push the frontier of the services trade literature forward.

1

Measuring the Cost of Barriers to Trade in Services

Philippa Dee, Kevin Hanslow, and Tien Phamduc

To what extent can the traditional tools of trade policy analysis be used to analyse the economic costs of barriers to trade in services?

Traditional analysis of trade barriers has focused primarily on the effects of tariffs. These are discriminatory taxes levied on foreign-produced goods at the border of a country.

The Heckscher-Ohlin (HO) framework is a standard framework in which tariffs have been analyzed (Heckscher [1919] 1949; Ohlin 1933). This framework assumes perfect substitutability between domestically produced and foreign goods of the same type, fixed endowments of primary factors of production, and perfect mobility of those factors between sectors within an economy. The framework has been extended to consider more than two goods and factors (Jones and Scheinkman 1977), the presence of a sector-specific factor of production (Mayer 1974; Mussa 1974), imperfect competition (Markusen 1981), increasing returns to scale (Melvin 1969) and product differentiation (Krugman 1979; Helpman 1981).

However, barriers to trade in services are unlike tariffs. They are typically regulatory barriers, rather than explicit taxes. They need not discriminate against foreigners. Indeed, barriers to market access are often designed to protect incumbent firms from *any* new entry, be it by domestic or foreign firms. And barriers to services trade are not restricted to affecting the *out-*

Philippa Dee is assistant commissioner, and Kevin Hanslow is director, at the Productivity Commission, Australia. Tien Phamduc is at the International Business School at Griffith University.

The views expressed in this paper are those of the authors and do not necessarily reflect those of the Productivity Commission. The authors are grateful for early discussions with Anne O. Krueger and comments from Kym Anderson, Chang-Tai Hsieh, Fukunari Kimura, and Will Martin.

put of services firms. One particularly important category of barriers to services trade—restrictions on foreign direct investment by service firms—affects the use of primary factors. These restrictions are recognized in the General Agreement on Trade in Services (GATS) under the World Trade Organization (WTO), since this agreement recognizes commercial presence as one of the modes by which services are traded.

To date, few papers of either a theoretical or an empirical nature have reviewed all these aspects of barriers to services trade. Some early papers largely dismissed concerns that the determinants of comparative advantage in services might differ from those in goods (Hindley and Smith 1984; Deardorff 1985). A few theoretical papers in the late 1980s examined some of the important characteristics of services, including knowledge intensity (e.g., Markusen 1989; Melvin 1989). This characteristic also featured in subsequent analysis of goods trade under imperfect competition (e.g., Grossman and Helpman 1991). However, those early theoretical papers did not look at the nature of barriers to services trade. Recently, a few empirical papers have examined the effects of removing barriers to trade in services. Many of these have failed to take account of barriers to commercial presence as an important category of barriers to trade in services (Brown et al. 1995; Brown, Deardorff, and Stern 1996; Hertel 1999; Nagarajan 1999). One seminal paper by Petri (1997) introduced a treatment of barriers to foreign direct investment in the services sector, but it failed to take into account barriers on the other modes of service delivery. Moreover, all empirical papers have suffered from a dearth of convincing empirical estimates of the incidence and economic significance of barriers to services trade.

A recent empirical paper by Dee and Hanslow (2000) sought to analyze the effects of removing barriers to services trade in a more comprehensive fashion.[1] The barriers included nondiscriminatory barriers to market access as well as discriminatory restrictions on national treatment. They included barriers to commercial presence as well as barriers to the other modes of service delivery. The focus of that paper was to compare the gains from liberalizing services trade with the gains from removing all post-Uruguay barriers to trade in agriculture and manufacturing. The paper also compared the gains from the total removal of barriers to services trade with the gains from several alternative approaches to partial liberalization. It identified significant second-best problems with some approaches to partial liberalization.

The purpose of this paper is to look more deeply at that analysis of services trade liberalization in order to assess the extent to which the traditional Stolper and Samuelson (1941) and Rybczynski (1955) results from the HO framework are still relevant in a more realistic model of services

---

1. Brown and Stern (2001) contains a services model that was developed independently and shares a number of conceptual and data features with the model presented here.

trade liberalization. In the process, the analysis examines whether and how the benefits of services trade liberalization are passed on to other sectors in the economy. Thus, the analysis tries to open up the "black box" of what is a rather complex general equilibrium model of services trade in order to gain insights into the sectoral results from that model in terms of more simple textbook treatments of trade policy analysis.

The structure of the paper is as follows. It first describes the model used—a multisector, multiregional computable general equilibrium model of world trade and investment. The theoretical structure of the model covers both foreign direct investment (FDI) and portfolio investment. The model's database contains estimates of FDI stocks and the activities of FDI firms, each on a bilateral basis. Thus, the model recognizes that both goods and services can be delivered via FDI as well as by conventional trade. The paper then looks at the size of the barriers to trade in services and the cost impost they impose on other sectors of the economy. This analysis uses the first of a comprehensive new set of estimates of barriers to services trade. To understand the general equilibrium effects of removing these barriers, the effects on each sector in selected economies are built up from a more restricted, partial equilibrium multicountry model. To this partial model are gradually added the resource constraints and income linkages associated with general equilibrium. It is as the resource constraints are added that the relevance of Stolper-Samuelson and Rybczynski effects can be analysed. The paper then briefly summarizes the implications of services trade liberalization for regional incomes. Finally, the paper identifies areas for further research.

1.1 The FTAP Model

The model is a version of the Global Trade Analysis Project model (GTAP; Hertel 1997) with foreign direct investment, known as FTAP. The treatment of FDI follows closely the pioneering work of Petri (1997). The FTAP model also incorporates increasing returns to scale and large-group monopolistic competition in all sectors. This follows Francois, McDonald, and Nordstrom (1995), among others, who adopted this treatment for manufacturing and resource sectors, and Brown et al. (1995) and Markusen, Rutherford, and Tarr (1999), who used similar treatments for services. Finally, FTAP makes provision for capital accumulation and international borrowing and lending. This uses a treatment of international (portfolio) capital mobility developed by McDougall (1993) and recently incorporated into GTAP by Verikios and Hanslow (1999). FTAP is implemented using the GEMPACK software suite (Harrison and Pearson 1996). Its structure is documented fully in Hanslow, Phamduc, and Verikios (1999). The model and its documentation are available at the Productivity Commission website at [http://www.pc.gov.au].

1.1.1   Theoretical Structure

The FTAP model takes the standard GTAP framework as a description of the *location* of economic activity and then disaggregates this by *ownership*. For example, each industry located in Korea comprises Korean-owned firms, along with U.S., Japanese, and other multinationals. Each of these firm ownership *types* is modeled as making its own independent choice of inputs to production, according to standard GTAP theory, and each firm type has its own sales structure.

On the purchasing side, agents in each economy make choices among the products or services of each firm type, distinguished by both ownership and location, and then among the individual (and symmetric) firms of a given type. Thus, the model recognizes the firm-level product differentiation associated with monopolistic competition. Firms choose among intermediate inputs and investment goods, whereas households and governments choose among final goods and services.

Agents are assumed to choose first among products or services from domestic or foreign locations, with a constant elasticity of substitution (CES) of 5. They then choose among particular foreign locations and among ownership categories in a particular location, both with a CES elasticity of substitution of 10. Finally, they choose among the individual firms of a particular ownership and location, with a CES elasticity of substitution of 15. With firm-level product differentiation, agents benefit from having more firms to choose among, because it is more likely that they can find a product or service suited to their particular needs. Capitalizing on this, Francois, McDonald, and Nordstrom (1995) show that the choice among individual firms can be modeled in a conventional model of firm types (not firms) by allowing a productivity improvement whenever the output of a particular firm type (and hence the number of individual firms in it) expands. However, because the substitutability among individual firms is assumed here to be very high, the incremental gain from greater variety is not very great, and this productivity-enhancing effect is not particularly strong (the elasticity of productivity with respect to output is $1/15 = 0.0667$).[2]

The first two choices, among domestic and foreign locations, are identical to the choices in the original GTAP model. They have been parameterized using values, 5 and 10, that are roughly twice the standard GTAP Armington elasticities. Two reasons can be given for doubling the standard elasticities. One is that only with such elasticities can GTAP successfully reproduce historical changes in trade patterns (Gehlhar 1997). The other is

---

2. The equivalent elasticity of productivity with respect to *inputs* is $0.0667/(1 - 0.0667) = 0.0714$, where this latter concept is used by Francois, McDonald, and Nordstrom (1995). The elasticities of productivity with respect to output and inputs are not equal because of the assumption of increasing returns to scale. Another reason that scale effects are not strong is that, with this nested structure, the economies of scale are regional rather than global.

that higher elasticities accord better with notions of firm-level product differentiation. Further calibration of the model to historical data using methods of maximum entropy (e.g., Liu, Arndt, and Hertel 2000) may provide a feasible means of refining the above estimates of firm-level substitution possibilities in the future.

The order of the first three choices, among locations and then among ownership categories, is the opposite of the order adopted by Petri (1997). The current treatment assumes that from a Korean perspective, for example, a U.S. multinational located in Korea is a closer substitute for a Korea-owned firm than it is for a U.S. firm located in the United States. Petri's treatment assumes that United States–owned firms are closer substitutes for each other than for Korean firms, irrespective of location.

There are two reasons for preferring the current treatment.

The first is that Petri's treatment produces a model in which multilateral liberalization of tariffs on manufactured goods produces large economic welfare losses, for most individual economies and for the world as a whole—an uncomfortable result at odds with conventional trade theory. The reason for the result is spelled out in more detail in Dee and Hanslow (2000).

The second reason for preferring the current treatment is that, in many instances, it accords better with reality. One of the distinguishing characteristics of services is that they are tailored each time to meet the needs of the individual consumer. Another characteristic is that they are often delivered face-to-face, sometimes making commercial presence (through FDI) the only viable means of trade. These characteristics taken together mean that service firms in a given location, irrespective of ownership, will tailor their services to meet local tastes and requirements and, thus, appear to be close substitutes, as in the current treatment.

Whereas the demand for the output of firms distinguished by ownership and location is determined as above, the supply of FDI is determined by the same imperfect transformation among types of wealth as in Petri (1997). Investors in each economy first divide their wealth between "bonds" (which can be thought of as any instrument of portfolio investment), real physical capital, and land and natural resources in their country of residence. This choice is governed by a constant elasticity of transformation (CET) semi-elasticity of 1, meaning that a 1 percentage point increase in the rate of return on real physical capital, for example, would increase the ratio of real physical capital to bond holdings by 1 percent. A bond is a bond, irrespective of who issues it, implying perfect international arbitrage of rates of return on bonds. However, capital in different locations is seen as different things. Investors next choose the industry sector in which they invest (with a CET semi-elasticity of 1.2). They next choose whether to invest at home or overseas in their chosen sector (with a CET semi-elasticity of 1.3). Finally, they choose a particular overseas region in which to invest (with a CET semi-elasticity of 1.4).

The less-than-perfect transformation among different forms of wealth can be justified as reflecting some combination of risk aversion and less-than-perfect information. It is important to note, however, that although the measure of economic welfare in FTAP currently recognizes the positive income contribution that FDI can make, it does not discount that for any costs associated with risk taking, given risk aversion. This is an important qualification to the current results and will be the subject of further research.

Although the chosen CET parameters at each "node" of the nesting structure may appear low, the number of nests means that choices at the final level (across destinations of FDI) are actually very flexible. For example, it can be shown that, holding total wealth fixed but allowing all other adjustments across asset types and locations to take place, the implied semi-elasticity of transformation between foreign destinations can easily reach 20 and can be as high as 60. The variation across regions in these implied elasticities comes about because of the different initial shares of assets in various regional portfolios.

The choice of CET parameters at each node was determined partly by this consideration of what they implied for the final elasticities, holding only total wealth constant. They were also chosen so that this version of FTAP gave results that were broadly comparable to an earlier version of GTAP with imperfect international (portfolio) capital mobility, for experiments involving the complete liberalization of agricultural and manufacturing protection (Verikios and Hanslow 1999). Imperfect capital mobility was also a feature of the GTAP-based examination of Asia-Pacific Economic Cooperation (APEC) liberalization by Dee, Geisler, and Watts (1996) and Dee, Hardin, and Schuele (1998). These parameters thus provide a familiar starting point from which refinements could be made in the future, possibly based on methods of maximum entropy.

In one respect, however, the current version of FTAP does differ from previous versions of GTAP with imperfect capital mobility. The GTAP variants assumed that capital was perfectly mobile across sectors, whereas FTAP has less-than-perfect sectoral mobility. Furthermore, the choice of sector is relatively early in the nesting structure, so that the implied elasticities guiding choice of sector, holding only total wealth constant, are relatively low (e.g., 1.2 in the United States). As a result, FTAP tends to exhibit the behavior that resources move less readily between sectors in a given region but more readily across regions in a given sector, although the differences are not dramatic. The current treatment is consistent with the idea that the knowledge capital often required to succeed in FDI, despite the difficulties of language and distance, is likely to be sector specific.

Petri's model assumed that total wealth in each region was fixed. In FTAP, although regional endowments of land and natural resources are fixed (and held solely by each region's residents), regional capital stocks can accumulate over time, and net bond holdings of each region can adjust to help finance the accumulation of domestic and foreign capital by each region's investors. The treatment of capital accumulation follows the original treatment of McDougall (1993) and was also used by Verikios and Hanslow (1999); Dee, Geisler, and Watts (1996); and Dee, Hardin, and Schuele (1998).

With this treatment of capital accumulation, FTAP provides a long-run snapshot view of the impact of trade liberalization, ten years after it has occurred. To the extent that liberalization leads to changes in regional incomes and saving, this will be reflected in changes to the capital stocks that investors in each region will have been able to accumulate. As noted, investors in each region are not restricted to their own saving pool in order to finance capital investment. They may also issue bonds to help with that investment, but only according to their own preferences about capital versus bond holding, and only according to the willingness of others to accept the additional bonds.

1.1.2 Model Database

The starting point for FTAP's database was not the standard GTAP database, because this includes measures of trade and investment barriers that are still to be eliminated under the Uruguay Round agreement. Instead, the starting point was an updated version of the GTAP database, following a simulation in which the barriers yet to be eliminated under the Uruguay Round had been removed. Such a database was provided by the work of Verikios and Hanslow (1999), under their assumption of less-than-perfect capital mobility.

Foreign Direct Investment Data

The Petri treatment of FDI requires the addition of data on bilateral FDI stocks and on the activity levels and cost and sales structures of FDI firms. The methods used to estimate such data were similar to those of Petri. Both APEC (1995) and United Nations (1994) provided limited data on FDI stocks by source, destination, and sector. These data were fleshed out to provide a full bilateral matrix of FDI stocks by source, destination, and sector, using RAS methods (Welsh and Strzelecki 2000). Thus, the individual bilateral estimates may be unreliable, although the more aggregate data match published totals. The resulting estimates are summarized in Dee and Hanslow (2000). The data were collected (and the model implemented) for nineteen regions and three broad sectors. The three sectors—primary (agriculture, resources, and processed food), secondary (other manufacturing), and tertiary (services)—correspond broadly to the three areas of potential trade negotiation in a new trade round. The intention is to use similar methods to produce a model with greater sectoral detail in the future.

One problem with such FDI data is that they distinguish FDI from portfolio investment according to whether the investor (or investing firm) has an

equity interest of 10 percent or more. This *ownership* share may not be sufficient to ensure *control* of an enterprise.[3] For some purposes, researchers
have instead considered affiliates that are majority owned—in which the
combined ownership of those persons individually owning 10 percent or
more from a particular country exceeds 50 percent. In the current context,
a better approach in the future may be to recognize explicitly the size of the
equity stake that different countries (including the local host) have in an enterprise, especially since some barriers to services trade are explicitly designed to control the extent of foreign ownership. This is an area for further
research.

The FDI stock data were used in turn to generate estimates of the output
levels of FDI firms. To do this, we estimated capital income flows by multiplying the FDI stocks by rates of return. These capital rentals were then
grossed up to get an output estimate for FDI firms, using ratios of capital
rentals to output from the GTAP database. Again, the resulting estimates
are similar to those in Petri (1997) and are summarized in Dee and Hanslow
(2000). A possible future refinement would be to use additional information
on the ratio of value added to output from U.S. and Japanese data on the
activities of offshore affiliates (e.g., Baldwin and Kimura 1998; Kimura and
Baldwin 1998). Petri (1997) shows how estimates obtained using different
methods can differ, sometimes widely. Nevertheless, experience shows that
models such as these are more sensitive to estimates of the extent of barriers to services trade than they are to estimates of the underlying services
trade and FDI flows.

The detailed cost and sales structures of FDI firms were assumed to be
the same as for locally owned firms and were obtained by prorating the
GTAP database. A subject for future research would be to make use of information on the true cost and sales structures of FDI firms, again using
available U.S. and Japanese data on the activities of offshore affiliates.

Estimates of Barriers to Services Trade

Estimates of existing barriers to services trade were injected into the
model's database, using the techniques of Malcolm (1998). The process is
documented in Hanslow et al. (2000).

The estimates of barriers to services trade were the first of a comprehensive new set of estimates, documented in Findlay and Warren (2000). The
general methodology of these studies is as follows.

- Qualitative information on barriers to services trade is converted to a
  quantitative index measure of trade restrictiveness, based on coverage

---

3. Another potential problem is that two or more countries can treat that same firm as a foreign affiliate. Although in some contexts this can lead to double counting, in the current context it does not because the FDI stock data have not been "grossed up" to account for other
owners (which could also include local joint venture partners).

and some initial judgments about the relative restrictiveness of the different sorts of restrictions.

- An econometric model is developed to measure the determinants of the economic performance (e.g., price, profit margin, cost, or quantity) of service firms in a given sector in different countries, taking account of all the factors that economic theory would suggest are relevant, including the index measure of trade restrictiveness.
- The economic model is used to estimate the determinants of economic performance. Wherever possible, the components of the trade restrictiveness index are entered separately so that the econometrics can reveal something about the relative weights attached to the separate components.[4]
- The results of the econometrics are used to calculate the effect of trade restrictions on performance. Where necessary, quantity or profit effects are converted to price or cost effects.

Estimates of barriers to trade in banking services along these lines were taken from Kaleeswaran et al. (2000), and estimates of barriers to trade in telecommunications services were taken from Warren (2000). The rates can be taken as indicative of post-Uruguay rates, because although the Uruguay Round established the architecture for services trade negotiations, it did not achieve much in the way of services trade liberalization (Hoekman 1995).

For modelling purposes, the barrier estimates were decomposed according to a two-by-two classification.

- The GATS framework distinguishes four modes of service delivery: via commercial presence, cross-border supply, consumption abroad, and the presence of natural persons. Accordingly, the FTAP model distinguishes barriers to establishment from barriers to ongoing operation. This is similar to the distinction between commercial presence and other modes of delivery, because barriers to establishment are a component of the barriers to commercial presence. Barriers to establishment are modeled as taxes on the movement of capital. Barriers to ongoing operation are modeled as taxes on the output of the service-providing firms.
- The GATS framework also distinguishes restrictions on market access from restrictions on national treatment. As noted above, the former are restrictions on entry, be it by locally owned or foreign-owned firms. In the FTAP model, they are treated as nondiscriminatory. Restrictions on national treatment mean that foreign-owned firms are treated less favorably than domestic firms. These restrictions are treated as discriminatory.

---

4. This is not possible where there is high multicollinearity between the various components, or where there is a lack of in-sample variation in some of the components.

Table 1.1   Classifying Barriers to Trade in Banking and Telecommunications Services

	Nondiscriminatory Barriers to Market Access	Discriminatory Derogations from National Treatment
	Barriers to Establishment	
Banking	Are there restrictions on the number of bank licenses?	Are there restrictions on the number of foreign bank licenses? Are there restrictions on foreign equity investment or requirements for foreigners to enter through a joint venture with a domestic bank? Are there restrictions on the permanent movement of people?
Telecommunications	One measure of restriction is actual number of competitors in fixed and mobile markets. Is there an enforced monopoly, partial competition or full competition in various fixed line markets and mobile market? What percentage of the incumbent fixed or mobile operator is privatised?	What percentage of foreign investment is allowed in competitive carriers?
	Barriers to Ongoing Operation	
Banking	Are there general restrictions on raising funds, lending, providing other lines of business, or expanding the number of banking outlets?	Are foreign banks restricted in raising funds, lending, providing other lines of business, or expanding the number of banking outlets? Are there restrictions on the proportion of foreigners on the board of directors? Are there restrictions on the temporary movement of people?
Telecommunications	Are there restrictions on leased lines or private networks? Are there restrictions on third party resale? Are there restrictions on connection of leased lines and private networks to the public switched telephone network?	Are there restrictions on callback services?

Source: McGuire and Schuele (2000) and Warren (2000).

The decomposition of trade barriers into this two-by-two classification follows the classifications used by Kaleeswaran et al. (2000) and Warren (2000). Table 1.1 shows how they classify barriers to trade in banking and telecommunications services. Note that in the banking sector, prudential regulations were not counted as trade barriers or included in the restrictiveness index. This was based on the recognition that they are designed to address a genuine market failure and the judgment that they are generally implemented in an appropriate fashion to that end. It is also consistent with the so-called "prudential carve-out" allowed for in the GATS.

Note also that in the banking study, horizontal (i.e., not sector-specific) restrictions on the permanent movement of people were counted as a barrier to establishment, and hence they were modeled as a barrier to the movement of capital. More properly, these restrictions should be modeled as a barrier to the movement of labor, but so far FTAP does not allow for international labor mobility. Similarly, horizontal restrictions on the temporary movement of people were counted as a barrier to ongoing operation, affecting both offshore affiliates and services delivered via "cross-border" trade, where the latter is broadly defined to include services delivered via the temporary movement of the consumer or the producer. In reality, the barriers affecting true cross-border trade are sufficiently different from those affecting trade involving temporary movement to warrant modeling them separately. These are areas for further research.

A simple average of the estimated price effects of barriers to trade in banking and telecommunications was taken as being typical of most services—all of the GTAP service categories of trade and transport; finance, business, and recreational services; and half of public administration, defense, education, and health. The remainder of public administration, defense, education, and health, along with electricity, water and gas, construction, and ownership of dwellings were assumed to be strictly non-traded (note that engineering services are part of business services, not construction). The resulting average estimates of barriers to trade in the tertiary sector would have been about 50 to 100 percent bigger had the banking and telecommunications estimates been taken as indicative of the whole of the services sector. A procedure for future research is to use the next version of the GTAP database, which will have more services-sector detail, to model barriers to each service separately, thus overcoming the extreme arbitrariness of these assumptions. In the meantime, the computational results should be treated as preliminary and interpreted with appropriate caution.

The resulting structure of post-Uruguay barriers to trade in services is summarized in table 1.2. Barriers to trade in primary (agricultural, resource, and processed food) and secondary (manufacturing) products are also shown for comparison purposes. Barriers to primary products are represented via a combination of taxes on imports, and subsidies (shown in table 1.2 as negative taxes) on exports and output. Unfortunately, at FTAP's three-sector level of aggregation, the actual taxes on primary exports and output are a combination of subsidies used for protective purposes, and taxes (e.g., excises on alcohol and tobacco) used for revenue raising. (Although the average taxes on primary output are not shown in table 1.2, they are all relatively small and mostly positive.) In future, using a database with greater sectoral detail will reduce the problems associated with "aggregation bias."

In the services sector, as noted above, barriers to establishment have been

Table 1.2   Tax Equivalents of Post-Uruguay Barriers to Trade and Investment (%)

	Imports		Exports		Domestic Output (tertiary)	Foreign Affiliates' Output (tertiary)	Domestic Capital (tertiary)	Foreign Affiliates' Capital (tertiary)
	Primary	Secondary	Primary	Tertiary				
Australia	1.69	7.30	0.65	4.81	0.00	0.69	0.62	14.79
New Zealand	1.16	4.51	-3.25	3.78	0.00	0.67	0.41	4.18
Japan	16.19	1.81	-8.12	4.41	3.59	4.75	0.33	3.01
Korea	12.95	6.61	-1.22	4.57	5.11	6.78	1.91	22.01
Indonesia	4.40	6.71	0.00	4.68	13.23	28.11	22.69	68.06
Malaysia	21.18	5.97	6.68	4.50	3.58	10.20	15.35	37.58
The Philippines	16.16	18.51	-0.10	4.80	8.38	22.65	7.40	54.28
Singapore	3.22	0.56	0.01	4.70	3.40	8.32	2.42	24.50
Thailand	12.12	14.81	-16.98	4.14	4.69	13.36	12.16	36.49
China	8.92	28.45	5.13	4.08	18.75	36.40	123.46	250.66
Hong Kong	0.00	0.00	0.00	9.91	1.39	2.36	1.35	5.41
Taiwan	27.31	5.63	-1.82	4.35	2.88	4.90	1.90	19.19
Canada	3.57	1.40	-0.43	3.54	0.25	1.67	0.53	6.11
United States	1.29	2.24	-0.02	4.26	0.07	1.08	0.00	3.83
Mexico	-1.50	2.99	1.89	5.23	2.17	5.59	0.68	12.99
Chile	6.76	10.26	0.02	4.36	2.97	4.11	14.15	20.36
Rest of Cairns[a]	3.82	13.39	6.30	4.49	0.98	5.55	7.19	19.45
European Union	3.17	1.13	-2.33	4.72	0.10	1.31	1.33	6.49
Rest of world	15.94	13.67	0.59	4.95	4.89	13.92	39.07	86.97

Source: FTAP model database.

[a]Rest of Cairns group: Brazil, Argentina, Colombia, and Uruguay.

modeled as taxes on capital. Barriers to ongoing operation may affect either FDI firms or those supplying via the other modes and have been modeled as taxes on the output of locally based firms (either domestic or foreign owned) and taxes of the same size on the exports of firms supplying via the other modes, respectively. The estimates of export taxes on services in the fourth column of table 1.2 are trade-weighted averages of the taxes on exports to particular destinations, where these are equal in turn to the taxes on foreign affiliates' output in the destination region, shown in the sixth column. These are modeled as taxes in the exporting region, rather than as tariffs in the importing region, to allow the rents created by the barriers to be retained in the exporting region. The issue of rents is addressed in more detail shortly.

The model also distinguishes restrictions on market access from restrictions on national treatment. The taxes on domestic capital and domestic output in table 1.2 represent the effects of restrictions on market access (affecting establishment and ongoing operation, respectively). The taxes on the capital and output of foreign affiliates are higher than the corresponding taxes on domestic firms, because they represent the effects of restrictions on both market access and national treatment.

The estimates in table 1.2 indicate that barriers to trade in services are generally at least as large as those on agricultural and manufactured products. Most economies have at least some significant barriers to trade in services. The only regions where barriers are low across the board are New Zealand, Japan, Hong Kong, Canada, the United States, and the European Union. However, this statement should be heavily qualified, because it is based only on estimates of barriers to banking and telecommunications. In the same vein, the estimates of overall barriers to services trade for China are very high, because the estimates of barriers to telecommunications services in China are particularly high, as they are in a number of other low-income developing economies. Estimates based on a broader set of services sectors are likely to produce less variation in overall estimates of services trade barriers across economies.

Barriers to trade in services have been modeled as tax equivalents that generate rents—a markup of price over cost—rather than as things that raise costs above what they might otherwise have been (e.g., Hertel 1999; Brown and Stern 2001). This decision was based on the way in which the price impacts of barriers to trade in banking and telecommunications services were measured. Kaleeswaran et al. (2000) measured the effects of trade restrictions on the net interest margins of banks, a direct measure of banks' markup of price over cost.[5] Warren (2000) measured the effects

---

5. Net interest margins—a measure of the difference between borrowing and lending rates of interest—can also be thought of as the "price" of financial intermediation services. The econometric model used to test the significance of barriers to trade in banking services was developed from an economic model of financial intermediation.

of trade restrictions on the quantities of telecommunications services delivered, and these were converted to price impacts using an estimate of the elasticity of demand for telecommunications services. Thus, Warren's estimates did not provide direct evidence of a markup of price over cost, but the relative profitability of telecommunications companies in many countries suggests that some element of rent may exist. By contrast, there is evidence that trade restrictions in sectors such as aviation raise costs (Johnson et al. 2000). As estimates of the effects of trade barriers in these sectors are incorporated into the model, it will be appropriate to treat some restrictions as cost-raising rather than as rent-creating.

One important implication of the current treatment is that welfare gains from liberalizing trade in services are likely to be understated, perhaps significantly. If trade restrictions create rents, then the allocative efficiency gains from trade liberalization are the "triangle" gains associated with putting a given quantum of resources to more efficient use. By contrast, if trade restrictions raise costs, the gains from trade liberalization include "rectangle" gains (qualified by general-equilibrium effects) from lower costs, equivalent to a larger effective quantum of resources for productive use.

Because barriers to services trade appear to be significant, and because they have been modeled as taxes, the rents they generate will be significant. A key issue is whether those rents should be modeled as being retained by incumbent firms, appropriated by governments via taxation, or passed from one country to another by transfer pricing or other mechanisms. In FTAP, the rents on exports have been modeled as accruing to the selling region, and those on FDI have been modeled as accruing to the region of ownership, after the government in the region of location has taxed them at its general property income tax rate. Despite this, the asset choices of investors are modeled as being driven by pretax rates of return. This is because many economies, in the developed world at least, have primarily destination-based tax systems. For example, if tax credits are granted for taxes paid overseas, investors are ultimately taxed on *all* income at the owning region's tax rate. Although such tax credits have not been modeled explicitly, their effect has been captured by having investors respond to relative pretax rates of return. Nevertheless, investor choices are also assumed to be determined by rates of return excluding any abnormal rent component. Investors would like to supply an amount of capital consistent with rates of return including abnormal rents, but they are prevented from doing so by barriers to investment. The amount of capital actually supplied is, therefore, that amount that investors would like to supply at rates of return excluding abnormal rents.

Thus, a portion of the rent associated with barriers to services trade is assumed to remain in the region of location in the form of property income tax revenue, whereas the remainder accrues to the region of ownership. Thus, liberalization of services trade could have significant income effects in both home and host regions as these rents are gradually eliminated. Dee and Hanslow (2000) show in detail how significant these effects are, relative to the allocative efficiency effects and other effects normally associated with trade liberalization.

A final point to note is that the model's database does not contain estimates of barriers to investment in agriculture and manufacturing, even though they are likely to be significant. It is unlikely that a new trade round would include negotiations on them. Nevertheless, their omission will affect the model's estimates of the effects of liberalization elsewhere, and the results need to be qualified accordingly.

1.2   The Cost Impact of Barriers to Trade in Services

Table 1.2 shows that the direct "tax equivalents" of barriers to trade in services are often significant, compared with the trade barriers expected to remain in agriculture and manufacturing after full implementation of the Uruguay Round. It also shows that barriers to services trade tend to be much higher in developing than in developed economies.

A priori, this does not mean that the services sectors in developing economies would suffer most from services trade liberalization. Because barriers to services trade are unlike tariffs, there are two key mechanisms by which the services sectors in developing countries could expand following services trade liberalization.

- Not all services trade barriers discriminate against foreign services suppliers, so the services sector could expand because of new domestic entry.
- Some services trade barriers restrict inward FDI, so the services sector could expand because of new foreign entry.

These mechanisms could be sufficient to offset the traditional mechanisms by which a protected sector can be harmed by removal of protection.

- Some services barriers discriminate against foreign services delivered cross-border, so the services sector could contract in the face of additional import competition.
- Services trade liberalization may benefit downstream using industries, and the services sector may lose out in the competition for domestic resources (e.g., labor).

Figure 1.1 examines the extent to which downstream using industries are likely to benefit from services trade liberalization. It shows the direct and indirect cost impost of domestic barriers to trade in services on all sectors in selected model regions, as calculated from the FTAP model database.

In general terms, the figure shows the direct and indirect input requirements needed to produce a unit of final demand in each sector. For example,

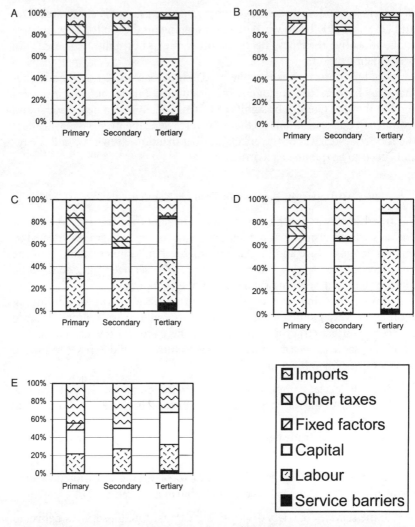

Fig. 1.1   Direct and indirect input requirements per unit of final demand: *A*, Japan; *B*, the United States; *C*, Korea; *D*, Taiwan; *E*, Hong Kong; *F*, Indonesia; *G*, Malaysia; *H*, the Philippines; *I*, Singapore; *J*, Thailand; *K*, China

a unit of processed food (a primary activity) sold to households might require inputs of unprocessed food (another primary activity), as well as packaging materials from the secondary sector. The packaging materials might again require inputs from forestry (a primary activity), along with electricity from the tertiary sector. Each of these direct and indirect inputs would have its own requirements for labor, capital, fixed factors (land and natural resources), and imported inputs, and these can be added up. Where

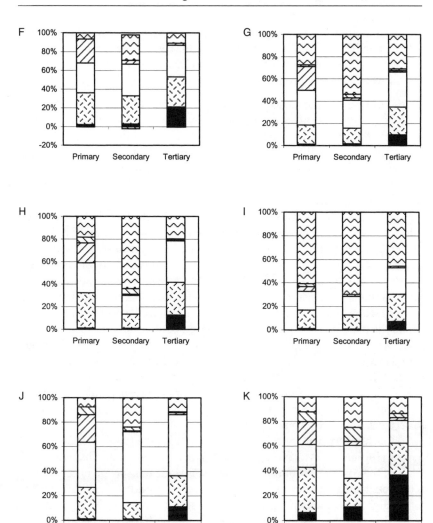

Fig. 1.1   (cont.)

the cost of the direct and indirect inputs is inflated by taxes, the direct and indirect tax contributions can also be calculated.

Thus, the direct and indirect cost impost of domestic barriers to services trade has been calculated by adding together the following:

- the output and capital taxes on direct and indirect services inputs, where those taxes represent the effects of domestic barriers to commercial presence (both establishment and ongoing operation); and
- the export taxes in the source region falling on direct and indirect im-

ported inputs, where these export taxes represent the effects of domestic barriers to cross-border services trade (where the term *cross-border* is interpreted loosely to include services traded via the temporary movement of the producer or consumer).

All other domestic taxes are collected in the contribution of "Other taxes," and all other taxes on imports (primarily tariffs) are included with the contribution of "Imports."

Figure 1.1 shows that, in every region shown, the greatest unit cost impost from services trade barriers falls on the services sector itself. This reflects two factors. First, the services sector experiences a direct taxing effect, whereas in other sectors the burden is indirect, through the higher cost of service inputs. Second, this effect is reinforced by the fact that in both developed and developing economies, the services sector itself tends to have a higher direct services input requirement than any other sector. Although other sectors may need service inputs, the greatest intensity of use of services is within the services sector itself. Thus, as will be seen, the benefits of services trade liberalization in many economies are concentrated within the services sector. This result is contrary to the normal effects of tariff removal, where the benefits are typically concentrated in other sectors.

Another feature of figure 1.1 is that in the economies with the highest per capita incomes (Japan, the United States, Korea, Taiwan, and Hong Kong), the cost impost of domestic services trade barriers on other sectors is minimal. Although these economies tend to be more service dependent, in terms of having higher direct service input requirements, their domestic barriers to services trade are also relatively low.

Somewhat surprisingly, in the economies with the lowest per capita incomes (Indonesia, Malaysia, the Philippines, Singapore, Thailand, and China), the cost impost of domestic services trade barriers on other sectors is not much greater. Only in China, where services trade barriers are particularly high, does the cost impost on other sectors approach 10 percent.[6]

By showing the cost impost of only domestic barriers to trade in services, figure 1.1 understates the potential first-round impact of multilateral liberalization of services trade. When barriers are removed globally, not only will domestic goods and services be cheaper, but so too will goods and services available in other economies. This benefit is likely to be significant in the highly import-intensive economies such as Korea, Taiwan, Hong Kong, Malaysia, the Philippines, and Singapore. Moreover, because the trade and transport services used to ship goods internationally will also be

6. The cost impost is estimated to be particularly high in China because its telecommunications market is particularly restrictive. When estimates of services barriers are incorporated for a broader range of services than banking and telecommunications, the overall cost imposts could differ from those shown here. Not only could the overall impost in China be lower, but the impost in developed countries could also be higher (since banking and telecommunications happen to be sectors in which developed countries are particularly liberal).

cheaper, there will be an additional cost reduction effect not captured in figure 1.1.

1.3   The Sectoral Effects of Removing Barriers to Trade in Services

1.3.1   Partial Equilibrium Effects on Sectoral Output

A useful way to understand the sectoral effects of removing barriers to trade in services is to start with a partial equilibrium framework and to gradually add the economy-wide constraints that distinguish a general from a partial equilibrium approach. This is a very useful technique of analysis, developed by Hertel (1997).

An initial partial equilibrium model is obtained by "turning off" the following parts of FTAP:

- *Factor supply constraints.* Each sector in each region can get all the labor and capital it needs at the going wage or rental price. Thus, the secondary and tertiary sectors in each region have horizontal supply curves (which nevertheless move downward as services barriers are removed). The primary sector continues to have an upward-sloping supply curve because fixed factors (land and natural resources) are still treated as being in fixed supply in each economy.
- *Income linkages.* Irrespective of what is projected to happen to factor prices and other variables, the model's measure of welfare is held fixed in each region. This "equivalent variation" is essentially a measure of net national product, or the real income accruing to the residents of each economy. In general equilibrium, it is affected not just by the amount of activity generated within a region, but also by net foreign interest and dividend payments associated with foreign borrowing and lending and with FDI.
- *The endogenous productivity and taste changes* associated with a love of variety. (In the full FTAP model, firms benefit from a wider choice of intermediate inputs in the same way that consumers benefit from a wider choice of final goods.)

In partial equilibrium, all the demand-side substitution possibilities of the full FTAP model are still in operation. Thus, for example, the demand for the output of the secondary sector in a region will depend on the following factors:

- how the cost (and hence price) of its output changes relative to the cost (and price) of output of secondary sectors in other economies, and how consumers and users in each region substitute between domestic and various imported sources of secondary output as a result of those relative price changes;

Table 1.3   Partial Equilibrium Effects on Selected Regions of Removing Global Barriers to
                     Trade in Services, by sector (%)

	Primary			Secondary			Tertiary		
	Q	Pd	Pm	Q	Pd	Pm	Q	Pd	Pm
Japan	−0.3	−2.4	−0.9	−1.4	−2.6	−3.9	−3.4	−2.1	−21.9
United States	−1.3	−0.7	−1.3	−7.3	−0.6	−3.3	−4.3	−0.4	−13.6
Korea	−0.1	−1.9	−1.0	2.3	−2.9	−3.1	−2.3	−3.5	−16.3
Taiwan	1.2	−1.2	−0.9	2.9	−2.3	−2.9	−4.5	−2.5	−14.7
Hong Kong	6.9	−1.1	0.1	15.2	−3.8	−3.9	−14.5	−5.2	−23.1
Indonesia	2.7	−0.5	−0.9	8.8	−3.6	−3.1	13.4	−12.1	−30.6
Malaysia	4.2	0.5	−0.9	6.9	−3.3	−2.9	0.3	−8.2	−21.3
The Philippines	1.2	−1.1	−0.9	2.9	−2.9	−2.9	8.1	−7.5	−27.9
Singapore	18.2	−1.6	−0.9	8.9	−3.5	−3.0	1.9	−6.7	−19.6
Thailand	3.1	0.6	−1.3	−5.3	−1.9	−3.0	0.3	−7.6	−21.9
China	36.6	18.1	−1.2	132.0	−10.2	−2.6	245.2	−27.9	−31.9

Source: FTAP model projections, partial equilibrium closure.
Notes: Q = domestic output quantity; Pd = domestic price; Pm = import price.

- how the cost (and hence price) of its output changes relative to the average price (across sources) of primary and tertiary output, and how domestic consumers and government substitute between the outputs of these different sectors as a result of these relative price changes;[7] and
- what the secondary input requirements are per unit of output in other sectors, and whether those other sectors are expanding or contracting.

Thus, even in the partial equilibrium model, the richness of substitution possibilities and interindustry linkages on the demand side make for a rather complicated story.

Because real incomes in each economy are assumed to be fixed, it would be expected that unless substitution effects dominate, the demand for, and hence output of, a commodity or service should increase whenever services trade liberalization reduces its price. And the only sector in which services trade liberalization would conceivably *not* reduce the price is the primary sector, where the return to the fixed factor could conceivably be bid up. Thus, the presumption is that services trade liberalization should reduce prices and increase output. Where this does not occur, it must be as a result of substitution effects.

Within the services sector itself, prices fall and output rises in the ASEAN economies and China (table 1.3). Note that although the prices of domestic services fall in these economies, the prices of imported services

7. In FTAP, as in GTAP, consumers and government are the only agents to substitute directly among different commodities. For intermediate and investment usage, different commodities (aggregated across sources) are used in fixed proportions.

fall by significantly more. Thus, substitution toward imports in these economies might suggest that services output should fall. Offsetting this, however, is an increase in exports of services from these economies. In the services sector, the price of a service import in the destination country can fall by significantly more than its output price in the exporting country. This is primarily because services trade liberalization involves removing the "export tax" equivalent of barriers to cross-border trade imposed by the destination country. Thus, although domestic services in the ASEAN region and China are disadvantaged relative to imports at home, when the same services are exported, their prices compare favorably with service exports from most other regions. (This is indicated indirectly by the fact that the domestic output price of services falls by more in ASEAN and China than in the other regions.) Thus, the services output expansion in ASEAN and China is primarily an export story.

In the higher per capita income economies, services output falls, despite a reduction in the domestic price, because of substitution toward imports. This is in accordance with the relative price movements shown in table 1.3.

The declines in the output of the secondary sector in Japan and the United States are because of substitution toward imports, especially in intermediate usage. For the other higher-income economies (Korea, Taiwan, and Hong Kong), the prices of domestic secondary output do not change greatly relative to secondary import prices, so the secondary-output expansions in these economies are primarily an export story. In ASEAN and China, the secondary-output expansions are because of both increased exports and substitution away from imports.

Although in the secondary and tertiary sectors the results are driven primarily by substitution among different sources of each commodity, in the primary sector it is possible to see the effects of each region's households' substituting among different commodities. This explains the slight falls in the output of the primary sector in Japan and Korea. In these economies, the prices of imported services fall significantly more than the prices of any other final commodity. Households tend to substitute toward imported services and away from everything else. Thus, primary output in these economies falls, despite the fact that the price of domestic primary output falls by more than its import price.

In the United States, the effect on primary-sector output of households' switching away from the primary sector in general is reinforced by substitution (in relative terms) toward primary imports.

In Taiwan, Hong Kong, Singapore, and Thailand, the expansion of the primary sector is primarily an export story. (The landed cost plus insurance and freight [c.i.f.] price of Thai primary exports falls, despite a slight increase in the domestic output price, because of cheaper international trade and transport services.) This can be confirmed by looking at more detailed model results not shown in table 1.3.

In Indonesia, Malaysia, the Philippines, and China, the switch by households away from the primary sector in general is offset by increased intermediate input demand, and some increase in export demand, for primary-sector output. The increased intermediate demand occurs despite an adverse relative price movement against imports (in all but the Philippines), because of interindustry linkages between the primary sector and the downstream secondary and tertiary sectors.

In summary, multilateral liberalization of services trade reduces domestic costs and prices across all economies, and the partial equilibrium sectoral effects are of three types.

- In economies such as those of Japan and the United States, where initial domestic services barriers are particularly low, domestic prices do not fall by much, and substitution toward cheaper imports leads to a reduction in output in all sectors of the economy. Real income can remain constant, however, because of the cheaper imports.
- At the other extreme, in the economies of the ASEAN region and China, where initial domestic services barriers are relatively high, domestic prices tend to fall significantly, and output in (almost) all sectors of these economies expands.
- In between are the economies of Korea, Taiwan, and Hong Kong, where initial domestic services barriers are moderate, but where all sectors are more trade exposed than in Japan and the United States. Thus, although the services sectors in these economies may not benefit from services trade liberalization, at least some of their other sectors benefit from cheaper domestic and imported inputs and thus gain an advantage on export markets.

1.3.2   General Equilibrium Effects on Sectoral Output

The partial equilibrium results of table 1.3 assumed that each sector in each economy could get any additional labor and capital at the going wage or rental price. The results also ignored the income implications of services trade liberalization.

In table 1.4, these effects are gradually reintroduced into the model. The first column reproduces the partial equilibrium results from table 1.3. In the second column, primary factor supply constraints are imposed. As in textbook models, aggregate supplies of capital and labor are assumed to be fixed, and these factors are treated as being perfectly mobile within each sector of the economy. In the third column, sectoral capital stocks are assumed to take the values they would in the full general equilibrium model. Thus, not only do aggregate capital stocks change in each economy, but capital is no longer perfectly mobile across sectors. Finally, the full general equilibrium results are presented. These incorporate not only the primary factor behavior of the full general equilibrium model, but also the associ-

Table 1.4   Partial and General Equilibrium Effects on Sectoral Output in Selected Regions of Removing Global Barriers to Trade in Services (%)

	Full Partial Equilibrium	Fixed Factors	Capital as in General Equilibrium	Full General Equilibrium
Japan				
Primary	−0.3	0.2	−0.3	−0.4
Secondary	−1.4	0.9	−0.5	−0.3
Tertiary	−3.4	−0.3	0.2	0.1
United States				
Primary	−1.3	2.4	0.4	0.6
Secondary	−7.3	1.9	0.2	0.6
Tertiary	−4.3	−0.7	−0.2	−0.4
Korea[a]				
Primary	−0.1	−0.5	−0.7	−0.8
Secondary	2.3	−0.4	−1.4	−1.6
Tertiary	−2.3	0.3	1.0	1.1
Taiwan[a]				
Primary	1.2	0.4	−0.1	0.1
Secondary	2.9	2.5	0.2	1.0
Tertiary	−4.5	−1.1	0.1	−0.2
Hong Kong[a]				
Primary	6.9	3.7	0.0	0.2
Secondary	15.2	9.0	−1.2	−2.2
Tertiary	−14.5	−2.1	0.4	0.6
Indonesia[a]				
Primary	2.7	−3.4	0.3	0.3
Secondary	8.8	−9.7	2.5	2.6
Tertiary	13.4	8.9	8.5	9.2
Malaysia[a]				
Primary	4.2	−1.1	0.0	0.1
Secondary	6.9	−1.8	0.2	0.1
Tertiary	0.3	3.3	1.5	1.5
The Philippines[a]				
Primary	1.2	−3.0	−1.9	−1.9
Secondary	2.9	−6.6	−2.6	−3.6
Tertiary	8.1	5.6	2.3	2.5
Singapore				
Primary	18.2	3.2	−4.0	−4.0
Secondary	8.9	−4.2	−5.6	−6.6
Tertiary	1.9	4.5	0.6	1.0
Thailand[a]				
Primary	3.1	0.7	−0.2	−0.1
Secondary	−5.3	−8.0	−0.7	−0.8
Tertiary	0.3	8.1	1.3	1.3
China[a]				
Primary	36.6	−6.5	−1.1	−0.2
Secondary	132.0	−16.7	4.5	2.5
Tertiary	245.2	43.7	28.7	32.5

Source: FTAP model projections, partial and general equilibrium closures.

[a]Aggregate capital stock projected to increase in general equilibrium closure.

ated income effects (including the net foreign income flows associated with FDI).

In broad terms, the imposition of factor supply constraints is the single most important step in taking the partial equilibrium sectoral results toward their general equilibrium values.

Even with factor supply constraints, the results for the tertiary sector in each region are qualitatively quite close to the partial equilibrium results:

- the services sectors in Japan and the United States are still smaller than in the absence of services trade liberalization;
- the services sectors in most other high-income economies are also projected to decline (Korea is the exception); and
- the services sectors in the ASEAN region and China still gain from services trade liberalization.

Now, however, the wage-rental ratios in each economy adjust to ensure that the induced output changes in other sectors do not lead to a violation of the overall primary factor supply constraints. Thus, the output of the primary and secondary sectors in Japan, the United States, Taiwan, and Hong Kong is now projected to rise to counteract the decline in their services sectors. In the ASEAN region, China, and Korea, output in many of the primary and secondary sectors is now projected to decline to offset the expansion of their services sectors.

One question is whether the changes in wage-rental ratios in the "fixed factors" version of the model are consistent with those predicted by the Stolper-Samuelson theorem. That theorem would predict that in the face of a decline in the relative price of services (induced by services trade liberalization) there would be a decline in the real return to the factor of production used relatively intensively in the services sector. In most economies, that factor is labor (see figure 1.1). Although the assumption of fixed factor supplies and perfect factor mobility is consistent with the assumptions of the Stolper-Samuelson theorem, there are many other assumptions in the "fixed factors" model that do not match the textbook Stolper-Samuelson assumptions exactly. It is nevertheless useful to see if the "fixed factors" model retains a Stolper-Samuelson flavor in the context of services trade liberalization.

Broadly speaking, the Stolper-Samuelson theorem would predict a decline in the wage-rental ratio in most economies; by contrast, the wage-rental ratio faced by producers in all economies in the "fixed factors" model is projected to rise. The reason is simple. Services trade liberalization includes liberalization of FDI in the services sector. The removal of taxes on service-sector capital leads to a direct and significant decline in capital rentals, relative to wages, because, with fixed capital supplies, the loss of rents from barriers to capital are borne directly by capital owners.[8]

---

8. The implication of this for regional incomes is not yet incorporated.

Thus, the nature of barriers to services trade leads to a significant departure from one of the standard textbook trade theorems.[9]

The results in the third column hint at the complexity of the capital supply story in the full FTAP model. Even though services trade liberalization involves removing taxes on service-sector capital, it is not always the case that cautious investors would invest more in those service sectors than they would if they viewed investment in any sector as being equally desirable (consistent with perfect sectoral capital mobility). In Japan, the United States, Korea, Taiwan, and Hong Kong, service-sector capital stocks are larger than in the "fixed factors" case, but in the other economies they are smaller. This demonstrates how the capital supply behavior in the FTAP model plays an important role in relocating capital across regions within a sector, as opposed to the textbook treatment of capital allocation across sectors within a region.

One question is whether the sectoral output responses associated with a change in aggregate capital stocks are consistent with those predicted by the Rybczynski theorem. That theorem states that if product prices are fixed (say, by "world prices"), an expansion in capital would lead to an expansion in the output of the product that uses capital relatively intensively and a contraction of the other product. Leamer and Levinsohn (1994, 7) give an insightful reinterpretation of the Rybczynski theorem:

> What is really at stake here is not the Rybczynski Theorem but rather its traveling companion, the Factor Price Equalization Theorem. These results together imply that factor supply changes . . . do not have much affect [sic] on factor prices because the potential affect [sic] on factor prices is dissipated by product mix changes in favor of the products that use the accumulating factor intensively.

Clearly, critical assumptions of the Rybczynski Theorem do not hold in the FTAP model. Products are imperfect substitutes, so that product prices are not "given" to any single region. As a result, relative factor prices can also change to absorb the impact of an increase in capital, so that it does not have to be absorbed by changes in product composition.

However, one would expect the FTAP model to display the same underlying economic forces that lead to the Rybczynski result under its special set of assumptions. This can be demonstrated in an intermediate simulation in which aggregate capital in each region moves as it does in general equilibrium but is still perfectly mobile between sectors (thus, each region still has a unique economy-wide wage-rental ratio). In this intermediate simulation,

---

9. The particular result depends on the assumption that barriers to trade in services create rents rather than raising costs. If barriers to services trade raise costs and hence move the production possibility frontier toward the origin, then services trade liberalization could raise the real returns to all factors of production, although the effects on relative factor returns could still be unclear a priori. Brown, Deardorff, and Stern (2000) also show how real returns to both factors can be raised by the additional gains from trade arising from increasing returns to scale, competition, and product variety, even when barriers are treated as tariff equivalents.

there is the expected relationship between the direction of movement of the capital stock and whether the wage-rental ratio is higher or lower than in the "fixed factors" version of the model. When the capital stock rises, the wage-rental ratio is higher than in the "fixed factors" case, and when the capital stock falls, the wage-rental ratio is lower.

The final column of table 1.4 incorporates the FTAP model's income linkages: real income in each region is no longer constant but reflects the induced changes in factor prices and international capital movements. Dee and Hanslow (2000) demonstrate that such income effects are crucial to the welfare implications of liberalizing trade in services, as will be seen shortly. However, table 1.4 shows that these income effects do not have strong additional effects on the sectoral distribution of gains from services trade liberalization.

General equilibrium models are often regarded as "black boxes," offering little chance of understanding what is inside. The above analysis suggests that because the structure of barriers to services trade is complex, the hardest part about understanding the effects of multilateral liberalization of services trade is understanding what happens in partial equilibrium.

The partial equilibrium results help to demonstrate how liberalization of services trade can differ from tariff removal. Barriers to services trade affect domestic new entrants as well as foreign suppliers, and the sector to benefit most in output terms from liberalization can often be the services sector itself.

The transition from partial to general equilibrium analysis also demonstrates how some of the standard textbook results fail to hold in the context of services trade liberalization. In particular, because services trade barriers affect the price of service-sector capital as well as service-sector output, the Stolper-Samuelson theorem fails to hold: the movement of relative factor prices is dominated by the removal of the barriers to capital movement. The Rybczynski theorem also fails to hold in its textbook form, but the underlying economic forces that lead to its result are still relevant.

1.3.3 General Equilibrium Welfare Effects

The first column of table 1.5 summarizes the effects of full liberalization of services trade on economic well-being in selected model regions (Dee and Hanslow [2000] present results for all model regions). As in GTAP, the measure of economic well-being is the equivalent variation—essentially a measure of the change in real income in each region, where the deflator is an index of the prices of household consumption, government consumption, and national saving. For FTAP, however, the relevant measure of national income is net national product—the income accruing to the residents of a region—rather than net domestic product—the income generated within the borders of a region. Thus, net domestic product is adjusted for the income earned on outward FDI, net of the income repatriated overseas from inward FDI, plus the income from net bond holdings.

Table 1.5   **Welfare Effects of Full Multilateral Liberalisation of Services Trade (absolute change in US$ millions)**

	Equivalent Variation	Contribution of Endowment Change to Equilibrium Variation	Contribution of Change in Real FDI Stocks to Equilibrium Variation	Contribution of Change in Real Bond Holding to Equilibrium Variation	Contribution of Change in Rents on FDI Capital and Output
Japan	4,130	–1,030	3,120	–2,978	–8,730
United States	–1,809	–5,713	2,665	1,708	–6,716
Korea	1,886	438	–5	39	123
Taiwan	–142	312	378	–583	–423
Hong Kong	5,896	102	7,829	–621	–8,211
Indonesia	2,470	7,158	–541	–4,519	530
Malaysia	1,015	367	–103	–168	585
The Philippines	1,236	164	–91	47	214
Singapore	–247	–1,071	–198	–108	1,049
Thailand	1,698	305	–24	–393	486
China	90,869	52,164	–12,649	–5,776	12,849

Source: FTAP model projections, general equilibrium closure.

Three of the selected economies are projected to have incomes lower than otherwise as a result of full multilateral liberalization of services trade—the United States, Taiwan, and Singapore. Dee and Hanslow (2000) show that in each case, the losses from multilateral liberalization of services trade would be more than offset by income gains accruing from multilateral liberalization of trade in agriculture and manufacturing. Nevertheless, the source of the income losses from multilateral liberalization of services trade warrants further investigation, especially for the United States, where the losses are projected to be significant.

Dee and Hanslow show that for agricultural and manufacturing liberalization, the welfare results are dominated by two things: the contribution of improvements in allocative efficiency, and the contribution of induced changes in the terms of trade. The model's regions are projected to experience positive income gains, or in a few cases small losses, as a result of these effects.

For services liberalization, changes in FDI patterns contribute several additional effects. First, FDI can lead to an expansion or contraction in the capital stock located within a region, leading to a positive or negative contribution to income generated within a region from this change in national endowments. Second, it can lead to changes in net FDI and net lending positions, with consequent changes in net foreign income flows accruing to residents. Third, it can induce changes in the returns earned on those net foreign asset holdings. An important example here is changes in the rents earned on FDI.

The second column of table 1.5 shows the contribution to real income from changes in real capital endowments. Generally, if capital endowments are higher than otherwise, real GDP will be higher than otherwise, and vice versa.[10] A major reason that Singapore is projected to lose slightly from services trade liberalization is that its capital stock is projected to be lower than otherwise.

However, a lower capital stock located domestically need not always lead to lower incomes for domestic residents. Earnings from higher outward FDI and higher lending abroad could offset it. The third column of table 1.5 shows the contribution to residents' real income from changes in real FDI stocks. The fourth column shows the contribution from changes in real bond holdings. Both also help to indicate the way in which changes in capital endowments are financed.

For example, Japan's capital stock is lower than otherwise, but it has a big increase in outward FDI. In fact, it also borrows (a negative change in bond holding) in order to finance its outward FDI. By contrast, China's increase in capital endowments comes partly from a large increase in inward FDI

---

10. For a few regions, real GDP can be higher than otherwise, even if endowments are lower than otherwise, because the endowments are used more efficiently.

and partly from additional foreign borrowing. Thus, the large projected increase in China's service-sector output and exports, noted above, comes as much from an expansion in foreign-owned service firms located in China as it does from an expansion in Chinese-owned service firms. The United States is projected to have a smaller capital endowment than otherwise, but this is offset to some extent by an increase in outward FDI and increased lending to other regions.

For a few regions, real incomes are affected not so much by changes in net asset positions, but by changes in returns on those assets. Although the details are not shown in table 1.5, Taiwan is projected to lose slightly from services trade liberalization, primarily because in the FTAP database it is a net creditor economy and is adversely affected by a small induced fall in real interest rates.[11]

A further source of change in asset returns is the change in rents generated by barriers to services trade. The last column of table 1.5 shows the income contribution to recipient countries of changes in the rents accruing to FDI, as barriers to services trade are eliminated. What is striking is the loss of rents to the main providers of outward FDI—Japan, Hong Kong, and the United States. In fact, the loss of rents to U.S. incumbent multinationals is more than sufficient to explain its overall projected income loss from multilateral liberalization of services trade. Note, however, that this result is sensitive to the assumption that all barriers to services trade are rent-creating rather than cost-raising.[12]

Generally, although induced changes in capital stocks—both those located domestically and those owned abroad—do not appear to play a major role in explaining the effects of multilateral services trade liberalization on sectoral output, they play a major role in explaining the effects on real regional incomes. Barriers to services trade affect capital movements as well as the output of services firms, so services trade liberalization can have a significant effect on the regional location and ownership of capital. The flow-on effects to regional incomes demonstrate another way in which liberalization of services trade can differ from tariff removal.

1.4 Agenda for Further Research

Much of the research agenda for further development of the FTAP model has been outlined already. It involves continuing to obtain estimates of the price impacts of barriers to services trade along the lines outlined in

---

11. Interest rates fall primarily because of an assumption that government saving *rates* are held constant. Growing revenues and saving *levels*, therefore, allow some government debt retirement.

12. Brown, Deardorff, and Stern (2000) show that if barriers are cost-escalating, then welfare effects are dominated by the movement of real physical capital. However, they have a more simple treatment of profit repatriation and debt service payments than here.

Findlay and Warren (2000), both for additional sectors and for additional modes of service delivery within a sector. The methodologies should in the process reveal whether the barriers are rent-creating or cost-raising. Such methods could also be used to estimate the price impact of barriers to FDI in agriculture and manufacturing. More sectoral detail needs to be incorporated into FTAP, to model the barriers to each service separately. More research is required to obtain more realistic output estimates and cost and sales structures for FDI firms and, if possible, a realistic initial allocation of rents. Additionally, the welfare measure in FTAP needs to be amended to take account of the costs of risk taking, given risk aversion.

In addition, some of the simplifying assumptions made during the original development of FTAP could now be relaxed, and the sensitivity of the results to these assumptions tested. One such assumption was the uniformity of behavioral parameters across sectors and regions. Although this reflected a deliberate research strategy, its importance could be tested using systematic sensitivity analysis (Arndt and Pearson 1996). The importance of data issues (e.g., the initial distribution of rents) and theoretical issues (e.g., investor behavior) could also be explored.

However, there is also scope for much more work using simple analytical models of services trade that better incorporate the features of services and the nature of the barriers to their trade. Insights of the sort available in Markusen, Rutherford, and Hunter (1995), for example, provide invaluable guidance to those attempting to build better empirical models of FDI and services trade.

References

Arndt, C., and K. Pearson. 1996. *How to carry out systematic sensitivity analysis via Gaussian Quadrature and GEMPACK.* GTAP Technical Paper no. 3. Purdue University: Center for Global Trade Analysis.
Asia-Pacific Economic Cooperation (APEC). 1995. *Foreign direct investment and APEC economic integration.* Singapore: APEC Economic Committee.
Baldwin, R. E., and F. Kimura. 1998. Measuring U.S. international goods and services transactions. In *Geography and ownership as bases for economic accounting,* ed. R. E. Baldwin, R. Lipsey, and J. D. Richardson, 9–48. Chicago: University of Chicago Press.
Brown, D. K., A. V. Deardorff, A. K. Fox, and R. M. Stern. 1995. Computational analysis of goods and services liberalization in the Uruguay Round. In *The Uruguay Round and the developing economies,* ed. W. Martin and L. A. Winters, 365–80. Washington, D.C.: World Bank.
Brown, D., A. Deardorff, and R. Stern. 1996. Modeling multilateral trade liberalization in services. *Asia Pacific Economic Review* 2 (1): 21–34.
———. 2000. CGE modeling and analysis of multilateral and regional negotiating options. Paper presented at conference on Issues and Options for the Multilat-

eral, Regional, and Bilateral Trade Policies of the United States and Japan. 5–6 October, University of Michigan.

Brown, D., and R. Stern. 2001. Measurement and modeling of the economic effects of trade and investment barriers in services. *Review of International Economics* 9 (2): 262–86.

Deardorff, A. 1985. Comparative advantage and international trade and investment in services. In *Trade and investment in services: Canada/U.S. Perspectives*, ed. R. Stern, 39–71. Toronto, Canada: Ontario Economic Council.

Dee, P., C. Geisler, and G. Watts. 1996. *The impact of APEC's free trade commitment.* Industry Commission Staff Information Paper. Canberra, Australia: Australian Government Publishing Service, February.

Dee, P., and K. Hanslow. 2000. *Multilateral liberalization of services trade.* Productivity Commission Staff Research Paper. Canberra, Australia: Ausinfo.

Dee, P., A. Hardin, and M. Schuele. 1998. *APEC early voluntary sectoral liberalization.* Productivity Commission Staff Research Paper. Canberra, Australia: Ausinfo.

Findlay, C., and T. Warren, eds. 2000. *Impediments to trade in services: Measurement and policy implications.* London: Routledge.

Francois, J. F., B. McDonald, and H. Nordstrom. 1995. Assessing the Uruguay Round. In *The Uruguay Round and the developing economies*, ed. W. Martin and L. A. Winters, 117–214. Washington, D.C.: World Bank.

Gehlhar, M. 1997. Historical analysis of growth and trade patterns in the Pacific Rim: An evaluation of the GTAP framework. In *Global Trade Analysis: Modelling and Applications*, ed. T. Hertel, 349–63. Cambridge: Cambridge University Press.

Grossman, G., and E. Helpman. 1991. *Innovation and growth in the global economy.* Cambridge: Massachusetts Institute of Technology Press.

Hanslow, K., T. Phamduc, and G. Verikios. 1999. The structure of the FTAP model. Productivity Commission, Canberra. Research Memorandum, December.

Hanslow, K., T. Phamduc, G. Verikios, and A. Welsh. 2000. Incorporating barriers to FDI into the FTAP database. Productivity Commission, Canberra. Research Memorandum.

Harrison, J. W., and K. R. Pearson. 1996. Computing solutions for large general equilibrium models using GEMPACK. *Computational Economics* 9 (2): 83–127.

Heckscher, E. [1919] 1949. The effect of foreign trade on the distribution of income. Reprinted in *Readings in the theory of international trade*, ed. American Economic Association. Philadelphia, Blakiston.

Helpman, E. 1981. International trade in the presence of product differentiation, economies of scale, and monopolistic competition: A Chamberlinian-Heckscher-Ohlin approach. *Journal of International Economics* 11:304–40.

Hertel, T. 1997. *Global trade analysis: Modeling and applications.* Cambridge: Cambridge University Press.

Hertel, T. 1999. Potential gains from reducing trade barriers in manufacturing, services, and agriculture. Paper presented at the 24th Annual Economic Policy Conference, Federal Reserve Bank of St. Louis. 21–22 October, St. Louis, Miss.

Hindley, B., and A. Smith. 1984. Comparative advantage and trade in services. *World Economy* 7:369–90.

Hoekman, B. 1995. Assessing the general agreement on trade in services. In *The Uruguay round and the developing economies*, ed. W. Martin and L. A. Winters, 327–64. Washington, D.C.: World Bank.

Johnson, M., T. Gregan, G. Gentle, and P. Belin. 2000. Modeling the benefits of increasing competition in international air services. In *Impediments to trade in services: Measurement and policy implications*, ed. C. Findlay and T. Warren, 119–51. London: Routledge.

Jones, R., and J. Scheinkman. 1977. The relevance of the two-sector production model in trade theory. *Journal of Political Economy* 85 (5): 909–35.

Kaleeswaran, K., G. McGuire, D. Nguyen-Hong, and M. Schuele. 2000. The price impact of restrictions on banking services. In *Impediments to trade in services: Measurement and policy implications*, ed. C. Findlay and T. Warren, 215–30. London: Routledge.

Kimura, F., and R. E. Baldwin. 1998. Application of a nationality-adjusted net sales and value-added framework: The case of Japan. In *Geography and ownership as bases for economic accounting*, ed. R. E. Baldwin, R. Lipsey, and J. D. Richardson, 49–82. Chicago: University of Chicago Press.

Krugman, P. 1979. Increasing returns, monopolistic competition, and international trade. *Journal of International Economics* 9:469–79.

Leamer, E., and J. Levinsohn. 1994. International trade theory: The evidence. NBER Working Paper no. 4940. Cambridge, Mass.: National Bureau of Economic Research, November.

Liu, J., C. Arndt, and T. Hertel. 2000. Estimating trade elasticities for GTAP: A maximum entropy approach. Paper presented at the Third Annual Conference in Global Economic Analysis. 26–29 June, Melbourne, Australia.

Malcolm, G. 1998. Adjusting tax rates in the GTAP data base. GTAP Technical Paper no. 12. Purdue University, Center for Global Trade Analysis.

Markusen, J. 1981. Trade and the gains from trade with imperfect competition. *Journal of International Economics* 11:531–51.

———. 1989. Trade in producer services and in other specialized intermediate inputs. *American Economic Review* 79 (1): 85–95.

Markusen, J., T. Rutherford, and L. Hunter. 1995. Trade liberalization in a multinational dominated industry. *Journal of International Economics* 38 (1–2): 95–117.

Markusen, J., T. F. Rutherford, and D. Tarr. 1999. Foreign direct investment in services and the domestic market for expertise. Paper presented at the Second Annual Conference on Global Economic Analysis. 20–22 June, Denmark.

Mayer, W. 1974. Short-run and long-run equilibrium for a small open economy. *Journal of Political Economy* 82:955–68.

McDougall, R. 1993. Incorporating international capital mobility into SALTER. SALTER Working Paper no. 21. Industry Commission, Canberra, Australia.

Melvin, J. 1969. Increasing returns to scale as a determinant of trade. *Canadian Journal of Economics* 2:389–402.

———. 1989. Trade in producer services: A Heckscher-Ohlin approach. *Journal of Political Economy* 97 (5): 1180–96.

Mussa, M. 1974. Tariffs and the distribution of income: The importance of factor specificity, substitutability, and intensity in the short and long run. *Journal of Political Economy* 82:1191–204.

Nagarajan, N. 1999. The millennium round: An economic appraisal. Economic Papers no. 139. Directorate General for Economic and Financial Affairs, European Commission, November.

Ohlin, B. 1933. *Interregional and international trade*. Cambridge, Mass.: Harvard University Press.

Petri, P. A. 1997. Foreign direct investment in a computable general equilibrium framework. Paper prepared for the conference on Making APEC Work: Economic Challenges and Policy Alternatives. 13–14 March, Keio University, Tokyo, Japan.

Rybczynski, T. 1955. Factor endowment and relative commodity prices. *Economica* 22:336–41.

Stolper, W., and P. Samuelson. 1941. Protection and real wages. *Review of Economic Studies* 9:58–73.
United Nations. 1994. *World investment directory: Latin America and the Caribbean.* New York: United Nations.
Verikios, G., and K. Hanslow. 1999. Modeling the effects of implementing the Uruguay Round: A comparison using the GTAP model under alternative treatments of international capital mobility. Paper presented at the Second Annual Conference on Global Economic Analysis. 20–22 June, Denmark.
Warren, T. 2000. The impact on output of impediments to trade and investment in telecommunications services. In *Impediments to trade in services: Measurement and policy implications,* ed. C. Findlay and T. Warren, 85–100. London: Routledge.
Welsh, A., and A. Strzelecki. 2000. Estimating domestic and foreign returns to capital for the FTAP model. Productivity Commission, Melbourne, Australia. Research Memorandum, May.

Comment   Fukunari Kimura

The research group of the Productivity Commission of Australia has conducted a series of great effort in quantifying the economic effects of liberalizing trade in services and has contributed to constructive discussion in a number of international academic and semi-academic forums. Admittedly, it is not at all easy to measure the magnitude of barriers to services trade as well as formulating a reasonable policy simulation model with rigorous theoretical framework. Nevertheless, it is crucially important to quantify possible effects of trade liberalization in order for policy makers to carry on a constructive discussion resisting various politico-economic pressures. The current paper presents a step forward in taking care of some of the features unique to trade in services vis-à-vis trade in goods.

Modes of Services Transactions

One of the novel features of this paper is the introduction of explicit treatment for modes of services transactions. The General Agreement on Trade in Services (GATS) of the Marrakesh Agreement defines four modes of services transactions: cross-border, consumption abroad, commercial presence, and natural persons. The past literature has tried to construct simulation models with a structure analogous to that for merchandise trade, but those models could only deal with the first mode, cross-border. The other three modes of services transactions require more sophisticated formulations in terms of factor movements across the national border, the place where factor services are inputted, and the nature of corresponding trade barriers.

Fukunari Kimura is professor of economics at Keio University.

The paper particularly puts emphasis on the most important mode of services transactions, that is, commercial presence. Service provision in this mode is initiated from the establishment of a local affiliate or a local branch through international capital movement (typically foreign direct investment [FDI]), and then services are produced in combination with local resources such as labor. The model traces the service supply structure by introducing international capital movement. Furthermore, the model distinguishes two types of barriers to services trade: barriers to establishing commercial presence and barriers to ongoing operation. These two roughly correspond to the concept of market access and national treatment in the table of concession of GATS. The former is modeled as taxes on the movement of capital, and the latter is formalized as taxes on the output of the service-providing firms.

Of course, the statistical measurement of such barriers, as well as the quantification of international capital movement, is not at all easy, and thus simulation results must be regarded as provisionary. However, the pioneering treatment of mode-3 service transactions will surely become a starting point to formalize services trade in future research.

Handling Capital

The authors call their model FTAP as a special version of GTAP with FDI. Foreign direct investment is different from simple international capital movements such as portfolio investment in an important way: it is accompanied by the movements of firm-specific assets such as technology and managerial know-how. Particularly in less developed economies, affiliates of foreign firms behave quite differently from local indigenous firms in terms of technology, managerial know-how, the pattern of purchases and sales, and the degree of exposure to foreign markets. Therefore, to seriously model FDI, we prefer to distinguish capital by the owner's nationality in addition to the location where capital services are used.[1]

Such expansion of dimension in policy simulation models raises a number of issues to be solved. One issue is the availability of statistical data. The pioneering work by Petri (1997) as well as that of Dee, Hanslow, and Phamduc basically relies on FDI flow data in estimating the magnitude of activities of affiliates of multinational enterprises (MNEs). However, we have a number of problems in this approach to data construction. First of all, we are not sure whether FDI data properly include reinvestment from retained earnings by affiliates abroad. The treatment of joint ventures is another problem. Moreover, the available figures are for investment flows, and thus we need uneasy transformation from flow to stock. After all, we have only capital stock estimates, which may not be a good proxy for the magnitude

---

1. See Baldwin and Kimura (1998) on the more detailed discussion on ownership, control, and location when considering FDI.

of activities of affiliates. We actually have some fragmental but direct information on activities of affiliates of MNEs. The Department of Commerce of the U.S. Government and the Ministry of International Trade and Industry (renamed the Ministry of Economy, Trade, and Industry in January 2001) of the Japanese Government, for example, compile ample activity data on U.S. and Japanese affiliates abroad. Some hosting countries such as Singapore and Malaysia include firm-nationality information in manufacturing censuses. Such information must be utilized to improve the quality of activity data in future research.

Another fundamental issue is the conceptual framework with which to introduce firm-specific assets in theoretically consistent models.[2] This paper makes an important contribution to the literature on this matter. In the model, investors first divide their wealth between "bonds" and "real physical capital," and then the former go to foreign portfolio investment with perfect arbitrage while the latter proceed to domestic physical investment and FDI with imperfect substitution. This treatment allows the introduction of firm-nationality-specific physical capital.

The paper includes an interesting discussion on the nesting structure of commodity demand. Petri (1997) sets the ownership of producers as a higher nest and then goes down to the location of producers as a lower-rank nest. In contrast, the present paper works with the opposite order. The former formulation is attractive if we think much of the existence of firm-specific assets. For instance, VCRs produced in a Sony plant located in Malaysia are closer substitutes for Japanese-made VCRs in Japan than they are for VCRs produced in a local indigenous plant located in Malaysia. Dee, Hanslow, and Phamduc, however, claim that services must meet local tastes and requirements and thus the location of production should come earlier than the nationality of producers. We obviously need more discussion on how to formulate the nationality of firms and the location of production.

Welfare Effects

The paper displays the simulation results of removing barriers to trade in services in a step-by-step, intuitive manner by starting from partial equilibrium effects and then explaining general equilibrium effects. The partial equilibrium results in which factor prices as well as domestic factor supplies are fixed look reasonable; with the removal of barriers to trade in services, countries originally with high barriers will gain the most, and those with low barriers will get hurt. Then factor prices are endogenously adjusted, and international capital movements are allowed in the general equilibrium, where the welfare gains are spread out to most of the countries.

---

2. Kimura and Tsutsumi (1998) list a number of conceptual issues to introduce firm-nationality-specific capital in computable general equilibrium (CGE) models.

Some uneasy results, particularly the negative welfare effects on the U.S. economy in the general equilibrium setting, should not be worried about too much. Rather, we must realize that the setting for FDI is crucially important in estimating the liberalization effects. If, for example, the U.S. service providers are more competitive than Japanese ones, the pattern of FDI may drastically change, resulting in different welfare impacts across countries. Firm-specific or firm-nationality-specific assets can also be a source of market power, which the symmetric constant elasticity of substitution (CES) nesting of product differentiation may not properly capture. Again, the key issue is how to formulate MNEs in simulation models.

Concluding Remarks

In summary, this paper makes a big step forward toward quantifying the cost of barriers to trade in services, which is crucially important in setting up a constructive policy discussion. The simulation results, however, have not reached the level of attracting very serious consideration from the nonacademic circle. A major task will be determining how to formulate MNEs in theoretically rigorous models. Although the paper makes a significant contribution to this subject, we have a number of things to settle in the future from both theoretical and empirical points of view. Another important task for us is to make primary statistical data collection of good quality. As for the FDI-related data, the best way to capture the activities of affiliates of MNEs is to collect information in the framework of host countries' establishment or firm censuses. Physical activities are, after all, much easier to capture than flows of financial transactions. As for service transactions, the balance-of-payments statistics cover only a small portion of trade in services. We must develop a statistical framework to capture various aspects of services trade covering four modes and possibly service contents in merchandise trade.

References

Baldwin, Robert E., and Fukunari Kimura. 1998. Measuring U.S. international goods and services transactions. In *Geography and ownership as bases for economic accounting*, ed. R. E. Baldwin, R. E. Lipsey, and J. D. Richardson, 9–36. Chicago: University of Chicago Press.

Kimura, Fukunari, and Masahiko Tsutsumi. 1998. The CGE modeling strategies with the nationalities of firms. Paper prepared for the conference Making APEC Work: Economic Challenges and Policy Alternatives. 13–14 March, Keio University, Tokyo, Japan.

Petri, P. A. 1997. Foreign direct investment in a computable general equilibrium framework. Paper prepared for the conference Making APEC Work: Economic Challenges and Policy Alternatives. 13–14 March, Keio University, Tokyo, Japan.

2

Shaping Future Rules for Trade in Services
Lessons from the GATS

Aaditya Mattoo

2.1 Introduction

The new round of services negotiations has begun, not with passionate intensity, but with a rather mechanical sense of "since we said we would, therefore we must." Although the lack of attention from those opposed to freer trade is cause for relief, the lack of conviction in supporters of new negotiations merits concern. The reason for both, however, is the limited impact so far of the General Agreement on Trade in Services (GATS). Creating a framework of rules in this difficult area was certainly an achievement, but the GATS has so far failed to deliver meaningful liberalization, and it has invariably been a step behind technological and regulatory developments in services. The agreement is generally perceived, not as a scourge of protection, but as a rather stodgy reaper of liberalization accomplished elsewhere.

In highlighting the limitations of the GATS, which is the main purpose of this paper, it is easy to understate what it has accomplished. In recognition of the fact that many services require proximity between consumers and suppliers, the agreement went beyond the traditional notion of trade (including only cross-border delivery) to encompass supply through the move-

Aaditya Mattoo is a senior economist in the Development Economics Research Group at the World Bank in Washington, D.C.

An earlier version of this paper was presented an NBER Conference on Trade in Services, Seoul, June 2000. It is part of the World Bank's research program on trade in services, supported in part by the United Kingdom Department for International Development. The paper has benefited from the comments of Rolf Adlung, Carsten Fink, Bernard Hoekman, Anne O. Krueger, Richard Snape, Arvind Subramanian, Chang Tai-Hsieh, and other conference participants. The views expressed here are those of the author and should not be attributed to the World Bank.

ment of both capital and labor.[1] The agreement also created a framework to deal with forms of protection more complex and less visible than tariffs. These include, first, a variety of quantitative restrictions, ranging from cargo sharing in transport services and limits on the number of (foreign) suppliers in telecommunications and banking to restrictions on the movement of service-providing personnel that affect trade in all services. Then there are numerous forms of discrimination against foreign providers, through taxes and subsidies as well as by allowing less favorable access to essential facilities such as ports, airports, or telecommunications networks. Finally, there is a subtle class of measures that are neither quotas nor explicitly discriminatory but nevertheless have a profound effect on services trade, that is, domestic regulations such as qualification and licensing requirements.

Gandhi said that it was pointless to dream of systems so perfect that human beings no longer need to be good. It is perhaps equally utopian to wish for international trade rules that can deliver liberalization without the willingness of governments. No doubt liberalization of services is primarily a challenge for domestic policy. Still, multilateral negotiations and agreements can help in four ways, by helping to achieve deeper liberalization through reciprocity-based market access negotiations; efficient protection and regulation through rules that favor the choice of superior instruments; credibility of policy through legally binding commitments; and a guarantee against discrimination through the most-favored nation (MFN) principle. How much has the GATS already delivered in these respects?

- The GATS has created an adequate framework to deal with explicit protection, but neither the negotiating momentum to reduce such protection nor the rules to ensure that it takes a desirable form.
- In dealing with the trade-impeding impact of domestic regulations, an admittedly difficult area, the agreement has achieved even less: the Uruguay Round provisions were weak, and only limited progress has been made in the last five years.
- More positively, many countries have taken advantage of the GATS to create a more secure trading environment by legally binding current levels of openness, and some have even precommitted to greater levels of future openness. However, the coverage of commitments for developing countries is limited, and in some cases commitments serve to protect the privileged position of incumbents rather than enhancing the contestability of markets.
- As befits a multilateral agreement, the GATS in principle prohibits a country from discriminating between its trading partners. The explicit

---

1. Developed country proponents of the GATS initially envisaged an inclusion only of capital movements, but developing country negotiators successfully pressed for the inclusion of labor movements also.

departures from this obligation, such as the exceptions for regional integration agreements and the exemptions listed by members, are well known. However, the difficulties in preventing implicit discrimination through domestic regulations and through the allocation of quotas have not been adequately appreciated.

- Finally, the GATS has so far done little to address the problem of private anticompetitive practices that fall outside the jurisdiction of national competition law, for example, in sectors like maritime and air transport. It has thus failed to reassure small countries that the gains from liberalization will not be appropriated by international cartels.

The rest of this paper develops the arguments presented above and provides suggestions on possible improvements not only in the rules of the agreement but also in the specific commitments made by countries and the negotiating methodology.[2] Where relevant, the paper draws upon the experience of the East Asian countries with the GATS. A basic tenet of the paper is that it is possible to make improvements in the GATS, and to make it a more effective instrument of liberalization, without fundamental structural changes, which are, in any case, of doubtful political feasibility.

2.2 Efficient Protection

The domestic political economic forces that lead to protection may also dictate that it is obtained through inefficient instruments. In goods trade, negotiations helped reduce protection, but ensuring that the efficient instruments of protection were chosen was the role of rules. Thus, General Agreement on Tariffs and Trade (GATT) rules broadly reflect the ranking of instruments suggested by economic theory: quotas are prohibited, tariffs are allowed but progressively negotiated down and bound, and production subsidies are permitted but subject to countervailing action under certain circumstances. The GATS rules on market access do not create a similar hierarchy. There are two basic rules: the market access provision (Article XVI) simply lists a set of measures, mostly different types of quotas, that cannot be maintained in scheduled sectors *unless prespecified;*[3] and the national treatment provision (Article XVII) prohibits any form of discrimination (including through subsidies) against foreign services and foreign ser-

---

2. The paper draws upon other research by the author, in particular Mattoo (2000a).

3. Article XVI stipulates that measures restrictive of market access that a WTO member cannot maintain or adopt, unless specified in its schedule, include limitations on (a) the number of service suppliers; (b) the total value of services transactions or assets; (c) the total number of services operations or the total quantity of service output; (d) the total number of natural persons that may be employed in a particular sector; (e) specific types of legal entity through which a service can be supplied; and (f) foreign equity participation (e.g., maximum equity participation). With the exception of (e), the measures covered by Article XVI all take the form of quantitative restrictions.

vice suppliers, again, *unless prespecified*.[4] Thus, in the services context, both the *level* and the *form* of protection are the outcome of negotiations between World Trade Organization (WTO) members.

2.2.1   Ranking Alternative Instruments

The question is: does economic theory in its current state suggest a hierarchy of instruments affecting services trade, and is it possible to create rules that favor a choice of superior instruments? The superiority of subsidies over trade restrictions is as valid for services as it is for goods,[5] and, in principle, tariffs are to be preferred to quotas for much the same reason as in the case of goods. However, there are at least three reasons that differences may arise. First, in some instances tariffs may not be easy to impose, and so the substitution of a more desirable policy instrument for a less desirable one may not be feasible. Second, some of the instruments that have a tariff-like effect in terms of inflicting costs on foreign providers (such as overly burdensome standards) are not, however, tariff-like in generating revenue. Finally, there are the numerous restrictions imposed on foreign direct investment and the movement of personnel that directly affect the market structure.

Consider each issue in turn. First, the difficulty of switching to fiscal instruments of protection has probably been exaggerated. As far as cross-border trade is concerned, the imposition of duties is probably most difficult—perhaps impossible, given the current state of technology—when a service is delivered electronically. In this case, however, other barriers to trade are also likely to be infeasible. Where quotas are feasible and maintained, as on cross-border trade in transport services, it is easy to conceive of tariff-type instruments: for example, a tax per passenger or unit of cargo carried by a foreign company. Moreover, the auction of a quota is analogous in economic effect to the imposition of a tariff. In the case of commercial presence, a number of fiscal instruments are possible, including entry taxes (or auctions of entry licenses), output taxes, and profit taxes. Ironically, the legal systems of many countries allow discrimination against foreigners through outright bans and entry quotas but make it difficult to impose discriminatory taxes.

---

4. Article XVII:1 states the basic national treatment obligation: "In the sectors inscribed in its Schedule, and subject to any conditions and qualifications set out therein, each Member shall accord to services and service suppliers of any other Member, in respect of all measures affecting the supply of services, treatment no less favourable than that it accords to its own like services and service suppliers." Unlike Article XVI, Article XVII provides no exhaustive list of measures inconsistent with national treatment. Nevertheless, Article XVII:2 makes it clear that limitations on national treatment cover cases of both de jure and de facto discrimination.

5. Both instruments encourage national production, the former by reducing the private costs of national producers and the latter by imposing a cost on foreign service providers. The latter is an inferior instrument because it leads to a deterioration in the price-quality mix that foreigners are able to provide local consumers. See also Hindley and Smith (1984), Hindley (1988), and UNCTAD and World Bank (1994) for a discussion of the economics of services trade.

Consider now the consequences of restrictive measures that increase foreign costs without generating revenue. In this case, part of the loss in consumer surplus is not offset by an increase in tariff revenue, so the loss in national and global welfare is much greater. Similarly, when quotas are imposed, their consequences for (national) welfare could be alleviated if the rents generated accrued domestically to importers or the government rather than to foreign exporters. However, the difficulties of intermediation in services suggest that quota-rents are more likely to be appropriated by exporters—or, more likely, quotas are likely to lead to socially wasteful administration costs and rent-seeking. Hence, one general conclusion is that if complete liberalization is not feasible, a shift from both quotas and non–revenue-generating measures to fiscal measures would lead to an increase in both national and global welfare.

A prohibition of quotas is unlikely to be politically feasible today. An intermediate step would be to build into GATS rules a legal presumption in favor of fiscal measures (see Deardoff 1994, 2000; Snape 1994; Hoekman 1996). The Uruguay Round Understanding on the Balance-of-Payments Provisions of the GATT 1994 provides a useful model. This understanding requires members to give preference to price-based measures and to use quotas only if price-based measures are inadequate, and the choice must be justified. In the GATS context, we would wish to see a shift from both quotas and wasteful discriminatory regulations to fiscal measures. Inducing a shift away from the former would require making the market access provision more stringent. Inducing a shift from latter has not been anticipated in the structure of the agreement and may be worth considering. In any case, greater flexibility in the national treatment provision (which prohibits all forms of discrimination) is not necessary, for, even if a country had committed to providing national treatment, then it is allowed to modify its commitments (under Article XXI) and switch instruments of protection—as long as the extent of protection does not increase.

2.2.2 Restrictions on Foreign Direct Investment

Restrictions on foreign investment assume particular significance in the case of services for which cross-border delivery is not possible, so that the price and quality of the service depend completely on the domestic market structure. Many developing countries, including some of those in East Asia, have been reluctant to allow unimpeded entry; instead, market access has been conceded either by allowing limited foreign entry or by increased foreign ownership of existing firms. Malaysia, the Philippines, and Thailand are among the countries that impose equity restrictions and restrictions on entry in key sectors like telecommunications and financial services, and many other East Asian countries have imposed one or the other type of restriction (see tables 2.1–2.3).

A central conclusion of the literature on privatization is that larger wel-

Table 2.1   East Asia Foreign Equity Participation, Degree of Competition, and Nature of Precommitment in Fixed Telecom Networks

Country	Limitations on FDI	Degree of Competition in Fixed Networks	Precommitment
Hong Kong	None	Oligopoly of 5 in domestic, monopoly in international	Will consider issuing more than the existing four licences for local fixed network services in June 1998.
Indonesia	GATS: 35%	Regional monopolies with scope for joint operating schemes	Policy review upon the expiry of the exclusive rights: exclusivity expires in 2011 for local service, in 2006 for long distance services, and in 2005 for international service.
Japan	20% in NTT and KDD	Full competition	Will increase foreign equity limits:
Korea	Variable: Facilities-based: 33%; KoreaTelecom: 20%; Resale-based: 0%	Full competition, phased in over several years	Facilities-based: 49% in 2001; Korea Telecom: 33% in 2001; Resale-based: 49% in 1999, 100% in 2001
Malaysia	GATS: 30%	Increasing competition; discretionary licensing	
The Philippines	GATS: 40%	Full competition; discretionary licensing	
Singapore	GATS: 73.99% (direct: 49%, indirect: 24.99%)	Monopoly	Oligopoly of 3 or more after April 2000
Thailand	Limited, in BTO arrangements	Monopoly, with some BTO arrangements	Will introduce revised commitments in 2006 when new law comes into force.

Source: Compiled by the author from GATS Schedules of Commitments.

fare gains arise from an increase in competition than from simply a change in ownership from public to private hands. Does the conclusion change when the change of ownership is from national to foreign hands? Foreign investment clearly brings benefits even in situations in which it does not lead to enhanced competition (e.g., there are entry restrictions). Foreign equity may relax a capital constraint, can help ensure that weak domestic firms are bolstered (e.g., via recapitalizing financial institutions), and can serve as a vehicle for transferring technology and know-how, including improved management. However, if foreign direct investment (FDI) comes simply because the returns to investment are artificially raised by restrictions on competition, the net returns to the host country may be negative (returns to the

Table 2.2   Market Access Commitments in Insurance in East Asia (direct includes life and nonlife)

			Limitations on Commercial Presence			
Member	Limitations on Cross-Border Supply	Limitations on Consumption Abroad	Legal Form	No. of Suppliers	Equity	Other
Brunei Darussalam	U	None excl. statutory ins.	Local registration	U (h)	U (h;	
Hong Kong	U	None excl. statutory ms.	S, B, or association of underwriters			
Indonesia	U	DL			100% of listed companies (G)	
Japan	Life: U Nonlife: for limited class; only with mode 3 for some services					
Korea	U except marine cargo and aviation insurance	U	S, B, joint ventures (but not with Korean licenses)		Restrictions on acquisition of existing firms; foreign portfolio invt only for listed stocks, ≤ 23% (h) LSO	
Malaysia	Life: U	Life: U	I	New: U	On incorporation of existing branches and for original owners: 51%; new participation in existing 30% (DLSO)	No branches for foreign > 50% (G)
The Philippines	Nonlife: DL U except for marine hull and marine cargo	Nonlife: DL U		DL	Acquisition or new: 51% (G)	

(continued)

Table 2.2   (continued)

Member	Limitations on Cross-Border Supply	Limitations on Consumption Abroad	Limitations on Commercial Presence			
			Legal Form	No. of Suppliers	Equity	Other
Singapore	U	None, excluding statutory insurance		New: U	Existing: 49% provided no foreign party is largest shareholder	
Thailand	U except for internal marine, aviation, and transit Nonlife for limited class	None Nonlife: none for limited class		DL	25%	

Source: Compiled by the author from GATS Schedules of Commitments.

Notes: B = branches; S = subsidiaries; h = restrictions in horizontal commitments; I = local incorporation required; None = commitment to impose no restrictions; No text = no restrictions, but reference to some regulations; U = unbound (no commitment); R = reciprocity condition or MFN exemption; DL = discretionary licensing or economic needs test; (D) LSO = (discretionary) limits on single ownership; G = grandfathering provisions.

Table 2.3   **Market Access Commitments in Banking in East Asia (acceptance of deposits and lending)**

Member	Limitations on Cross-Border Supply	Limitations on Consumption Abroad	Limitations on Commercial Presence					
			Legal Form	No. of Suppliers	Equity	Number of Operations (branches)	Value of Transactions or Assets	Other
Hong Kong	U	None	Deposits: S or B	DL for acquisition of locally inc bank		For banks, including overseas, max offices = 3 (G) Lending: none		For S, 10 years as authorized institution
Indonesia	None	None	New: I, joint venture (G of old B)	New: U	Acquisition of existing: 49% (G)	2 B/auxiliary office		
Japan	U	None						
Korea	U	U		Only branches of top 500 banks; unclear on S	Restrictions on acquisition of existing firms; foreign portfolio investment only for listed stocks, and ≤ 23% (h) LSO		Ceilings on foreign currency loans	
Malaysia	Deposits: U Lending ≥ RM25m only with mode 3	None		New: U	Existing: 30% (G) DLSO	U for B and ATMs of commercial banks		
PNG (o)	None	None	None					

(*continued*)

Table 2.3 (continued)

Member	Limitations on Cross-Border Supply	Limitations on Consumption Abroad	Legal Form	No. of Suppliers	Equity	Number of Operations (branches)	Value of Transactions or Assets	Other
					Limitations on Commercial Presence			
The Philippines	U	None	Single form of presence + local incorporation: DL	DL, R	Acquisition or new: 51% (G)	10 new B (1995–2000), individual max = 6	30% max foreign share of total assets	
Singapore	U	None		Deposits new: U Lending: none	Deposits: 40% LSO Lending: none	Deposits: 1 office (incl. ATM) Lending: off Premise ATM: U	Lending local currency to non-residents: DL	
Thailand	U	U	I or B	S: U B: DL	Acquisition of existing: 25% (limitations on individual ownership) DL on > 25%	Existing banks with a B before 1995: 2 additional BS (G); new Bs: DL		

Source: Compiled by the author from GATS Schedules of Commitments.
Notes: See table 2.2.

investor may exceed the true social productivity of the investment). To some extent the rent appropriation may be prevented by profit taxation or by holding competitive auctions of licenses or equity, but the benefits of competition would still not be obtained.[6]

Entry restrictions are becoming harder to justify in the face of growing evidence of the benefits of competition.[7] Why then do we observe such widespread restrictions on entry? While it is possible to construct special models of market and/or regulatory failure where entry barriers enhance welfare (Laffont 1998), there are usually more prosaic reasons for the barriers. First, restrictions generally aim to protect the incumbent suppliers from immediate competition for infant industry–type reasons, to facilitate "orderly exit," or simply due to political economy pressures. The result is protection not only of national firms but also of foreign incumbents—as in the case of foreign telecom monopolies in Hong Kong, foreign insurance companies in Malaysia, and, most strikingly, the bilateral agreements in air transport. Other instruments, such as discriminatory subsidies or taxes, could be better targeted. Monopolistic or oligopolistic rents are also sometimes seen as a means to help firms to fulfill universal service obligations through cross-subsidization. However, governments are increasingly devising means of achieving these objectives without sacrificing the benefits of competition: for example, by imposing universal services obligations on new entrants or asking for competitive bids for subsidies to serve unprofitable areas. In some cases, a form of "investment pessimism" exists, leading to the belief that promises of oligopoly rents are necessary to attract new investment. However, it is not clear why the market structure needs to be determined by policy, unless there are some initial investments the benefits of which may be appropriated by rivals. Finally, governments may seek to raise revenue (or rents for politicians and bureaucrats) by auctioning monopoly or oligopoly rights. This amounts to indirect appropriation of consumers' surplus. However, the static and dynamic inefficiencies consequent upon lack of competition would still exist.

Ideally, multilateral rules should make it difficult for governments to resort to trade restrictions to pursue objectives that are better achieved through other means. In each of the cases mentioned above, entry restric-

---

6. It is also difficult to provide an economic rationale for foreign equity restrictions. The incentive to transfer technology or otherwise to improve performance is bound to be less for foreign investors if they will only receive a fraction of the gain. It would, therefore, be optimal to allow full foreign ownership to prevent dilution of incentives and extract potential rents through the initial sale price. However, political concerns about foreign control probably account for the broad ownership restrictions in countries like Malaysia and the Philippines and in the incumbent firm in the telecom sector in Japan.

7. In Latin America, for example, countries that granted monopoly privileges to telecom operators of six to ten years to the privatized state enterprises saw connections grow at 1.5 times the rate achieved under state monopolies but only half the rate in Chile, where the government retained the right to issue competing licenses at any time.

tions are at best a second- or third-best instrument to achieve the objective in question, but they are chosen because of constraints such as the inability to raise revenue without economic or political cost. It will probably be difficult and not necessarily desirable to outlaw completely barriers to entry. However, it may be possible to create a legal presumption against such barriers by requiring that a country that imposes them demonstrate that they are necessary—in the sense that more appropriate instruments are not feasible. This idea is developed below.

2.3   Credibility through GATS Commitments

It is well known that the freedom to change one's mind can be a nuisance. The GATS offers a valuable mechanism to make credible commitments to policy. Failure to honor these commitments would create an obligation to compensate those who are deprived of benefits, making the commitment more credible than a mere announcement of liberalizing intent in the national context. Governments can bind current policy or commit themselves to implement liberalization at a future date.

Unfortunately, with some exceptions, not much was made of this opportunity. In general, countries made limited commitments, excluding many sectors and many modes (see Hoekman 1996; Adlung 2000). The larger East Asian economies did bind a certain level of access in segments of major services sectors like business, communication, finance, transport, and tourism, but few countries made commitments in sectors like distribution, education, and environmental and health services. Several countries in the region bound at less than the status quo, at least with respect to some aspects of their regimes. The Philippines, for example, did so with respect to foreign equity participation in commercial banks, binding at 51 percent when domestic law allows 60 percent. Korea also stopped short of reflecting in its GATS offer all the present and future liberalization commitments made at the Organization for Economic Cooperation and Development (OECD).[8]

2.3.1   Precommitment

One reason governments may be reluctant to liberalize immediately is a perceived need to protect the incumbent suppliers from competition—either because of infant industry–type arguments or to facilitate "orderly exit." The failure of infant industry policies in the past, and the innumerable examples of perpetual infancy, is attributable in part to the inability of a

---

8. Furthermore, under the terms of the International Monetary Fund agreement, the de facto regime with respect to foreign capital is already more liberal than the GATS offer. For instance, president Kim Dae-Jung was quoted as saying that "from now on there is no need for discrimination between indigenous and foreign capital. We are living in an era where foreign investment is more important than foreign trade" (*Financial Times,* 29 December 1997).

government to commit itself to liberalize at some future date and hence to confront incumbents with a credible deadline. One way of overcoming the credibility problem is for governments to make binding commitments under the GATS to provide market access at a precise future date.

In general, the use of the GATS as a mechanism for lending credibility to future liberalization programs has been disappointing. The telecommunications sector is an exception, however. In this sector, several East Asian governments are among those who have taken advantage of the GATS to strike a balance between their reluctance to unleash competition immediately on protected national suppliers and their desire not to be held hostage in perpetuity either to the weakness of domestic industry or to pressure from vested interests (table 2.1). Singapore and Korea have bound themselves to introduce competition at precise future dates. Indonesia and Thailand are among the countries that have made weaker commitments. Greater use needs to be made of the GATS in this respect, for there is growing evidence that reform programs that are believed are more likely to succeed.

2.3.2 Grandfather Provisions

A particularly perverse use of commitment from an economic point of view is the inclusion of grandfather provisions in the financial services schedules of some countries under negotiating pressure. The issue arose because domestic law, pertaining to foreign ownership, branching, and other rights, had changed since foreign firms first established commercial presence. For instance, Malaysia began to implement its indigenization policy after several fully foreign-owned firms were already operating in its market. The home countries of the firms were unwilling to see a dilution of what they saw as "acquired rights," whereas Malaysia was unwilling to grant the same rights to new entrants. The negotiated solution was for Malaysia to commit to preserve the rights of incumbents while offering inferior terms to new investors (see table 2.4). Where differences in ownership and legal form affect firm performance, new entrants have been placed at a competitive disadvantage. Thus, the triumph of moral over economic reasoning has meant that the GATS was used to make markets less contestable.

2.4 Regulatory Disciplines

Most of the key regulatory challenges must necessarily be addressed at the national level, and, even more than in the case of other policies, there are limits to what should and can be addressed at the multilateral level. Still, there are likely to be benefits from strengthened multilateral disciplines on domestic regulations. First of all, such disciplines are needed to enable exporters to address regulatory barriers to their exports in foreign markets. For instance, unless disciplines are developed to deal with licensing and

Table 2.4   **Grandfather Provisions in GATS Schedules on Banking and Insurance Services in East Asia**

Country	Provision
	Foreign Equity-Related
Indonesia	Banking and insurance: Share ownership of foreign services suppliers is bound at the prevailing laws and regulations. The conditions of ownership and the percentage share of ownership as stipulated in the respective shareholder agreement establishing the existing individual joint venture shall be respected. No transfer of ownership shall take place without the consent of all parties in the joint venture concerned.
Malaysia	Banking: Entry is limited to equity participation by foreign banks in Malaysian-owned or controlled commercial and merchant banks with aggregate foreign shareholding not to exceed 30 per cent, but the thirteen wholly-foreign owned commercial banks are permitted to remain wholly-owned by their existing shareholders.
	Insurance: New entry is limited to equity participation by foreign insurance companies in locally incorporated insurance companies with aggregate foreign shareholding not to exceed 30%. Foreign shareholding not exceeding 51% is also permitted when (i) existing branches of foreign insurance companies are locally incorporated, which they are required to be by 30 June 1998, and (ii) for the existing foreign shareholders of locally incorporated insurance companies which were the original owners of these companies.
The Philippines	Insurance and banking: New investments of up to 51% of the voting stock, but existing investments of foreign banks will be maintained at their existing levels.
	Legal Form-Related
Hong Kong	Banking: The condition that branches of foreign banks are allowed to maintain offices in one main building and no more than two additional offices in separate buildings, does not apply to banks incorporated outside HKSAR licensed before May 1978 in respect of fully licensed banks and before April 1990 in respect of restricted licence banks.
Indonesia	Banking: Existing branches of foreign banks are exempted from the requirement imposed on new entrants to be in the form of locally incorporated joint venture banks.
Malaysia	Insurance: Branching is only permitted for direct insurance companies with aggregate foreign shareholding of less than 50 per cent but companies are permitted to maintain their existing network of branches. (See also foreign equity-related provision above.)
Thailand	Banking: While the establishment of new branches is subject to discretionary licensing, existing foreign banks which already had the first branch office in Thailand prior to July 1995 will each be permitted to open no more than two additional branches.
	General
The Philippines	Insurance: Limitations in market access listed in the specific insurance sub-sectors do not apply to existing wholly or majority foreign-owned authorized insurance/reinsurance companies as of the entry into force of the WTO Financial Services agreement.

Source: Compiled by the author from GATS Schedules of Commitments.

qualification requirements, market access commitments in areas like financial and professional services will have only notional value. Furthermore, the development of such disciplines can play a significant role in promoting and consolidating domestic regulatory reform. The telecommunications negotiations, which led to the early institution of independent regulators in many countries, provide an example of this possibility. Finally, there is a class of problems that must necessarily be addressed at the multilateral level: the problem of international cartels in sectors like maritime transport.

2.4.1 The Case for a Horizontal Approach

One of the ironies of the GATS is that among its weakest general provisions are those dealing with domestic regulations.[9] The reason is not difficult to see: it is extremely difficult to develop effective multilateral disciplines in this area without seeming to encroach upon national sovereignty and unduly limiting regulatory freedom. Nevertheless, it is desirable and feasible to develop horizontal disciplines for domestic regulations (see also Feketekuty 2000). Such a generic approach is to be preferred to a purely sectoral approach for at least three reasons: it economizes on negotiating effort, leads to the creation of disciplines for all services sectors rather than only the politically important ones, and reduces the likelihood of negotiations' being captured by sectoral interest groups. It is now widely recognized that the most dramatic progress in the European Union single-market program came from willingness to take certain broad cross-sectoral initiatives. In the World Trade Organization (WTO) context, the experience of the accountancy negotiations shows the propensity for single-sectoral negotiations on domestic regulations to produce a weak outcome: although a valuable "necessity test" was instituted, the elaboration of disciplines on measures such as qualification requirements was disappointing.

Even if a horizontal approach is desirable, is it feasible? The diversity of services sectors, and the difficulty in making certain policy-relevant generalizations, would seem to favor a sector-specific approach. However, even though services sectors differ greatly, they have much in common in terms of the underlying economic and social reasons for regulations. Focusing on these reasons provides the basis for the creation of meaningful horizontal disciplines. The economic case for regulation in all services sectors arises essentially from market failure attributable primarily to three kinds of problems: natural monopoly or oligopoly, asymmetric information, and externalities (see table 2.5).

---

9. The relevant provision (Article VI) requires members not to apply licensing and qualification requirements and technical standards so as to undermine market access commitments in a manner that "could not reasonably have been expected" when the specific commitments were made. This provision may provide a defense against new restrictions but could be interpreted to mean that old regulations whose persistence could reasonably have been expected cannot be challenged.

Table 2.5   A Multilateral Approach to Dealing with Domestic Regulations

Market Failures	Services sectors	Multilateral Approach
Monopoly/oligopoly	Network services: transport (terminals and infrastructure), environmental services (sewage) and energy services (distribution networks)	Generalize key disciplines in telecom reference paper to ensure cost-based access to essential facilities, be they roads, rail tracks, terminals, sewers or pipelines
Asymmetric information	Intermediation and knowledge based services: financial services, professional services, etc.	Nondiscrimination and generalization of the "necessity" test. Use the test to create a presumption in favor of economically efficient choice of policy in remedying market failure.
Externalities	Transport, tourism, etc.	
Social objectives Universal service	Transport, telecommunications, financial, education, health	

2.4.2   Dealing with Domestic Monopolies

Market failure due to natural monopoly or oligopoly may create trade problems because incumbents can impede access to markets in the absence of appropriate regulation. Because of its direct impact on trade, this is the only form of market failure that may need to be addressed directly by multilateral disciplines. The relevant GATS provision—Article VIII, dealing with monopolies—is limited in scope. As a consequence, in the context of the telecom negotiations, the reference paper with its competition principles was developed in order to ensure that monopolistic suppliers would not undermine market access commitments (Tuthill 1997). It might be possible to generalize these principles to a variety of other network services, including transport (terminals and infrastructure) and energy services (distribution networks), by ensuring that any major supplier of essential facilities provides access to all suppliers, national and foreign, at cost-based rates.[10]

2.4.3   Other Sources of Domestic Market Failure

In all other cases of market failure, multilateral disciplines do not need to address the problem *per se*, but rather to ensure that domestic measures to deal with the problem do not serve unduly to restrict trade. (The same is true for measures designed to achieve social objectives.) Such trade-restrictive effects can arise from a variety of technical standards, prudential regulations, and qualification requirements in professional, financial, and numerous other services, as well as from the granting of monopoly rights to complement universal service obligations in services like transport and

10. Even though it would be extremely difficult to determine what cost-based rates are, the provision should at least make it possible to challenge the more egregious departures.

telecommunications. The trade-inhibiting effect of this entire class of regulations is best disciplined by complementing the national treatment obligation with a generalization of the so-called "necessity" test. This test leaves governments free to deal with economic and social problems provided that any measures taken are not more trade restrictive than necessary to achieve the relevant objective. The test is already applied to technical barriers to trade in goods and is part of the recently established "pilot" disciplines for the accountancy sector. It might make sense to go beyond the GATT precedent and to use the test to create a presumption in favor of economically efficient choice of policy in remedying market failure and in pursuing noneconomic objectives (Mattoo and Subramanian 1998). For instance, in the case of professionals like doctors, a requirement to requalify would be judged unnecessary, because the basic problem, inadequate information about whether they possess the required skills, could be remedied by a less burdensome test of competence.

The necessity test is generally seen as an additional discipline on nondiscriminatory measures. It has not been recognized that without some such test it would be difficult to apply even the fundamental disciplines of national treatment (Article XVII) and MFN (Article II), for it would be impossible to determine if a measure is in effect nondiscriminatory.[11] Both articles prohibit discrimination between *like* services and *like* service suppliers, but likeness itself is not easy to establish. If a doctor is a doctor, a regulation that imposed any additional burden on a doctor trained in Country A (abroad) than on a doctor trained in Country B (at home) would violate Article II (Article XVII). If a doctor trained in one country is deemed not to be "like" a doctor trained in another country, then the disciplines contained in the articles would simply not apply. The former interpretation may be unduly stringent and politically unsustainable; the latter is unduly permissive and would open the door to all manner of regulatory protection. The necessity test would seem to be the perfect solution. Countries are not prevented from imposing additional qualification and training requirements, but these should not be more burdensome than necessary, in the sense described above.

To conclude, the arguments in this section must not be taken to mean that there is no need for sector-specific work. Such work is necessary and should involve consumers, industry, and regulators, to help determine how best to deal with asymmetric information and differences in standards between countries in particular sectors. However, the application of a necessity test is necessary today because harmonization and mutual recognition are not meaningful alternatives—even though they can play a role at the regional or plurilateral level. The pessimism with regard to harmonization is based

---

11. There is no explicit mention of the necessity test in the national treatment and MFN provisions.

on the absence of widely accepted international standards in services. Where such standards exist, as in banking or maritime transport, meeting them is seen as a first step toward acceptability rather than as a sufficient condition for market access. With regard to mutual recognition agreements (MRAs), it would seem that even in strongly integrationist Europe, despite a significant level of prior harmonization, the effect of MRAs may have been limited by the unwillingness of host country regulators to concede complete control (Nicolaidis and Trachtman 2000).

2.4.4 Competition Policy: The International Dimension

The procompetitive rules developed for the basic telecommunications sector were designed to protect the rights of foreign suppliers. Is there a need for broader competition policy disciplines in the GATS to protect the interests of consumers more directly? Article IX of the GATS deals with "certain business practices of service suppliers, other than those falling under Article VIII, [that] may restrain competition and thereby restrict trade in services." However, its disciplines are weak and require little more than consultation and information sharing.

There may be a need to strengthen these disciplines. Consider one particularly important example. Maritime transport costs have a profound influence on international trade. Their persistent high level has been attributed not only to restrictive trade policies but also to private anticompetitive practices such as rate-binding agreements, primarily but not exclusively of the maritime conferences (Hummels 1999). The high incidence of such agreements is due to the fact that the United States, the European Union, and many other countries exempt shipping conferences from antitrust regulation on the grounds that they provide price stability and limit uncertainty regarding available tonnage. In the case of routes serving the United States, the exemption from antitrust law is compounded by the Federal Maritime Commission's (FMC) role in helping to police price-fixing arrangements.[12]

A recent econometric analysis suggested that although public restrictions adversely affect maritime transport costs, private anticompetitive practices have an even stronger impact.[13] Thus, it would seem that even though there has been an erosion in the power of conferences due to the entrance

12. The 1984 U.S. Shipping Act required all ocean carriers to file their rates with the FMC and publish their rate and schedule information. Secret discounting on filed rates was until recently considered illegal. The FMC was authorized to ensure, through the imposition of fines, that the filed rates were actually charged. The rationale for these measures was ostensibly to protect small shippers from being disadvantaged by their inability to extract discounts from shipping companies. Certain other aspects of U.S. maritime policy are discussed in section 2.5.

13. Fink, Mattoo, and Neagu (2000) estimate that the breakup of conference and other price-setting agreements leads to a more dramatic reduction in transport prices (38 percent) than restrictive cargo allocation policies (11 percent). The estimated potential savings from the elimination of both could be as high as one billion U.S. dollars on goods carried to the United States alone. Francois et al. (1996) and Francois and Wooton (1999) also find significant costs of public and private restrictions on trade in maritime transport services.

in the market of efficient outsider shipping companies and of a certain tightening in the law, collusive arrangements have not disappeared.[14] As recently as May 2000, the European Commission imposed fines on shipping lines serving the East Asian and U.S. routes and on those serving the transatlantic route for collusive pricing that went beyond the scope of the exemptions that had been granted.

What are the implications for policy? The negotiations on maritime transport were the only post–Uruguay Round services negotiations that completely failed. This failure implied an unfortunate loss of political momentum for reform of domestic policies and, less obviously, a lost opportunity to develop procompetitive rules. To some extent, an effort was made to develop rules that would ensure nondiscriminatory access to port services.[15] However, these rules, concerned primarily with ensuring market access, did little to protect consumers from the anticompetitive practices of international cartels. An international initiative is needed, because these practices cannot be adequately addressed only through national competition policy, given the weak enforcement capacity of small states. A further reason for developing a first best international response to these practices is to prevent recourse to an inferior national response: recall that the costly cargo-sharing schemes imposed by many developing countries were primarily a reaction to the perceived power of conferences.

One approach would be to deal with the problem by creating sector-specific competition rules, as has been attempted in basic telecommunications services under the GATS. However, if there is sufficient evidence that anticompetitive practices also affect other services sectors, such as air transport and communications, there may a need to strengthen the general GATS disciplines, that is, Article IX, dealing with anticompetitive business practices.[16] This would serve to reassure small countries in particular that the gains from liberalization will not be eroded by collusive pricing.

2.5 The Most-Favored Nation Principle

The GATS and its MFN obligation came into effect before WTO members were willing to eliminate completely discriminatory measures in services trade. The agreement therefore had to strike a difficult balance between creating meaningful multilateral disciplines and accommodating discrimi-

---

14. A recent change in U.S. regulation regarding international shipping, notably the Ocean Shipping Reform Act (OSRA) of 1998, allows for the confidentiality of key terms (prices are included in this category) in contracts between shippers and carriers but preserves the antitrust immunity of the rate-setting conference system.

15. In some respects, the approach to port services, which can be seen as "essential facilities" often controlled by "major" or monopoly suppliers, was analogous to the approach to basic telecommunications networks established in the regulatory principles referred to above.

16. It is also conceivable that these issues could be addressed as part of broader competition policy disciplines in the WTO.

natory trade practices. The challenge to multilateral disciplines posed by the explicit departures from the MFN obligation, such as the exceptions for regional integration agreements and the MFN exemptions listed by members, are widely recognized. However, the difficulties arising from less visible, implicit discrimination have not been adequately appreciated.

2.5.1   The Scope and Significance of Explicit Departures from Most-Favored Nation

Consider the explicit exemptions first. Around 380 MFN exemptions have been listed by some seventy members, with many members listing several exemptions in the same sector (see Mattoo 2000b, table 1). Nearly two-thirds of the exempted measures are to be found in communication services and in transport services. One reason specified for these measures is the existence of sector-specific preferential regional agreements, or other bilateral or plurilateral agreements. For instance, in audiovisual services, more than half of the exemptions mention promotion of common (regional) culture as a motive for limiting access to joint programs to finance and diffuse audiovisual works; and in maritime transport, nearly half the exemptions are by developing countries for measures implementing the provisions of the United Nations Convention on a Code of Conduct for Liner Conferences.[17] The other reason cited for exemptions is a unilaterally imposed reciprocity condition, which specifies that a member is willing to guarantee access to its market only to those members who provide it with access to their markets. These are particularly significant in air transport services and financial services.[18]

In cases in which the exemptions coexist with specific commitments (as in financial services)[19] or legitimize preferences that do not greatly affect the pattern of trade (as in cross-border supply of land transport services), there is probably not much cause for concern. Most-favored nation exemptions would seem to matter most, and be most difficult to eliminate, in sectors like audiovisual services and maritime transport, where few specific commitments have been made and discriminatory practices seem to be empirically important.

Perhaps even more important than the MFN exemptions that have been listed are those that did not need to be. The Annex on Air Transport specif-

17. These provisions, in principle, divide 80 percent of the liner trade on a traffic route between the shipping companies of the two states at each end, leaving only 20 percent for shipping companies of other nationalities. Full implementation of this rule is apparently rare, and third-country ships usually have access to a larger share of the market. Many members chose to maintain MFN exemptions despite the suspension of the obligation for the sector.

18. The exemptions listed for air transport services pertain to the services falling within the scope of the GATS, that is, repair and maintenance, selling and marketing of air transport services, and computer reservation system (CRS) services.

19. Market access guaranteed under specific commitments must be extended on a nondiscriminatory basis to all trading partners, even if an MFN exemption has been sought. The MFN exemption can provide legal cover only for better treatment for some trading partners than provided for in the specific commitments.

ically excludes the complex network of bilateral agreements on air traffic rights from GATS rules.[20] Thus, a sector that is in urgent need of liberalization remains fragmented into cozy duopolies, and prospects for progress at the multilateral level are dim. One source of hope is the increasing agreement among WTO members to push for the liberalization of a cluster of services related to tourism. Excluding air transport from this initiative would be like leaving the Prince of Denmark out of a certain play.

The U.S. exemption in maritime transport was more like Banquo's ghost: it was not explicitly listed—because the MFN obligation was suspended for the sector—but had a completely disruptive effect on the negotiations.[21] The United States did not believe that the quality of its trading partners' market-opening commitments justified giving up its right to take retaliatory action against foreign restrictive practices. One way of making progress in the current round is to bundle transport negotiations together and focus on the liberalization of multimodal transport, a central concern of U.S. industry. Also, the development of competition disciplines, along the lines suggested above, would help to address the anticompetitive practices that the United States believes impede access to foreign markets.

The other main departure from MFN is the provision (Article V) for economic integration agreements, which allows any subset of WTO members to liberalize trade in services among themselves under certain conditions. This provision is broadly modeled on the corresponding provision in the GATT. The agreements that have been notified so far include those establishing the North American Free Trade Agreement (NAFTA), the European Communities, and their member states; their agreements with the Slovak Republic, Hungary, Poland, the Czech Republic, Romania, Norway, Iceland, Liechtenstein, and Bulgaria; and agreements between Canada and Chile and between Australia and New Zealand.[22]

There is an important question of whether regional and other preferential agreements are capable of achieving deeper and more effective liberalization of services, and whether they would contribute to or impede further multilateral liberalization. A discussion of the role of these agreements and the possible reform of the relevant GATS rules is beyond the scope of this paper (see Stephenson 2000). The paper concentrates on

20. International air transport services are for the most part governed by arrangements negotiated under the Chicago Convention (i.e., the International Air Services Transit Agreement, done at Chicago, 7 December 1944).

21. The original U.S. MFN exemption for maritime transport services reserved the "right to investigate and take action against foreign carriers to address adverse or unfavourable actions affecting United States shipping or United States carriers in United States oceanborne commerce and the cross trades between foreign ports."

22. A related exception from the MFN rule, for the movement of natural persons, is permitted by Article V bis of the GATS. This allows countries to take part in agreements that establish full integration of labor markets. The only such agreement notified so far is the one involving Denmark, Finland, Iceland, Norway, and Sweden.

the narrower issue of the economic and legal implications of discrimination in services, where the instruments of protection are domestic regulations and quotas.

2.5.2 Discrimination through Domestic Regulations and Quotas: Economic Considerations

The consequences of discrimination between trading partners through taxation (or duties) are well understood. Does discrimination through domestic regulations and quotas raise new analytical issues from the economic and legal point of view?

When tariffs are the instruments of protection, the costs of trade diversion for the importing country may be an important deterrent to preferential liberalization agreements. Despite the increase in consumers' surplus from any liberalization, governments may nevertheless be averse to such agreements because the displacement of high-tariff imports from third countries by low or no-tariff imports from preferential sources implies lost revenue. The same reasoning also applies to other regulations that imply a transfer from foreign suppliers to domestic interest groups. However, the situation is different when the protectionist instrument is a regulatory barrier that imposes a cost on the exporter without yielding a corresponding revenue for the importing government or other interest group. There is then no cost to the country granting preferential access because there is no revenue to lose. The same is true in the case of quotas when the rents were either dissipated or appropriated by foreign suppliers. Therefore, in these cases, preferential liberalization is necessarily welfare enhancing for the importing country—as well as for the exporting country that obtains improved access.[23]

However, if third countries supply the market in question, they lose because prices decline due to increased sales from the preferred source. The impact on global welfare depends on the nature of regulatory measure. If it generates no revenues or rents, then global welfare will increase. In effect, exempting some suppliers from the measure reduces their costs and leads to a reduction in price in the importing country. The gain to consumers from any decline in price is necessarily greater than the loss to a subset of suppliers. This suggests that multilateral rules should take a more tolerant view of preferential arrangements like recognition agreements that help eliminate wasteful duplication (e.g., of training) and are therefore global welfare-enhancing. However, we must not lose sight of the fact that nonpreferential liberalization would enhance welfare even more because the service would be supplied by the most efficient locations.

---

23. This reasoning does not take account of the fact that there may be greater spillover benefits (e.g., relating to technology) arising from trade with certain partners than with others.

2.5.3   Legal considerations

Recognition Agreements

Recognition agreements are like sector-specific preferential arrangements and can have similar trade-creating trade-diverting effects. Their result may well be to create trade according to patterns of mutual trust rather than the pattern of comparative advantage. The interpretation of the GATS provision on recognition (Article VII) is, therefore, likely to be of considerable importance. The provision attempts to strike a difficult balance. On the one hand, it is permissive and allows a member at any point of time to recognize the standards of one or more members and not of others. On the other hand, it seeks to ensure that this freedom is not abused by prohibiting the use of recognition as a means of discrimination and requiring a member who enters into a recognition agreement (RA) to afford adequate opportunity to other members to negotiate their accession to such an agreement or to negotiate comparable ones. In this respect, Article VII mandates an openness vis-à-vis third countries in a way that Article V, dealing with economic integration agreements, does not.[24]

How can it be established whether acceptance of some standards and not others is discriminatory? The approach discussed with regard to domestic regulations is also applicable here. Making distinctions between services and service suppliers in the pursuit of certain domestic policy objectives, such as to ensure the quality of professional services, financial stability, and competitive market conditions, is economically sensible. It would, therefore, be desirable to allow members the legal freedom to pursue such objectives but to discipline the exercise of such freedom by ensuring that the choice and level of instruments is not more burdensome than necessary—with economic efficiency considerations playing a role in this assessment. The text of Article VII does not contain an explicit necessity test, but, as in the case of the MFN and national treatment obligations, it is difficult to see how the provision can be given meaningful content without the inclusion of such a test.

Nondiscriminatory Allocation of Quotas

One central legal issue in the GATS, which has received surprisingly little attention, is how quotas are to be allocated in a manner consistent with the

---

24. Article V on integration agreements does not explicitly preclude RAs, and several countries (such as Australia and New Zealand) have chosen to notify their RAs under this provision. It would seem desirable to establish that Article VII, with its desirable nondiscriminatory and open-ended nature, overrides Article V of the GATS as far as RAs are concerned. This interpretation would help to generalize the liberalizing impact of RAs, for although an RA amounts to an acceptance of likeness vis-à-vis suppliers from a particular country, it also defines the appropriate standard of treatment vis-à-vis suppliers from other countries.

nondiscrimination obligation. In the past, this was not a major issue because commitments reflected the status quo, and the quotas, particularly with regard to service suppliers, were descriptions of the existing market structure.[25] In the future, however, as genuine liberalizing commitments are made, the nondiscriminatory allocation of quotas is bound to be an important issue. For instance, it has been reported that China, as part of its accession negotiations, promised the European Union that its firms would be granted a specific number of licenses in the insurance sector. How is this assurance to be reconciled with the MFN obligation?

The goods precedent offers limited guidance. General Agreement on Tariffs and Trade Article XIII, on the "nondiscriminatory administration of quantitative restrictions," requires aiming at a distribution of trade approximating the shares that countries might be expected to have in the absence of such restrictions or supplied during a previous representative period. In the services context, the requirement to replicate historical shares may have no relevance if there was no previous foreign presence, or it may perpetuate historical discrimination if previous quotas were allocated to favored suppliers.[26]

More appropriate candidates for a nondiscriminatory allocation of quotas would seem to be first-come, first-served rule (e.g., a large number of work permits are being issued) or a system of auctions to the highest bidder (e.g., a few telecom licenses are being issued). Neither rule would necessary lead to distributions that "might be expected to obtain in the absence of such restrictions."[27] It would seem, therefore, that the rules for ensuring nondiscriminatory allocation of quotas under GATS would need to look beyond the GATT precedent. It is possible that a less elaborate variant of the disciplines in the Agreement on Government Procurement, designed to ensure competitive tendering on a nondiscriminatory basis, will need to be considered.

2.6   Reviving Reciprocity?

Reciprocity has been a central principle governing GATT/WTO negotiations: one country reduces its level of protection in return for a reciprocal

25. Thus, when Bangladesh committed to "four licenses issued" in cellular telephony, the ambiguity in the choice of tense was not an accident: the licenses in question had already been issued.

26. In the Bananas Case, the European Union's method of allocating import licenses for bananas from certain sources was found to be inconsistent with Article II because it reallocated quotas and quota rents away from the importers who traditionally imported from these sources (see paragraphs 7.350–7.353 of the Panel Report). In a sense, the panel's reasoning followed the logic of GATT Article XIII.

27. It is obvious that first-come, first-served favors the proximate. Auctions would give the relatively efficient producers larger shares than they would have obtained in the absence of quotas (when quotas are set at below unrestricted trade levels). Jackson (1997, p. 140), however, notes that first-come, first-served and auctions would seem to fulfill the MFN obligation, and refers to the Article XIII reliance on historical patterns as a "quasi" MFN principle.

reduction by its trading partner.[28] While reciprocity-based negotiations are widely credited with the substantial reduction in levels of protection achieved in goods trade, it is surprising that the limited application of the principle has not conversely been seen as the reason for the disappointing results in services trade.

The GATS had a deliberately symmetric structure. In principle, there was scope for developed and developing countries to exploit their modal comparative advantage: improved access for capital from developed countries being exchanged for improved temporary access for individual service providers from developing countries. In practice, there was little political will to improve access for foreign individuals (except for the limited class of skilled intracorporate transferees), and a trade-off between modes of delivery simply did not take place. Moreover, even the negotiating links across services sectors and between services and goods sectors do not seem to have been particularly fruitful. Thus, the GATS commitments reflect for the most part the existing levels of unilaterally determined policy rather than liberalization achieved through a reciprocal exchange of "concessions."

It might well be that reciprocity cannot and will not play a major role in services trade. Services liberalization could for the most part be undertaken unilaterally, and the GATS would be important only in preventing its reversal—that is, in its credibility role (see Hoekman and Messerlin 2000). Indeed, for countries that are either determined to liberalize or determined to protect, negotiations are not important. However, for countries in the middle ground, who are open to reform but whose ability to implement reform is constrained by domestic opposition, multilateral negotiations can be useful. Many developing countries are today in this situation, and a wider application of the principle of reciprocity may deliver greater liberalization and more balanced outcomes.

2.6.1 Facilitating Reciprocity across Modes

A collective commitment to the use of appropriately designed formulas offers the best chance of linking different modes of delivery.[29] Such formulas can also help overcome concerns about free-riding that arise in an MFN-based system. But is it technically feasible to link concessions across

28. This emphasis on achieving a "balance of (liberalizing) concessions" has led to the perception of WTO negotiations as a mercantilist process driven by political forces that nevertheless leads to the desirable outcome of reduced levels of protection. In effect, reciprocity serves to mobilize the support of export interests to counterbalance the protectionist interests of import-competing firms and workers. In an important recent paper, Bagwell and Staiger (1999) show that reciprocity can be seen as neutralizing the adverse terms of trade effects associated with unilateral reductions in protection, and it therefore leads to greater liberalization.

29. Developing countries have resisted this option, preferring the use of a request-and-offer approach. Their reluctance stems from defensive considerations and a belief that they would be obliged to concede excessively high levels of openness if a formula approach were adopted.

modes (see Sapir 1998 and Thompson 2000)? One simple option is to take advantage of the current political pressure for accelerated liberalization in selected sectors, such as environmental services. This approach could be accepted on the condition that there was no gerrymandering: that is, all countries would liberalize access in all modes, including the movement of individuals. Environmentalists and environmental service exporters could then be relied on to counter the opposition of employees and individual suppliers in the domestic environmental industry. Furthermore, with severe shortages of skilled labor in the United States and Europe and the powerful constituency of high-technology companies lobbying for relaxation of visa limits, the prospects for serious intermodal trade-offs—such as obtaining labor movement in return for allowing greater commercial presence for foreign service providers—are now greater.

An alternative way of creating a link between modes is by requiring each country to provide increased "foreign labor content entitlements" to its domestic firms in relation to the country's increased exports of services (Mattoo and Olarreaga 2000a). Entitlements would be global rather than bilateral, and the extent and pattern of use would be determined by sound economic considerations of modal comparative advantage. Some of the social and political difficulties could be overcome by clarifying that liberalization is only with respect to temporary movement of service suppliers and does not imply migration. Establishing clear links between increased exports and increased foreign labor content entitlements may also help make the political case. The presence of foreign workers would be seen as a direct consequence of increased opportunities for export abroad, and also as contributing to the increased competitiveness that makes it possible to exploit these opportunities.

2.6.2   Reciprocity within Modes across Sectors

It would be wrong to suggest that reciprocity must necessarily take an intermodal form. There may, for instance, be scope for cross-sectoral reciprocity in the same mode. Trade in electronically delivered products—falling within the scope of cross-border supply—is of growing importance and offers an increasingly viable alternative to the movement of individuals. If the United States can supply financial and audiovisual services to the Philippines electronically, the Philippines in turn can supply software development and data processing services to the United States. Fortunately, most electronic commerce is already free of barriers, and so the main concern should be preventing the introduction of new barriers if they ever become technically feasible. Members of the WTO have so far focused on prohibiting the imposition of customs duties on electronically delivered products. Because the bulk of such commerce concerns services, open trading conditions are more effectively secured through deeper and wider commitments under the GATS on cross-border trade regarding market access (which would preclude quantitative restrictions) and national treatment (which would preclude all forms of discriminatory taxation).[30] One possible formula would be for all members to agree that no restrictions would be imposed on cross-border delivery, either of all services or of a bundle whose composition could be negotiated.

2.6.3   Remedying the Hold-Back Problem through a Credit Rule

One undesirable aspect of an emphasis on reciprocity is that it creates the temptation to hold back from unilateral liberalization. This is why most economists view reciprocity with suspicion This hold-back problem can be overcome, however, by rules which create an *ex ante* assurance (at the end of a round of negotiations) that credit would be given in future rounds of negotiations for unilateral liberalization undertaken between rounds. The impulse to liberalize unilaterally then need not be inhibited by the fear of loss of negotiating coinage. The proposed rule is different from the demands for credit that are typically made at the *beginning* of a new round of negotiations. The acceptance of such demands would have only a distributional effect, favoring those who have already undertaken liberalization, and the granting of such credit relies on the unlikely generosity of those who have not liberalized. The proposed *ex ante* assurance of credit rule has three virtues:[31] it would help induce or enhance liberalization in some countries between negotiating rounds; more striking, it could also lead to deeper levels of multilateral liberalization and force other countries to go further than in the absence of a rule; and, most important, such a rule does not rely on altruism to be generally acceptable.

Article XIX:3 of the GATS requires that in each future round "modalities shall be established" for the treatment of liberalization undertaken autonomously by members since previous negotiations. In principle, this is precisely the type of *ex ante* assurance of credit that would be desirable. However, the nebulousness of the provision and the postponement of the establishment of modalities suggest that in practice the provision may provide little more than a basis for *ex post* demands for credit. One way of giving the rule operational content is by establishing that any agreed liberalizing formula would be applied not to current actual levels of protection but to the levels bound in the previous round of negotiations.[32]

---

30. There is considerable scope for an improvement in commitments. For instance, in software implementation and data processing, of the total WTO membership of over 130, only 56 and 54 members, respectively, have made commitments; only around half of these commitments guarantee unrestricted market access, and a similar proportion guarantees unqualified national treatment. It is particularly striking that in the core banking services in which around 75 WTO members have made commitments, about one-third of the developing countries guarantee unrestricted cross-border supply, whereas only one out of the ten developed countries does so.

31. The alternative rules are discussed more fully in Mattoo and Olarreaga (2000b).

32. This suggestion was in fact contained in a proposal from Brazil submitted just before the Seattle Ministerial.

2.7 Conclusions

"Reveal and bind all trade-restricting measures." "Make national treatment a general obligation." It would be tempting to make such clear and powerful proposals, but it would not be realistic or useful. The GATS is here to stay in its present form, and radical reform will not occur in this round of negotiations—nor, probably, in the next. Those who think that this is unduly conservative need only take a closer look at the negotiations in the Working Party on GATS rules. The results of five years' work on subjects such as safeguards, subsidies, and government procurement are no more tangible than the emperor's new clothes. In defense of the GATS, however, it must be recognized that it took about a half century to reduce goods barriers, and there is still a distance to go. Additionally, of course, agriculture was exempted from the GATT negotiations until the Uruguay Round and continues to pose special problems. Thus, a fairly long time horizon is probably needed to achieve meaningful liberalization in services because of the difficulty in overcoming resistance by deep-rooted vested interest in many countries.

It seemed more constructive, therefore, to take a close look at the existing provisions of the agreement and make precise proposals on how they can be improved. The main conclusions are the following:

- Wasteful regulations and entry restrictions pervade services trade. Unlike the GATT, the GATS has created no hierarchy of instruments of protection, although the analysis here suggests that the ranking of instruments in the case of both goods and services is similar. Although it may not yet be politically feasible to impose the same hierarchy as in goods, an attempt should nevertheless be made to create a legal presumption in favor of instruments (such as fiscal measures) that provide protection more efficiently.
- Greater advantage must be taken of the valuable opportunity offered by the GATS to lend credibility to reform programs by committing to maintain current levels of openness or by precommitting to greater levels of future openness.
- Multilateral rules on domestic regulations can play an important role in promoting and consolidating domestic regulatory reform, even when they are primarily designed to prevent the erosion of market access commitments. It would be desirable to generalize the application of procompetitive principles developed for basic telecommunications to other network-based services sectors and the application of the "necessity test" instituted for accountancy services to regulatory instruments in all sectors.
- Anticompetitive practices could be important in sectors like maritime, air transport, and communication services. Because these practices

cannot be adequately addressed through national competition policy, given the weak enforcement capacity of small states, GATS rules in this area must be strengthened.

- Explicit departures from the MFN rule matter most in sectors like maritime transport, audiovisual services, and air transport services—which have been excluded from GATS disciplines. Progress will not be easy, but bundling sectoral negotiations together (e.g., in transport) may help. It is also necessary to develop rules to ensure the nondiscriminatory allocation of quotas and to maintain the desirable openness of the GATS provision covering mutual recognition agreements.
- If the GATS is to advance the process of services liberalization beyond levels undertaken independently and lead to more balanced outcomes from the developing country point of view, then reciprocity must play a greater role in negotiations. This may be facilitated by devising negotiating formulas that establish credible links across sectors (both goods and services) and across modes of delivery. To overcome a possible hold-back problem, it is necessary to provide credible *ex ante* assurance of negotiating credit for unilateral liberalization.

Finally, three sets of issues have been neglected by this paper: How can the provisions of the GATS and the schedules of commitments be made clearer and more accessible? What rules should be developed for safeguards, subsidies, and government procurement? What form do preferential agreements in services take, and how can GATS rules for such agreements be improved? Some work has been done in each of these areas, but there is need for much more research (see Sauvé and Stern 2000).

References

Adlung, Rudolf. 2000. Services trade liberalization from developed and developing country perspectives. In *GATS 2000: New directions in services trade liberalization*, ed. Pierre Sauvé and Robert M. Stern, 112–31. Washington, D.C.: Brookings Institution.

Deardoff, Alan. 1994. Market access. In *New world trading system: Readings*, 57–63. Paris: Organization for Economic Cooperation and Development.

———. 2000. Tariffication in services. Paper presented at *Conference on Issues and Options for Multilateral, Regional, and Bilateral Policies of the United States and Japan*. Universisty of Michigan, Ann Arbor, MI, 5–6 Novermber. Available at [http://www.Fordschool.umich.edu/rsie/conferences.html].

Feketekuty, Geza. 2000. Assessing and improving the architecture of GATS. In *GATS 2000: New directions in services trade liberalization*, Pierre Sauvé and Robert M. Stern, 85–111. Washington, D.C.: Brookings Institution.

Fink, Carsten, Aadity Mattoo, and Cristina Neagu. 2000. Trade in international maritime services: How much does policy matter? Washington, D.C.: World Bank. Mimeograph.

Francois, Joseph, Hugh Arce, Kenneth Reinert, and Joseph Flynn. 1996. Commercial policy and the domestic carrying trade: A general equilibrium assessment of the Jones Act. *Canadian Journal of Economics* 29 (1): 181–98.

Francois, Joseph, and Ian Wooton. 1999. Trade in international transport service: The role of competition. *Center for Economic Policy Research Discussion Paper no. 2377*. London: Center for Economic Policy Research.

Hindley, Brian. 1988. Service sector protection: Consideration for developing countries. *World Bank Economic Review* 2:205–24.

Hindley, Brian, and Alasdair Smith. 1984. Comparative advantage and trade in services. *World Economy* 7:369–90.

Hoekman, Bernard. 1996. Assessing the general agreement on trade in services. In *The Uruguay Round and the developing countries*, ed. W. Martin and L. A. Winters, 88–124. Cambridge: Cambridge University Press.

Hoekman, Bernard M., and Patrick A. Messerlin. 2000. Liberalizing trade in services: Reciprocal negotiations and regulatory reform. In *GATS 2000: New directions in services trade liberalization*, ed. Pierre Suavé and Robert M. Stern, 487–508. Washington, D.C.: Brookings Institution.

Hummels, David. 1999. Have international transportation costs declined? University of Chicago, Graduate School of Business, Manuscript.

Jackson, J. H. *The world trading system: Law and policy of international economic relations.* Cambridge, Mass.: MIT Press.

Laffont, Jean Jacques. 1998. Competition, information, and development. Paper presented at *Annual Bank Conference on Development Economics.* April 20–21, Washington, D.C.

Mattoo, Aaditya. 2000a. Developing countries and the new round of GATS negotiations: Towards a pro-active role. *World Economy* 23:471–89.

———. 2000b. MFN and the GATS. In *Most-favored nation treatment: Past and present*, ed. Thomas Cottier and Petros Mavroidis, 51–99. Ann Arbor, Michigan: Michigan University Press.

Mattoo, Aaditya, and Marcelo Olarreaga. 2000a. Reciprocity across modes of supply: A negotiating formula. *World Bank Policy Research Paper*, forthcoming.

———. 2000b. Should credit be given for unilateral liberalization in multilateral negotiations? *World Bank Policy Research Paper*, forthcoming.

Mattoo, Aaditya, and Arvind Subramanian. 1998. Regulatory autonomy and multilateral disciplines. *Journal of International Economic Law* 1:303–22.

Nicolaidis, Kalypso, and Joel P. Trachtman. 2000. From policed regulation to managed recognition: Mapping the boundary in GATS. In *GATS 2000: New directions in services trade liberalization,* ed. Pierre Suavé and Robert. M. Stern, 241–82. Washington, D.C.: Brookings Institution.

Sapir, A. 1998. GATS 1994–2000. *Journal of World Trade* 33:51–66.

Suavé, Pierre, and Robert M. Stern, eds. 2000. *GATS 2000: New directions in services trade liberalization.* Washington, D.C.: Brookings Institution.

Snape, Richard. 1994. Services and the Uruguay Round. In *The new world trading system: Readings.* Paris: Organization for Economic Cooperation and Development.

Stephenson, Sherry. 2000. GATS and regional integration. In *GATS 2000: New directions in services trade liberalization*, eds. Pierre Suavé and Robert M. Stern, 509–42.

Thompson, Rachel. 2000. Formula approaches to improving GATS commitments. In *GATS 2000: New directions in services trade liberalization*, ed. Pierre Suavé and Robert M. Stern, 473–86.

Tuthill, Lee. 1997. The GATS and new rules for regulators. *Telecommunications Policy*, 23 (November): 783–98.
UNCTAD and World Bank. 1994. Liberalizing international transactions in services: A handbook. New York: *United Nations*.

Comment   Anne O. Krueger

This is an excellent paper and well worth careful reading. It addresses the issues and challenges that confront successful multilateral negotiations to liberalization of trade in services, yet neither gives way to despair nor insists upon politically infeasible solutions (such as national treatment with no barriers to foreign services or services providers).

Efforts to liberalize trade in services are increasingly important as services trade increases even more rapidly than trade in goods, paralleling the increasing share of modern services in the economies of the industrialized countries. However, the difficulties of negotiating liberalization are formidable: whereas barriers to trade in goods were traditionally border measures (tariffs, quotas, etc.) that were reasonably comparable across goods (one could always estimate the tariff equivalent of quotas and surcharges on imports), trade in services is restricted through a variety of measures—licensing of financial intermediaries, imposition of quality and standards regulation of services providers, regulations governing foreign direct investment in service sectors in which domestic presence is essential to trade, and restrictions on access (as with airplane landing rights and control over port access).

As Mattoo points out, there is good reason to believe that a global multilateral deal could be struck between capital-exporting industrialized countries who want access to developing country markets for their capital-intensive services and labor-abundant developing countries, who could profitably export unskilled labor–intensive services if temporary entry were granted to foreign workers for work on such services as construction.

A key issue, as Mattoo recognizes, is how much can be achieved by cross-sectoral negotiations. Some observers, noting the different types of barriers to entry in different service sectors, have concluded that sector-specific negotiation (such as happened in telecoms) is the only way forward. However, the difficulties with this are well known and deserve repetition here. The genius of the General Agreement on Tariffs and Trade and the World Trade Organization has been in achieving cross-sectoral tying in such a way that countries were willing to give up protection of some items in return for re-

Anne O. Krueger is first deputy managing director of the International Monetary Fund and a research associate of the National Bureau of Economic Research.

duced trade barriers (tariffs) in other sectors. Clearly, there are some activities in which liberalization is desirable, with or without a quid pro quo, that will take place unilaterally. In other cases, protection has such strong political backing that it will not be removed even with multilateral bargaining. In between, however, lie a host of protectionist measures that would or could be reduced given sufficient incentives (in the form of reciprocal reduction of other services barriers to the country's exports). Most of these in-between cases—which is what the multilateral tariff negotiations enabled for trade in goods—can be traded off only against reduced protection in other sectors.

Hence the need for cross-sectoral negotiations. Some things, Mattoo notes, can be done. Finding a least-cost way to achieve domestic objectives and showing the necessity of a barrier can help. However, Mattoo seems to despair of finding any uniform metric to enable formula-based negotiations (such as was finally possible in agriculture with the development of the producer-subsidy-equivalent measure). This, it seems to me, is too pessimistic. To be sure, we do not have such a metric yet, but the tariff equivalent of existing protection on maritime services, barriers to entry of insurance or legal firms, and others may be estimable. Surely, as economists, we should keep trying to find acceptable ways of quantifying these barriers.

Finally, I wonder whether things are quite as bleak for services as Mattoo paints them. We do have the General Agreement on Trade in Services, and some barriers are listed and therefore bound. Perhaps in a next round countries could agree to list all remaining service barriers and to bind them to those (or even lower) levels. Further, there might be an agreement to eschew imposing barriers on new services. Perhaps we could even begin to develop a regime for some temporary movement of workers to provide services in exchange for removal or relaxation of ceilings on the percentage ownership of foreigners in some or all service industries.

That said, there are a great many constructive suggestions in Mattoo's paper. Let us hope that they, at least, can influence the next round of multilateral trade negotiations.

Regulating Services Trade
Matching Policies to Objectives

Richard H. Snape

3.1 Introduction

Analysis of the relationship between objectives and optimum policies has a long history, at least back to Tinbergen, Haberler, Meade, and others in the early 1950s. Bhagwati and Ramaswami, and then Harry Johnson, Max Corden, and Bhagwati developed it further in a trade context, in Johnson's case in his classic "Optimal Trade Intervention in the Presence of Domestic Distortions" (1965) and for Corden most completely in his *Trade Policy and Economic Welfare* (1974).

The general message is that the policy instrument should be targeted closely on the policy objective, minimizing by-product distortions. In the context of international trade policy, domestic "market failures" are best addressed by domestic policies, whereas international trade policies are best reserved for objectives associated with international trade itself. In the latter category, as far as national economic objectives are concerned, a country's ability to affect its terms of trade is usually regarded as the principal reason for using a trade policy instrument. Although second-best considerations may modify these conclusions, they still provide a good starting point.

As far as noneconomic objectives are concerned, similar considerations apply. To the extent that such objectives are associated with trade as such (such as to satisfy isolationist objectives), trade policies would be the efficient means to pursue the objective. To the extent that they are essentially

Richard H. Snape is deputy chairman of the Australian Productivity Commission and emeritus professor of Economics, Monash University, Australia.

The views expressed in this paper are those of the author and should not be attributed to the Productivity Commission (except where they are actually attributed to the commission by the author), nor to the Australian government. The author is grateful to Geraldine Gentle, Greg McGuire, Ralph Lattimore, and the seminar discussants for comments and assistance.

domestic (for example, to expand production of cars in Australia) trade policies are not optimum policies, in the absence of second-best or administrative cost considerations.

Most of this analysis has been in relation to trade in goods, where the goods move between countries but the factors of production and the consumers or users of the goods do not. More recently, appropriate regulation of services trade has been addressed, within countries and internationally, particularly in the context of the negotiation of the General Agreement on Trade in Services (GATS) and subsequently.

In this paper the policy-objective link is addressed for services. First, the value of generic, rather than industry- or sector-specific, policies is emphasized. International aviation is referred to to show the problems that can emerge from sector-specific policies in the context of (bilateral) international agreements. Various objectives for the regulation of services are then considered. These are objectives associated with foreign investment and establishment, consumer protection, social and cultural matters, and access to the services of "essential facilities." In all of these the desirability of defining the true objectives clearly and of adopting consistent policies across sectors and, where appropriate, across international agreements, is highlighted. Extensive reference is made to some of the reports of the Australian Productivity Commission to illuminate these issues.

3.2 Generic Policies

The rules of the General Agreement on Tariffs and Trade (GATT) are mostly framed in forms that cover all goods (although there are exceptions for agriculture and for clothing and textiles). In the GATS the general rules are quite modest in comparison. The GATS is a framework agreement with special provisions being negotiated for each sector. Much of the activity by the members in the GATS has been sector specific, with little or no attempt to develop cross-sector rules or trade-offs.

In some quarters there have been recent attempts to develop generic, or horizontal, disciplines for services at the domestic and multilateral level (Mattoo 2000, 483–7). As Mattoo points out, "a generic approach is to be preferred to a purely sectoral approach for at least three reasons: it economises on negotiating effort, leads to the creation of disciplines for all services rather than only the politically important ones, and reduces the likelihood of negotiations being captured by sectoral interest groups" (484). In addition, a generic approach helps to ensure that the same criteria and policies are applied for different products and industries to address the same policy objective. Provided the policies are well chosen, this reduces the distortions of resource allocation and choice. However, the case here applies not only to services: it applies to goods as well.

Further, despite the many differences between some forms of goods trade and services trade, and the different forms of regulation, the arguments for generic or horizontal policies that apply to policies for goods and to policies for services also apply to policies across both goods and services. Where there are objectives that relate to both goods and services, there is a strong case for applying generic policies—nationally and internationally—that embrace both.

As an example of what can develop when services policy is not generic but is highly industry specific, international aviation stands out. A very restrictive framework of bilateral aviation agreements has developed, and the trade-offs are all within aviation. There is no "nondiscrimination" rule, as there is under both GATT and GATS: indeed, discrimination is of the essence.

For the last fifty-odd years, international aviation has been conducted within the context of bilateral reciprocity, based on protection of national designated flag carriers. International aviation (except ground handling and similar services) is explicitly excluded from the provisions of GATS. There are now more than three thousand of these bilateral agreements worldwide. They are based on "freedoms": that is, nothing is allowed unless it is explicitly permitted. The agreements typically specify the number of seats that can be offered by the designated airlines of the two parties to the agreement for flights between the two countries and whether they may pick up traffic en route and fly beyond the parties. They rarely allow carriage within the foreign partner. They may also contain provisions relating to fares, which may range from notification to governments to fare approval or control by governments. Some agreements provide for the sharing of revenue between the airlines of the two parties. Typically the agreements (or other legislation) limit the foreign ownership of airlines.

Partners to agreements may effectively veto foreign takeovers of airlines in their partner countries by refusing to recognize an airline as a designated national airline of the partner country, for the exercise of the rights under the air services agreement. (This was threatened by the United States when Aerolineas Argentinas was taken over by Iberian Airlines; in response to the threat, the United States received concessions from Argentina.) Thus, the ownership provisions are at the core of the system.

There has been substantial liberalization of air service agreements by many countries in recent years, but still on a bilateral basis. The United States has entered into more than thirty bilateral "open-skies" agreements, which involve the removal of most of the restrictions on capacity and routes but still prohibit foreign airlines from carrying domestic passengers within the United States, and they do not ease the tight restrictions on the national ownership of U.S. airlines.

Within this context, the Productivity Commission[1] was asked by the Australian government to recommend the best policy for the Australian people

---

1. The Productivity Commission, an independent statutory agency, is the Australian government's principal advisory body on microeconomic policy and regulation. Its reports, together with submissions made to it in the course of its inquiries, are available without charge on its website: http://www.pc.gov.au.

as a whole.[2] Industry protection as such was not an objective within the government's Terms of Reference for the inquiry, nor is it within the matters that the commission is required to consider by its act.

The commission caused surprise in some quarters by not recommending unilateral liberalization (Productivity Commission 1998). The reason was that in the tight bilateral system, unilateral liberalization could not be guaranteed to increase the traffic between Australia and other countries. The trading conditions have to be agreed upon by the parties at both ends of a flight (and with intermediate countries, if there are any). Thus, the terms of trade (here meaning the terms of aviation trade) are negotiated between the parties, and no country is a "small country" in the usual economic sense of being unable to affect its terms of trade.

The commission recommended that the Australian government try to negotiate an open plurilateral club of open-skies agreements and, better still, a liberal multilateral agreement for aviation under the GATS. However, within the constraints of the bilateral system the commission's main recommendation was that the government attempt to negotiate agreements that are as unrestricted as possible with the bilateral partners. The policy would offer unrestricted capacity, routes (including intermediate and beyond points), fares, code sharing, number of designated airlines, and ownership as a basis for designation. (A policy of unilateral unrestricted open skies, within overall negotiated capacity, was recommended for all airports with the exception of Sydney, Melbourne, Brisbane, and Perth, on the grounds that negotiating power was limited for all airports except these four.) Domestic cabotage was to be negotiated bilaterally. The government has accepted these recommendations, with the exception of that relating to cabotage.

Lessons that can be drawn from this inquiry relate to the danger of going down the path of product-specific reciprocity (Snape 2001). Even if industry protection is not a part of the objective of the policy framework (and even if it were, there are likely to be other and better means by which to pursue it), the ability to pursue the general interest is limited by the stance of foreign partners to the bilateral agreements.

Generic principles and agreements are, of course, of no use for those industries for which their application is excluded. However, there would appear to be no reason in principle that generic principles and regulations (international and national) could not be applied to all forms of international transport (of goods, people, and services), whether it is provided in physical form, by wire, or by the electromagnetic spectrum. The regulation could

2. The commission's act requires it to have regard (inter alia) to the need "to improve the overall economic performance of the economy through higher productivity in the public and private sectors in order to achieve higher living standards for all members of the Australian community" and "to reduce regulation of industry . . . where this is consistent with the social and economic goals of the Commonwealth Government."

then address the basic national objectives, rather than being specific to particular modes of transport. If the objective is industry protection as such, there are more efficient means by which to achieve it than through a morass of bilaterally restrictive agreements, which in many cases provide few incentives for economy efficiency.

How international aviation would develop in a multilateral, nondiscriminatory aviation world is a matter of conjecture. Hub and spoke systems could well develop, as in the United States, together with airline mergers that would in part replace the current alliances. General competition policy, national and across jurisdictions, would then have a higher profile in aviation. Such generic policies could be expected to promote competition, whereas the current bilateral system has at its roots the restriction of international competition.

3.3   Objectives for Services Regulation

Services are regulated to pursue a variety of objectives, economic and noneconomic. Among the former are problems associated with asymmetric information (including consumer protection and prudential requirements), monopoly (including natural monopolies), public goods and externalities, protection of intellectual property, and the improvement of the terms of foreign trade and investment. There are also technical matters (e.g., the scarcity of the radio frequency spectrum) that are of economic importance. Among the noneconomic objectives are distributional matters including universal availability, cultural and social objectives, and national ownership. Of course, there are also straight-out industry protection objectives for many governments.

3.3.1   Foreign Investment and Establishment

Of the national economic objectives, only the terms of trade and investment, and some externalities (in particular in the form of technology transfer), would seem to be matters on which policy would lead necessarily to a consideration of measures that would discriminate between foreigners and nationals. However, there would seem to be no difference in principle here between policies for goods and for services. The fact that many services could not be traded without establishment would not of itself imply that different principles should be applied to investment in the production of goods and investment in the production of services. Many goods also might be traded only (or more efficiently) if a local presence is established for assembly or distribution.

If there are externalities associated with some investments but not others (for example, knowledge transfer), then these differences may lead to differential treatment of different investments. However, that does not necessarily lead to different treatment of investments in goods production in general and investments in services production in general. There seems to be no reason to assume that as a general proposition investment in services is more or less a source of knowledge transfer or of other externalities than investment in goods production. Similarly, if national ownership is a government objective, there would seem to be no reason to discriminate on a general basis between goods and services (although, of course, governments may decide such ownership is more important for some industries than others).

Thus, there would appear to be a case for generic rules to apply to investments in both goods and services domestically as well as in international agreements. However, the fate of the proposed Multilateral Agreement on Investment suggests that such a broad approach at the multilateral level will be some time away. In the meantime, the multilateral rules governing investment related to goods trade are quite weak; for services, apart from the requirement for nondiscrimination, their strength under the GATS depends mainly on the specific commitments undertaken by members for particular sectors.

3.3.2   Consumer Protection

As to asymmetric information or consumer protection, there is much that is applicable to both goods and services. Just as many services are regulated for consumer protection, many goods are subject to regulations aimed at the same type of objective—for example, safety specifications for motor vehicles. Generally it is better to specify requirements in terms of performance rather than inputs: for example, that tractors should not turn over on forty-five-degree slopes, rather than that they should have a tonne of ballast at wheel level. Since 1947, Article III of the GATT has stated that regulations and requirements affecting the internal sale of products should not be applied so as to afford protection to domestic production. However, this has not prevented the adoption of unique national requirements.

The Agreement on Technical Barriers to Trade (as well as the Agreement on the Application of Sanitary and Phytosanitary Products) addresses such matters for goods.[3] The agreement requires that technical barriers "are not applied in a manner which would constitute a means of arbitrary or unjustifiable discrimination among countries where the same conditions prevail or a disguised restriction on international trade" (Preamble), and that foreign (WTO member) products should receive treatment that is no less favorable than like products of national origin (Article 2.1). Technical regulations should not create unnecessary obstacles to international trade and

---

3. The emphasis of the Agreement on Sanitary and Phytosanitary Products is on harmonization of international standards, but allowing for different standards when there is scientific justification that is related to the protection of human, animal, or plant life or health. Measures should not discriminate among members of the WTO where identical or similar conditions prevail.

should not be more trade restrictive than necessary to fulfill a legitimate objective. These objectives are national security, prevention of deceptive practices, and protection of human health or safety, of animal or plant life or health, or of the environment (Article 2.2). Adoption of international standards and standards based on performance are encouraged.[4] These provisions apply to all goods, whether or not they have been subject to negotiated trade barrier reductions and commitments.

In services, performance standards designed for the protection of consumers are often difficult to apply: medical practice is the classic example. (Ex ante rather than ex post protection is generally preferred) Thus, domestic regulations frequently specify required qualifications for service suppliers, and professional titles (such as architect, doctor, lawyer, and university) are reserved by legislation for those persons or institutions with the qualifications deemed appropriate. Although the emphasis on the qualifications of providers rather than on performance makes it difficult to specify generic rules across professions at the national as well as at the international level, there is scope for generic rules in multilateral agreements in regard to the foreign treatment of national qualifications.

In the GATS, Article VI.4 provides for the Council of Trade in Services (through bodies it may establish) to develop any necessary disciplines to ensure that measures relating to qualifications, technical standards, and licensing requirements do not constitute unnecessary barriers to trade in services. It specifies that such disciplines should be based on objective and transparent criteria and are not more burdensome that necessary to ensure quality, and that licensing procedures are not in themselves a restriction on supply. Such provisions would cover all services. Until such disciplines are developed, GATS provides that in those sectors for which there are specific commitments by members, these principles should apply. Little progress appears to have been made to develop these disciplines of general application for services. This is in contrast to goods, where the Agreement on Technical Barriers to Trade has general application.

As well as these provisions, Article VII of GATS provides for mutual or unilateral recognition of qualifications acquired abroad. Importantly, it provides that such recognition should not be applied in a manner that would discriminate between members in whose countries similar qualifications or experience have been obtained.

It would seem desirable to have a common set of principles for the recognition of standards, qualifications, and licensing requirements covering goods and services. Although there are two agreements covering the same issue, one for goods, the other for services (and the latter only for scheduled

---

4. Many product standards fall outside these provisions (the screw pitch of nuts and bolts, for example), whereas many that may be covered (for example, different safety specifications for motor vehicles and electrical connections) may inhibit trade but do not appear to be constructed for this purpose.

services), there are possibilities of inconsistencies even if the intent is the same.

In three recent Productivity Commission inquiries related to services—gambling, broadcasting, and architects—consumer protection has been an important issue. In the broadcasting inquiry, the regulation of content was addressed, in particular pornography and other material deemed by Parliament to be unsuitable for broadcasting. There are many issues here, including whether the rules for the Internet should be akin to those for books and magazines or to those for free-to-air (or subscription) broadcasting. (Technological convergence could suggest that all electronic platforms be treated in a similar manner, although there are also arguments that would distinguish between one-to-one dissemination and one-to-many.)

During the course of the inquiry, the government introduced regulation of online content. The regulations attempt to prohibit objectionable material hosted on Internet sites in Australia. They also attempt to require Australian Internet service providers to prevent Australians from obtaining access to material found to be objectionable (PC 2000a, 482–3).

On gambling, among other things the commission was asked to report on "the social impacts . . . , the cost and nature of welfare support services . . . the effects of regulatory structure . . . the implications of new technologies (including the internet), including the effects of traditional government controls on the gambling industries. . . ."

While emphasizing that there were substantial consumer benefits (entertainment, etc.) accruing from gambling, the commission also drew attention to the risks and costs arising from problem gambling. Lack of adequate information regarding the "odds," or price of gambling, and addictive behavior (together with the design of gambling machines that encouraged, or at least did not discourage, addictive behavior) were seen as the main problems. (There were also problems with government-legislated restrictions on competition in the supply of gambling facilities, complex and inconsistent regulation, and what some commentators have described as addiction by governments to the revenue from gambling.)

The commission saw avenues of regulation that would provide greater information and protection for consumers and support for problem gamblers. The development of Internet gambling raises questions, some of which were not unlike those associated with broadcasting, particularly in the international dimension. This involves taxation as well as consumer protection issues. Within Australia, the tax question is being solved by an agreement between jurisdictions that the tax revenue should be repatriated to the (Australian) jurisdiction of the gambler. Implementation requires licensing and enforcement. Internationally, such agreements would be much more difficult to achieve.

As with broadcasting of pornography, a main question was not whether online activities should be restricted, but the extent to which they could be controlled (and by whom). The commission saw a reasonable objective as being to reduce demand for and access to unlicensed gambling sites. Licensing is much easier to enforce for domestic sites than internationally, and just as unlicensed organized gambling is illegal in many countries and there are penalties on both suppliers and customers, so it could be with domestic gambling sites.

International sites pose more difficulty; they could offer better odds if taxes are not levied, although they would not be popular if the payment of winnings was in doubt or unenforceable. Blocking international sites probably is technically feasible, but the cost effectiveness can be questioned particularly if it were to be attempted on a general basis rather than in response to complaints. Blocking, other than in response to complaints, would be even more difficult for pornography: a search engine found about 7,000 "Internet gambling" sites but about 5 million sites that combined "XXX" and "sex" (PC 1999, 18.44). In broadcasting, the commission recommended that there be a review of the pornography policies when they have been in operation for one year.

A lesson from gambling and broadcasting is that addressing consumer protection objectives for services traded electronically is becoming more and more difficult. The consequence of e-commerce for international trade and trade rules is being investigated by many, including Drake and Nicolaidis (2000): regulation is likely to become increasingly difficult in some areas, and in some less necessary. (Of course, technological changes may lead to increased concentration and make the case for regulation stronger in some areas.) The principles of efficient regulation imply that generic rather than industry-specific platform-specific regulation should be sought as far as possible. The case for technological neutrality is strengthened substantially by rapid, and unforeseeable, technical change and by the impossibility of predicting the consequences. Here, however, as in some other service areas (for example, telecommunications), technology and other factors may require that the generic give way to the specific at some level. Good regulation would seem to call for the presumption to be for generic rules and principles, with the onus to be on justification for departures from them.

The Terms of Reference for the inquiry into architects specified that the commission report on "the preferred option for regulation, if any, of the architectural professional in Australia." In its report the commission took the view that there was no strong consumer protection case for retaining the legislated protection of the title "architect" and that the relevant legislation should be repealed (PC 2000b). It took the view that if there were public benefit in providing a stamp of approval, then that could be given by a professional body, as it is in many other professions, such as engineering and accounting. Building regulations are such that safety considerations are not relevant for the legislated restriction on the use of the term *architect*.

International considerations were raised in the course of the inquiry. Article VII of the GATS, and its provisions for mutual recognition, were referred to above. If other jurisdictions require government certification of architects for them to practice in their country, would the lack of government certification in Australia imply that Australian architects would be disadvantaged? Of course, the mutual recognition provisions of Article VII do not require that certification must be in the same form in the mutually recognizing jurisdictions: professional body recognition could suffice. On the other hand, if government certification were required for export of architectural services, it could be provided for those who wished to export. The commission saw no consumer or other reason that it should be required for all architects.

3.3.3   Social and Cultural Objectives

Social and cultural objectives are high on the agenda for many governments. Such matters are, of course, particularly difficult to define or measure, but so too are love and beauty, but few would deny their existence and importance. For the commission's inquiry into broadcasting, the Terms of Reference required it to "advise on practical courses of action to improve competition, efficiency and the interests of consumers in broadcasting services. In doing so, the Commission should focus particular attention on balancing the social, cultural and economic dimensions of the public interest and have due regard to the phenomenon of technological convergence to the extent that it may impact upon broadcasting markets." The existing act was referred to in the Terms of Reference as seeking "to protect certain social and cultural values, including promoting a sense of Australian identity, character and cultural diversity; encouraging plurality of opinion and fair and accurate coverage of matters of national and local significance. . . ."

At present there are tight limits to foreign equity investments in free-to-air and subscription television and in newspapers, but no limits for radio. There are rules that restrict cross-media investments: for example, the same enterprise cannot own both a newspaper and a television station in the same broadcast licence area. There are also rules for commercial free-to-air television regarding minimum overall Australian content, children's programs and documentaries, and Australian advertising content, and agreed industry codes for Australian music on commercial radio.

The commission did "not attempt to evaluate or comment on the social and cultural objectives of content regulation. Rather it takes the stated social and cultural objectives as given and . . . attempts to clarify them and to consider whether the existing policies address them effectively" (PC 2000a, 379). It attempted to distinguish between those policies that were essentially industry protection and those that addressed the social, cultural, and diversity objectives. The rapidly changing technology had implications for its recommendations as well as for more general regulation of broadcasting.

The commission concluded that the case for restricting foreign investment on the grounds that foreigners would be less likely to promote Australian culture was at best weak. It was not regarded as an appropriate policy instrument for this objective, an objective that is also addressed by the content quotas. More important was that diversity of media ownership (and hence of sources of information and content) was more likely to be promoted by treating foreign investment in the media in the same manner as foreign investment in other industries, that is, by not prescribing any limits. Recognizing the media concentration can be a problem, that the current cross-media rules were rapidly becoming obsolete through technical change (the rules do not cover the Internet or subscription television), and that this technological change could in the future multiply the sources of information and comment greatly, the commission recommended that the Trade Practices Act should be amended immediately to include a media-specific public interest test that would apply to all proposed media mergers. Thus social, cultural, and political dimensions of the public interest would be considered, in addition to the standard economic questions attending mergers. The commission also recommended that after regulatory barriers to entry in broadcasting had been removed and spectrum became available for new broadcasters, and after the repeal of restrictions on foreign investment, the cross-media rules should be removed.

For the content rules, the commission made the judgment that the requirement that advertising be 80 percent Australian was essentially aimed at industry assistance and that any valuable cultural or social "Australianness" would be likely to be met by advertisers, in their own interests, as a means to engage their (Australian) audience. (In any case, the 80 percent minimum has been exceeded regularly by about 10 percentage points.) The commission also decided that the children's, documentary, and Australian drama requirements were targeted to the social and cultural objectives more than is the advertising quota. In part the case for the quotas arises from the public-good nature of free-to-air television and Hotelling-type considerations.[5] The 55 percent overall Australian requirement was regarded as being much less targeted to social or cultural (or Hotelling-type and public-good) considerations.

More important however, was the new technology. The commission took the view that it was better to target the social and cultural objectives directly, rather than through particular broadcast platforms. It rejected the view that as new broadcast platforms develop, the content rules should be extended to them according to their degree of "influence."[6] Such a policy

5. This refers to the tendency, when there are few commercial free-to-air broadcasters (financed by advertising), for all of them to focus on the mass market, with none catering for thinner demand, even though individuals in these parts of the market may have strong content preferences.

6. The supposed extent of influence of various forms of broadcasting is an important criterion for the degree of content regulation in the existing Broadcasting Act.

would at best be one of "catch-up." Further, it recognized that regulating international electronic traffic will become increasingly difficult as its volume increases and as technologies develop. Instead, it recommended that to ensure that the social and cultural objectives of broadcasting continue to be addressed in the future digital media environment, the government should commission an independent public inquiry into Australian audiovisual industry and cultural policy, to be completed by 2004. Following this review, but prior to the final switch-off of analog television services, a new framework of audiovisual industry and cultural policy should be implemented. It recommended that the inquiry be based on the government's competition principles: that is, that regulations that restrict competition should be retained only if the benefits to the community as a whole outweigh the costs and if the objectives can be met only through restricting competition.

The aim of the commission in these recommendations was to encourage policies that were targeted on objectives and that were generic. The intention was that the policies not be platform-specific, both with regard to cultural and social objectives of audiovisual services (and perhaps the performing arts in general) and with respect to diversity of sources of information and content.

3.3.4   Access to Essential Facilities

Access to essential facilities has been a burning issue since the early days of negotiation of the GATS. It is also a pressing topic of policy in competition policy within many Organisation for Economic Cooperation and Development (OECD) and other countries. It is a matter that may apply to services rather more than to goods, although there would be no point in removing tariffs on imported goods if the imported goods did not have access to ports, unloading facilities, and internal transport.[7]

It was argued during the negotiation of the GATS that there was no point in removing frontier barriers to, say, telecommunications, if access could not be obtained to (monopoly) domestic distributions systems. Consequently, Article XVI (Market Access) was negotiated. For sectors in which market access commitments are made, it provides that there should be no limitations on suppliers (number, output, etc., or legal structure), or on foreign capital, unless the restrictions are specified in the member's schedule. Article VIII provides that for services subject to specific commitments, monopoly suppliers should not abuse their monopoly power, cross-subsidize

---

7. Article III.1 and III.4 of the GATT has some relevance here. For goods production there is also the matter of access to essential raw materials. This was much in the minds of the negotiators of the original GATT, with the backdrop of the trade policies of the 1930s. It is reflected in the Preamble of GATT1947: "developing the full use of the resources of the world," and in the provisions of Article XI (General Elimination of Quantitative Restrictions) that apply to exports as well as imports, although this article in fact allows for export prohibitions on "products essential to the exporting party."

other activities, or act in a manner that discriminates among foreign members of the WTO or is inconsistent with specific commitments of the member.

In the course of GATS negotiations on telecommunications, a reference paper was developed to establish principles for interconnection—without distinction between domestic and foreign telecommunication suppliers. It provided for interconnection to be supplied on terms no less favorable than for the owner's own like services or for those for nonaffiliated service providers. Mattoo (1999, 22) proposes that the same principles should be applied to other network services (e.g., transport terminals, energy services, and sewage). It could also be extended to embrace all forms of internal distribution. In this way horizontal or generic rules could be developed that cover goods and services, although the specifics of application would have to be different for each industry.

There is some tension between requiring access to essential facilities and incentives to invest where, for reasons of economies of scale or scope, it is efficient for there to be only one network supplier. If access to upstream or downstream competitors is required (or feared), then an investment in the essential facility may not be undertaken or may be truncated, even though the investment may be in the general interest. International competition rules need to allow for such tensions, as do domestic competition rules. The analytical problems here are by no means settled in principle, while the policies stemming from them depend on circumstances. Thus, although objectives could be agreed under the GATS or elsewhere, international rules to achieve them are probably best framed to allow discretion for national authorities in pursuing the objectives and for cooperation among these authorities.

3.4 Conclusion

A lesson that one can draw from the above is not surprising: that it is best to use policy structures that are attuned to objectives and to seek generic policies in this context. Trade policies are appropriate for trade objectives, social and cultural policies are appropriate for social and cultural objectives, investment policies are appropriate for investment objectives, competition policies are appropriate for competition objectives, and so on. Policies attuned to particular trade, investment, social and cultural, and competition (and so on) objectives should extend beyond specific industries and beyond specific forms of production, consumption, and trading.

This then leads to the development of consistent domestic and international rules for cross-border trading of goods *and* services, and to the conclusion that these rules should address cross-border trading alone and not such other issues as conditions of competition that apply to both domestic and foreign enterprises. Competition rules (for goods and services) are best designed generically to address all forms of competition, whether it is from domestic or foreign sources, although they will require adaptation to specific industries. Similarly, investment laws (domestic or international) are best designed to address investment consistently, whether it be for goods or services production.

Again, social and cultural objectives are best pursued by policies that do not discriminate between the platforms on which a particular service may be disseminated. As in many areas, it is consumption (and perhaps production) of cultural services (and goods) from which the cultural benefits may flow, not the platform of dissemination nor international trade as such. (Messerlin and Cocq [1999] discuss some of the problems in the European Union of attempting to assist cultural industries through the platforms of dissemination.)

In the multilateral context there is the question of how to go from here to there. A multilateral agreement on investment is off the agenda for the time being; an agreement for competition policy is also some distance away. The WTO reference paper on principles for telecommunications interconnection has been referred to above and could provide a basis for a generic policy on access to essential facilities, for incorporation into a competition agreement. The OECD secretariat is exploring a "cluster" approach for sector services that are closely related in terms of either production interconnections or the manner of regulation, and this could lead some way in the direction of generic policies.

A key issue is to contain the use of policies to address objectives for which the policies are not well suited. A prime example of this practice is the use of international trade policies to improve labor standards or the environment. However, perhaps most important is for countries' domestic policies to align objectives and policies consistently and efficiently.

References

Corden, W. Max. 1974. *Trade policy and economic welfare.* Oxford, England: Clarendon Press.

Drake, William J., and Kalypso Nicolaidis. 2000. Global electronic commerce and GATS: The Millennium Round and beyond. In *GATS 2000: New directions in services trade liberalization,* ed. Pierre Sauvé and Robert M. Stern, 399–437. Washington, D.C.: Brookings Institution.

Johnson, Harry G. 1965. Optimal trade intervention in the presence of domestic distortions. In *Trade, growth, and the balance of payments: Essays in honor of Gottfried Haberler,* 3–34. Amsterdam: Rand McNally and North-Holland.

Mattoo, Aaditya. 1999. Developing countries in the new round of GATS negotiations: From a defensive to a pro-active role. Paper prepared for the conference *Developing Countries and the Millennium Round.* September, Geneva, Switzerland.

————. 2000. Developing countries in the new round of GATS negotiations: Towards a pro-active role. *The World Economy* 23 (4): 471–89.

Messerlin, Patrick A., and Emmanuel Cocq. 1999. Preparing negotiations in services: EC audiovisuals in the Millennium Round. Paper prepared for the World Services Congress, *Services: Generating Global Growth and Opportunity*. November, Atlanta, Ga.

Productivity Commission (PC). 1998. *International air services*. Report no. 2, Canberra, Australia: Ausinfo. Available at [http://www.pc.gov.au/inquiry/airserv/finalreport/index.html].

————. 1999. *Australia's gambling industries*. Report no. 10 (3 volumes). Canberra, Australia: Ausinfo. Available at [http://www.pc.gov.au/inquiry/gambling/finalreport/index.html].

————. 2000a. *Broadcasting*. Report no. 11. Canberra, Australia: Ausinfo. Available at [http://www.pc.gov.au/inquiry/broadcst/finalreport/index.html].

————. 2000b. *Review of legislation regulating the architectural profession*. Report no. 13. Canberra, Australia: Ausinfo. May. Available at [http://www.pc.gov.au/inquiry/architects/finalreport/index.html].

Snape, Richard H. 2001. Reciprocity in trade agreements. In *Trade, development, and political economy: Essays in honour of Anne O. Krueger*, ed. Deepak Lal and Richard H. Snape, 141–56. London: Palgrave.

Comment   Takatoshi Ito

This paper is a useful survey on policy issues on service trade. As the title suggests, the author forcefully argues that matching policy objectives closely with policy tools is important. When policy objectives are given, it is better to use domestic subsidies, taxation, or regulations rather than international trade policies. I support this principle. The author, citing studies and reports done by the Productivity Commission, elaborated the principle with real-world examples in Australia, frequently referring to the General Agreement on Trade in Services (GATS).

In general, policy interventions in the market are called for when there are market failures—that is, when optimal resource allocation is not achieved through the market mechanism. Market failures arise from asymmetric information, externalities, public goods, scale economies, uncertainty, lack of complete markets, and other reasons. Cases have been explored in the literatures of advanced microeconomics and industrial organizations. Although the author is not explicit in this characterization, the author makes careful examinations of market failures in the context of services, as opposed to goods, and to the international aspects. It is interesting to examine what kinds of modifications are needed when well-known results in the domestic context are applied to international trades.

Takatoshi Ito is a professor at the Institute of Economic Research at Hitotsubashi University, Tokyo, and a research associate of the National Bureau of Economic Research.

Many services sectors have their own characteristics and particular policy requirements, even in the domestic context. It is important and interesting to explore how these services—such as aviation, banking and other financial services, telecom, legal service, medical service, and the like—are treated in international services trade policies.

Let me take up a particular example that the author highlights in the paper, namely, international aviation, that is a remarkable example of failure in the services trade framework. International aviation is governed by a network of bilateral agreements, rather than a comprehensive multilateral treaty; discrimination, rather than equal treatment, is the norm; domestic routes are typically prohibited for foreign airlines. The author explains a recommendation of the Productivity Commission to the government. The recommendation was "to negotiate an open plurilateral club of open-skies agreements and, better still, a liberal multilateral agreement for aviation under the GATS." An interesting part of the recommendations is that unilateral unrestricted open skies was not recommended for airports in Sydney, Melbourne, Brisbane, and Perth, on the ground that "negotiating power was limited for all airports except these four."

Now let me raise several issues that are inherent in international aviation. First, history matters. International aviation has been largely governed by a series of bilateral agreements and a cartel-like multilateral agreement (e.g., what used to be airfare agreements under the International Air Transport Association) on capacity controls and airfare controls. The United States, which is usually a harbinger of liberalization in many other areas, is not completely innocent in the field of international aviation, although the United States has recently been pushing for "open skies." The United States still imposes a restriction in cabotage (foreign carriers operating the domestic routes) and a restriction on investment in domestic carriers (ceiling in percentage of ownership). These restrictions effectively protect domestic carriers.

It is difficult to scratch all bilateral agreements in order to establish a comprehensive multilateral aviation treaty. The past bilateral agreements have typically favored particular airlines as "incumbent" carriers. As deregulation proceeded, the dividing line of the common interest often lies between incumbents and newcomers. It is well known that in the United States-Japan air service negotiation of the late 1980s to 1990s, the interests of United and Northwest have more common elements with Japan Airlines than American and Delta.

Second, production of airline services is characterized by technology of scale economies, primarily due to large fixed costs and network externality. Even in the domestic markets of small advanced countries and emerging-market economies, only one or two airlines would be fit to survive under a completely deregulated environment. When deregulation is extended to international services, scale economies would imply that some countries would lose domestic airlines (or "flag carriers"). Situations in Europe after its establishment of a single market vividly illustrate the point. Therefore, promoting liberal international aviation may result in a loss of domestically owned airlines. When the number of airlines declines domestically, regionally, and globally, there may be a danger of monopolistic pricing. The key to preventing monopolistic pricing is to guarantee free entry (and with affirmative action–like encouragement) and to promote contestable markets. A relevant question here is whether contestability is easier or more difficult to obtain in international airline markets compared to domestic airline markets.

Third, landing slots are scarce resources. When capacity constraints are binding, competition policy relying on deregulation and contestability may not work. A standard answer in economics for such a situation is an auction of scarce resources. However, auctions may further aggravate a problematic aspect of "scale economies." There are similarities to telecom frequency auctions. Unless landing spots of destination cities' airports are open or auctioned, unilateral auctions may not enhance national welfare of domestic consumers and companies.

Fourth, it has become difficult to define national welfare. Cross-border airline alliances—such as Star Alliance, One World, and Skyteam—have developed strong common business strategies, such as code sharing, through check-ins, and sharing airport lounges. Dividing lines of interest may be drawn between different alliances rather than across national borders. In this decade, Japanese airlines—Japan Airlines and All Nippon Airways—are aligned with U.S. airlines—American and United, respectively—in accordance with business affiliations. The nationality of airlines, or even the incumbent-newcomer classification, is no longer relevant in companies' welfare. The national policy goal becomes complicated, although maximizing consumers' welfare remains most important.

As shown above, a particular case such as international aviation can be studied carefully to analyze important factors in examining international service trade.

Let me raise some other issues that will be deemed important in the near future. Markets are changing, and regulatory considerations often have to chase the reality. First, e-commerce is spreading very quickly. One can purchase books, music, and financial products on-line. These will enhance competition internationally. Therefore, local physical stores will be under pressure. What used to be considered "nontradables" are now under international competitive pressure. Building transparent and equitable rules on these e-commerce transactions, including taxation rules, will be important. Second, loss of nationality is imminent from international alliances such as in international aviation and telecom companies. Other examples may follow. These will enhance the case for multilateral competition policy. Third, competition between different sectors may complicate competition policy.

Now, cable TV companies, local telephone companies, and mobile phone companies will be competing in similar products, such as in the market of "last mile" in local telephone service. Competition policy had to be reconsidered in this environment of greater contestability. Monopoly in one technology does not guarantee rents. This also calls for a multilateral framework for competition policy across industries that were traditionally different.

In sum, the paper is an excellent survey on the issues surrounding international service trade. My comments are meant to elaborate on some of the issues raised in the paper. International aviation was used to illustrate factors that should be considered in coming up with policy prescriptions, and some of the issues that arise from recent information technology are raised in arguing a case for multilateral competition policy.

Comment   Edwin L.-C. Lai

This is a very interesting and stimulating paper. It tells us of the chaotic nature of the regulation of trade in services nowadays. It also makes us think about why trade in services is so much more regulated than trade in goods.

The main arguments and points being made are these: (a) generic rules should be applied across sectors; (b) countries should adopt policies that achieve the objectives directly rather than indirectly, be they economic, cultural, or social; and (c) it is desirable to have a set of generic principles that govern trade in both goods and services.

These are intellectually compelling and valid points. They are based on sound economic principles. The author has given good examples to illustrate how to apply generic rules to sectors of similar nature: for example, (a) transport should include physical (aviation), wire (internet, phone services), and electromagnetic wave (broadcast); (b) qualifications, technical standards, and licensing requirements should all be governed by one set of rules; (c) the "principle of interconnection" in telecommunications should be applied to all network services, such as transport terminals, energy sectors (electricity transmission), and sewerage.

It is hard to establish a set of generic principles for all services. If there is such a set, it is probably very general. Because of the heterogeneity of services, there should be different sets of generic principles for different groups of sectors. The process of grouping the sectors and tailor-making rules for each group of sectors can be a really challenging task because of the complex differences among the services. In any case, this would seem to be a necessary first step.

Edwin L.-C. Lai is associate professor of economics at the City University of Hong Kong.

Should we adopt a systematic approach by first categorizing services according to their nature and then establishing a set of generic rules for each category? Or should we adopt an ad hoc approach—for example, begin with some sectors for which agreements have been reached, and then try to apply generic rules to sectors of similar nature? This would be a crucial question to be answered by the World Trade Organization (WTO).

In order to establish generic rules for services liberalization, we would also need to understand the fundamental differences between trade in goods and trade in services. According to the WTO, there are four modes of supply of services: (a) cross-border supply; (b) consumption abroad; (c) commercial presence; and (d) movement of natural persons. Trade in goods occurs mainly in mode (a), and to a lesser extent in mode (c) and (d) in the case of foreign direct investment (FDI) in manufacturing. Mode (b) is usually absent in the goods sector. On the other hand, all modes of supply are common in trade in services.

Moreover, the quality of a service usually depends on the qualification and competence of the personnel delivering the service. Examples are lawyers, engineers, medical doctors, and so on. In the absence of an international certifying body, individual countries have to have the right to determine which foreign qualifications to recognize. The challenge is to establish generic rules that disallow illegitimate discrimination against foreign entities yet allow for legitimate discrimination against foreign countries that have lower standards in the certification of personnel. Hence, harmonization of qualifications across countries would seem to be more complicated than harmonization of standards for goods.

Finally, the delivery of certain services, such as accounting and legal services, requires that the personnel understand the institutions and culture of the country. This necessity can serve as a barrier to trade in these sectors. Whether or not such culture-related knowledge should be part of the qualification requirement is a matter of debate. It has to be resolved before generic rules are set up in these sectors.

4

Taiwan's Accession into the WTO and Trade in Services
A Computable General Equilibrium Analysis

Ji Chou, Shiu-Tung Wang, Kun-Ming Chen, and Nai-Fong Kuo

4.1 Introduction

Starting in the mid-1980s, the implementation of a whole range of large-scale tariff reductions was undertaken by the Taiwanese government as a result of the country's huge trade surpluses and subsequent pressure from its trading partners for the opening up of its domestic market. With the realization of the importance of its participation in international economic institutions, Taiwan applied to join the General Agreement on Tariffs and Trade (GATT) in 1990 and subsequently the World Trade Organization (WTO) in 1995. The ensuing bilateral negotiations between Taiwan and WTO members have since provided further impetus for trade liberalization within Taiwan. The average nominal tariff rate has been reduced from 20.6 percent in 1987 to 8.25 percent in 1998, while the average effective tariff rate has also been reduced from 7.02 percent to 3.13 percent during the same period. Along with the liberalization of the agricultural and manufacturing sectors, the liberalization of Taiwan's trade in services has become one of the major issues in bilateral negotiations between Taiwan and the WTO members. Taiwan has made numerous commitments with regard to the liberalization of its trade in services and has also recently begun deregulation of the country's service industries in pursuit of its goal of becoming an Asia-Pacific Regional Operations Center (APROC).

Due mainly to the limitations of data, previous studies regarding the effects of trade liberalization in Taiwan have focused largely upon com-

Ji Chou is a research fellow at the Chung-Hua Institution for Economic Research. Shiu-Tung Wang is a professor at National Taiwan Ocean University. Kun-Ming Chen is an associate professor at National Chengchi University. Nai-Fong Kuo is an associate professor at Shih Hsin University.

modity trade; however, any assessment of trade liberalization that fails to consider trade in services may well underestimate its potential impact. The purpose of this study is therefore twofold. First of all, we apply Hoekman's method to the compilation of tariff equivalents for Taiwan's service sector. This is followed by the application of a global trade analysis project (GTAP) model, a multiregional computable general equilibrium model developed at Purdue University, which provides the means of analyzing the effects of trade liberalization in services as well as commodities. This paper considers technology spillovers—from the developed countries to developing countries—through the imports of intermediate inputs, capital goods, and services, while also investigating the potential differences resulting from WTO accession by both Taiwan and mainland China.

The remainder of this paper is organized as follows: Estimates of barriers to trade in services in Taiwan are carried out in the next section, including a comparison with barriers to service trade existing in other countries. This is followed in section 4.3 by presentation of the structure of the GTAP model and the extension of the model for adoption in this study. The simulation design and simulation results are discussed in the penultimate section, with conclusions being drawn in the final section.

4.2   GATS and Barriers to Trade in Services

4.2.1   GATS and Its Disciplines

Within the world economy, the service sector has invariably played an important role in developed countries, particularly in terms of business and financial services and of trade, transportation, and telecommunications services. As table 4.1 shows, of the total global output in 1995, on average, 58.1 percent was attributable to the service sector. Hong Kong had the highest share with more than half of its total output being generated from transportation services on its assumption of an entrepôt role for China.[1] The shares in Taiwan and Korea were between 40 percent and 50 percent, whereas in the developing countries of China and South Asia, the shares were below 40 percent. In comparison with manufacturing products, however, the proportion of exports or imports in international trade was relatively low (see tables 4.2 and 4.3). On the one hand, this may reflect the fact that a major proportion of service output is for domestic use (or nontradable), whereas on the other hand it may indicate that there are many barriers to trade in services.

---

1. Singapore's share of services would be similar to that of Hong Kong because it also plays an entrepôt role for Malaysia and Indonesia. To simplify the number of countries for multinational general equilibrium analysis in the next sections, we group Indonesia, Malaysia, the Philippines, Singapore, and Thailand as the Association of Southeast Asian Nations (ASEAN) in this study.

Table 4.1   Country/Regional Composition of Products in 1995, by Sector (%)

| Country/Region | Agriculture and Mining | Non-heavy Manufacturing | Heavy Manufacturing | Service | | | | | | Total Sector |
				Subtotal	Utilities	Construction	Trade, Transport and Telecommunications	Business and Financial Services	Social and Personal Services	
Taiwan	4.0	22.2	27.5	46.3	2.1	5.0	13.4	12.8	13.0	100.0
China	16.4	26.5	23.6	33.6	3.6	7.1	12.6	5.8	4.5	100.0
Hong Kong	2.8	12.4	8.0	76.9	1.2	7.3	52.9	11.6	3.9	100.0
Japan	3.0	19.6	19.4	58.0	4.0	9.9	18.7	20.4	5.0	100.0
Korea	7.1	25.0	24.0	44.1	2.9	10.0	11.0	12.7	7.5	100.0
ASEAN	13.0	22.3	21.9	42.8	2.0	8.5	18.6	9.4	4.3	100.0
South Asia	25.9	22.1	12.8	39.3	5.1	6.9	12.4	5.7	9.2	100.0
Australia and New Zealand	8.1	14.9	9.7	67.2	1.9	6.7	21.9	22.6	14.1	100.0
United States	3.8	14.8	16.3	65.2	2.7	7.4	18.2	24.9	12.0	100.0
Canada	7.1	17.8	17.8	57.3	3.6	7.7	19.5	22.1	4.4	100.0
European Union	3.1	16.8	17.3	62.8	2.0	6.5	15.2	26.9	12.2	100.0
Rest of world	13.2	22.5	15.7	48.5	3.7	6.5	15.6	13.4	9.3	100.0
Average	5.9	18.4	17.4	58.1	2.9	7.5	16.8	21.2	9.7	100.0

Source: GTAP data bank.

Table 4.2   Country/Regional Composition of Exports in 1995, by Sector (%)

Country/Region	Agriculture and Mining	Non-heavy Manufacturing	Heavy Manufacturing	Service						Total Sector
				Subtotal	Utilities	Construction	Trade, Transport and Telecommunications	Business and Financial Services	Social and Personal Services	
Taiwan	0.8	31.1	61.0	6.9	0.0	0.0	3.8	1.6	1.5	100.0
China	5.0	46.1	40.8	8.1	0.2	0.0	4.3	3.0	0.6	100.0
Hong Kong	0.3	20.9	20.1	58.6	0.1	1.0	52.3	4.3	0.9	100.0
Japan	0.3	11.7	76.1	11.7	0.0	0.0	8.1	3.5	0.1	100.0
Korea	0.7	27.4	55.3	16.2	0.0	0.6	10.2	3.0	2.4	100.0
ASEAN	9.7	25.6	48.5	16.3	0.0	0.1	12.1	3.5	0.6	100.0
South Asia	18.1	55.6	12.6	13.7	0.0	0.0	3.5	1.0	9.2	100.0
Australia and New Zealand	30.2	28.6	19.2	22.2	0.0	0.1	13.8	3.5	4.8	100.0
United States	6.5	20.9	47.2	25.1	0.0	0.0	10.4	11.3	3.4	100.0
Canada	13.2	29.7	47.4	9.7	0.4	0.0	5.4	3.2	0.7	100.0
European Union	4.3	32.0	45.2	18.6	0.2	0.6	7.9	7.4	2.5	100.0
Rest of world	29.1	25.7	27.6	17.6	0.1	0.7	10.7	2.8	3.3	100.0
Average	9.4	27.7	44.8	18.1	0.1	0.4	9.4	5.8	2.4	100.0

Source: GTAP data bank.

Table 4.3   Country/Regional Composition of Imports in 1995, by Sector (%)

Country/Region	Agriculture and Mining	Non-heavy Manufacturing	Heavy Manufacturing	Service						Total Sector
				Subtotal	Utilities	Construction	Trade, Transport and Telecommunications	Business and Financial Services	Social and Personal Services	
Taiwan	8.2	24.1	53.0	14.6	0.0	0.0	4.6	3.5	6.5	100.0
China	6.7	36.4	48.8	8.2	0.0	0.2	5.2	2.3	0.5	100.0
Hong Kong	6.8	29.6	46.7	16.9	0.4	2.6	7.3	5.6	1.0	100.0
Japan	22.5	27.5	25.9	24.0	0.0	0.0	18.2	5.7	0.1	100.0
Korea	18.3	21.2	45.4	15.1	0.0	0.0	7.6	4.0	3.5	100.0
ASEAN	7.6	19.8	59.1	13.3	0.0	1.2	4.8	5.4	1.9	100.0
South Asia	16.0	32.3	41.9	9.8	0.0	0.0	6.0	0.2	3.6	100.0
Australia and New Zealand	4.6	26.3	47.4	21.7	0.0	0.1	15.3	5.1	1.2	100.0
United States	9.3	23.8	52.6	14.3	0.1	0.0	6.5	5.9	1.8	100.0
Canada	5.7	24.1	55.9	14.3	0.0	0.0	7.8	5.9	0.6	100.0
European Union	9.1	31.6	41.4	17.9	0.2	0.3	8.2	6.4	2.8	100.0
Rest of world	8.5	32.4	42.8	16.1	0.2	0.7	9.5	3.7	2.0	100.0
Average	10.0	29.0	44.5	16.5	0.1	0.4	8.6	5.3	2.1	100.0

Source: GTAP data bank.

Prior to the Uruguay Round of GATT talks, trade in services had not been brought under multilateral disciplines, but the issue was subsequently addressed by the General Agreement on Trade in Services (GATS), which came into force in January 1995, with its key provision of nondiscrimination. This is reflected in the application of most-favored nation (MFN) status and national treatment rules. Although MFN is a general obligation for market access, GATS contains an annex that allows countries to invoke exemptions. The coverage for each GATS member is subject to a negative list that applies to all services, apart from those listed by each member. Most-favored nation exemptions are in principle expected to last no longer than ten years and are subject to renegotiation in subsequent trade-liberalization rounds of talks, the first of which must take place within five years of the agreement's entry into force. National treatment is defined as treatment no less favorable than that accorded to like domestic services and service providers.

The introduction of a market access commitment reflects one of the distinguishing characteristics of service markets—the fact that their contestability is frequently restricted by nondiscriminatory measures.[2] Because national treatment and market access are not general obligations within the context of GATS, the commitment schedules are crucial in determining the extent of market access opportunities resulting from the agreement; therefore, specific commitments form the core of GATS. Each member of GATS first decides which sectors will be subject to these disciplines and then decides, for each of these sectors, which measures violating market access or national treatment are to be kept in place.

The limitations and exceptions must be specified under four general modes of supply: (a) cross-border supply of a service without the physical movement of supplier or consumer; (b) consumption abroad, that is, provision involving movement of the consumer to the country of the supplier; (c) a commercial presence in the territory of another country; and (d) provision of services requiring the temporary movement of natural persons.

In addition to specific commitments, countries may also make horizontal commitments. These are usually laws and policies that restrict the use of a mode of supply to foreigner suppliers, independently of the sector involved.

Hoekman's Method

By mid-1994, more than sixty GATS members had submitted MFN exemptions. For the purpose of evaluating the restrictions of trade in services,

---

2. In general, there are six types of restrictions that are prohibited: limitations on the number of service suppliers allowed, on the value of transactions or assets, on the total quantity of service output, on the number of employees, on the total quantity of service output, on the type of legal entities through which a service supplier is permitted to operate, and on the participation of foreign capital.

Hoekman's (1995) study counted the specific commitments of each GATS member. This paper extends the Hoekman study to the case of Taiwan, both before and after its accession into the WTO.

Because commitment schedules are constantly subject to change, any direct comparison between Taiwan's case and that of other areas may be misleading, because the timing for compilation of the schedule is inconsistent; nevertheless, information provided by the study is still worthy of reexamination. As table 4.4 shows, generally speaking, the number of commitments in developed countries was quite high, usually over 300, with the notable exception of New Zealand, which had only 276 in 1995. The number of commitments for China, Indonesia, the Philippines, India, and Pakistan was less than 200; Hong Kong was just 200; Malaysia, Thailand, and Singapore were over 200; and in Korea there were already in excess of 300 in 1995. The number of commitments for Taiwan prior to WTO accession was 329, and this is expected to rise to 401, close to the Japanese and Australian levels of 1995 and even higher than the levels of other developed areas in the same

Table 4.4   **Number of Scheduled Commitments**

Country/Region	No. of Commitments
Taiwan	
Pre-WTO Accession	329
Post-WTO Accession	401
China	196
Hong Kong	200
Japan	408
Korea	311
ASEAN	
Singapore	232
Malaysia	256
Thailand	260
The Philippines	160
Indonesia	140
South Asia	
India	132
Sri Lanka	8
Bangladesh	4
Pakistan	108
Australia	412
New Zealand	276
United States	384
Canada	352
European Union	392

Sources: Data on Taiwan are obtained from the Schedule Concerning Commitments on Trade in Services, 1999; remaining data are obtained from Hoekman (1995).

Note: The maximum number of commitments is 620 (i.e., 155 activities), multiplied by four modes of supply.

year such as the United States, Canada, New Zealand, and the European Union (EU).[3]

In addition, to summarize the number of commitments, Hoekman developed a three-category weighting method to quantify GATS schedules, examining all GATS schedules and allocating a weight to each possible schedule entry.[4] That Hoekman's method reveals valuable information about restrictions on services trade for different countries is undeniable, and it seems that, so far, no more plausible replacement has emerged.

Brown et al. (1995) and Robinson, Wang, and Martin (1999) used Hoekman's data as their initial policy parameters to carry out trade liberalization simulations in their multinational computable general equilibrium (CGE) studies. However, Hertel (2000) argued that the idea of modeling protection with revenue-raising tariff equivalents from the work of Brown et al. was inappropriate. Although partially solving the problem with the view that the liberalization of restrictions on trade in service reduces the effective import price of services in the domestic market, Hertel conceded that his approach could succeed in capturing only a small part of the whole story of service-sector reform. Furthermore, as noted by Dee, Hanslow, and Phamduc (chap. 1 in this volume), many studies have failed to take account of barriers to commercial presence as an important category of barriers to trade in services in general. Dee used the FTAP model to distinguish *barriers to establishment* from *barriers to ongoing operations* and came up with more convincing results; however, the sector classification in the FTAP model is still too highly aggregated for this study to consider adopting.

Barriers to Trade in Services

Table 4.5 provides Hoekman's calculations of tariff equivalents for trade in services, along with the case of Taiwan calculated for this study, both pre- and post-WTO accession. The results seem consistent with the results in table 4.4, that is, that developed countries tend to show lower tariff equivalents. There are, however, some variations between sectors. The tariff equivalent rates in the sectors of construction, wholesale and retail distribution, and business and financial services are relatively low in developed countries, whereas transportation and telecommunications represent the sectors with high tariff equivalent ratios in all regions, because cabotage, air transportation proper (as opposed to ground services), post services, and basic telecommunications are all sectors in which access tends to be prohibited by most countries and where maximum tariff equivalent ratios are therefore assigned (see table 4A.1).

High-income countries (or regions) such as the EU, Australia, and New

3. The EU is counted as one member of GATS.
4. Readers may refer to Hoekman's paper in the 1995 World Bank Working Paper version (Hoekman 1995) and in a shorter version in the 1996 volume edited by Martin and Winters (Hoekman 1996).

Tables 4.5   Tariff Equivalents for Single-Digit ISIC Service Sectors

			ISIC Sector			
Country/Region	5: Construction	6: Wholesale and Retail Distribution	7: Transport Storage and Communications	8: Business and Financial Services	9: Social and Personal Services	6 + 7: Trade, Transport and Communications
Taiwan						
Pre-WTO Accession	10.0	10.2	134.6	24.3	26.2	85.7
Post-WTO Accession	10.0	10.2	82.2	21.4	25.6	53.9
China	25.0	34.5	150.9	29.9	42.4	105.2
Hong Kong	32.0	30.9	134.7	27.9	43.4	93.9
Japan	5.0	4.5	87.8	17.6	31.1	55.1
Korea	16.0	21.4	141.0	24.3	40.5	94.0
ASEAN	21.2	32.5	135.0	29.1	39.5	94.7
Singapore	12.0	33.2	130.2	25.3	34.8	92.1
Malaysia	10.0	33.6	128.2	24.7	35.3	91.1
Thailand	28.0	31.8	148.7	31.7	39.7	102.8
The Philippines	40.0	30.0	120.3	30.9	43.8	84.8
Indonesia	16.0	33.6	147.5	33.0	43.8	102.8
South Asia	37.0	34.4	151.6	38.8	41.4	105.6
India	34.0	34.5	148.7	37.1	41.2	103.9
Pakistan	34.0	34.1	151.7	36.0	36.4	105.5
Australia and New Zealand	8.5	11.4	140.0	16.0	30.6	89.5
Australia	12.0	8.6	141.1	13.2	25.1	89.1
New Zealand	5.0	14.1	139.0	18.7	36.1	89.9
United States	5.0	4.5	119.0	10.6	31.0	74.0
Canada	6.0	8.6	119.6	13.8	39.4	76.0
European Union	10.0	9.5	140.8	15.7	22.9	89.3
Rest of World	30.4	28.6	137.4	33.2	39.3	94.7

Sources: Details on Taiwan are from this study; details on the remaining countries are obtained from Hoekman (1995).

Zealand demonstrate high tariff equivalents in these sectors. Life insurance services are another highly protected area, but the aggregated services sector may well dilute its influence to some extent.

There are many limitations in Hoekman's method, as pointed out by Warren and Findlay (1999). First of all, it does not distinguish between barriers, in terms of their impact on the economy, with minor impediments receiving the same weighting as an almost complete refusal of access. Second, in many cases the coverage of the GATS schedule does not give an accurate picture of the actual barriers that are in place. Third, there is some evidence to suggest that nations with liberal policies left some services unbound so as to maintain a retaliatory capability in future market access negotiations. Therefore, some industries that are recorded within the ratio as impeded may well be open, at least to suppliers from some economies. Warren and Findlay also proposed a number of suggestions aimed at partially improving Hoekman's method; however, his data are the only estimates currently available to provide an initial consistent basis for analyzing the impact of services trade liberalization across countries. Furthermore, tariff equivalents for different periods will reveal the openness of the service sector in the specific country. In the case of Taiwan, restrictions on trade in services in transportation, storage, and communications are relatively high prior to WTO accession, as shown in table 4.5; however, these will be further rationalized following WTO accession, with particular emphasis on the area of telecommunications. In the basic telecommunications services, no restrictions will be applied to cross-border supply and consumption abroad. In terms of commercial presence, foreign suppliers will basically be allowed to enter into the market with some restrictions.[5] The presence of natural persons in the basic telecommunications services is unbound, except as indicated in the horizontal section. In addition, further liberalization of the horizontal commitments has been undertaken as follows:

1. Foreign businesses and individuals may directly invest in Taiwan with respect to portfolios in companies whose shares are listed in the securities market of Taiwan, without previous ceiling limitations;

2. The duration of business visits is extended from sixty days to ninety days; and

3. Intracorporate transferees may stay for a two-year initial period renewable each year, compared to the previous period of stay of no more than three years.

5. A service supplier should be a company limited by shares incorporated in Taiwan. Direct investment in a service supplier by non-Taiwanese persons cannot exceed 20 percent. The aggregate of direct and indirect investment by non-Taiwanese persons in service suppliers other than Chunghwa Telecom is limited to 60 percent. The aggregate proportion of shares held by non-Taiwanese persons in Chunghwa Telecom, including both direct and indirect investment, cannot exceed 20 percent.

4.3   Model Specifications

This paper employs the GTAP model of Hertel (1997), which has been widely used to analyze the impact of trade liberalization; however, in order to assist in determining the gains from import-embodied technology spillover, an equation that links the technology transfer and imports of capital goods, intermediate inputs, and services is also incorporated into the model.[6] One distinguishing feature of this model is that there is an international shipping sector that acts as an intermediary between the supply of, and demand for, international transport services, the demand for transport services being generated by international trade. These services account for the difference between free on board (f.o.b.) and cost plus insurance and freight (c.i.f.) values for a particular commodity shipped along a specific route. The supply of these services is provided by means of a Cobb-Douglas production function in which the service exports from each region are used as inputs. The production functions associated with these services are commodity or route specific. This formulation allows us to introduce commodity or route technical changes in international transport services. For instance, following Taiwan's accession into the WTO, it is highly likely that the current trade regulations between Taiwan and mainland China will be removed; thus, Taiwan's accession into the WTO will generate an efficiency increase in the international transport services for trade between Taiwan and mainland China.

Another distinguishing feature of this model is that there is a global bank that enables us to endogenize the balance of trade, so that it is unnecessary to adjust regional investment in line with regional changes in saving. The existence of this bank ensures that the global demand for saving equates to the global demand for investment in the postsimulation equilibrium. The global bank uses receipts from the sale of a homogeneous saving commodity to the individual regional households in order to purchase shares in a portfolio of regional investment goods. The size of this portfolio will adjust to accommodate changes in global saving. There are two alternative mechanisms for the allocation of investment across regions. One formulation assumes that investors behave in such a way that any changes in regional rates of return are equalized across regions; the other formulation assumes that the regional composition of capital stocks will not change in the postsimulation equilibrium. The first formulation would seem to be more appropriate in determining the efficiency gains from trade liberalization, particularly in trade in services, which often involves capital movement between regions; thus, the first formulation is adopted in this paper.

It is suggested that there are two distinct types of gains from trade liberal-

6. Values of elasticity parameters used in the model are all based on the original GTAP database developed at the Center for Global Trade Analysis, Purdue University. Readers who are interested in the data should refer to McDougall, Elbehri, and Truong (1998).

ization: (a) gains from more efficient utilization of resources, and (b) gains from technology spillover effects via expansion in the importation of capital goods, intermediate inputs, and services (see Robinson, Wang, and Martin 1999, 4–5). In order to quantify the first effect, we specify a macro closure that allows average rental rate adjustment to ensure that global capital is fully utilized and efficiently allocated, not only across sectors within each region, but also across different regions in the world, as discussed earlier. In order to capture the technology spillover effect of trade liberalization, following Robinson, Wang, and Martin (1999), we introduce an equation that links total factor productivity (TFP) to the imports of capital goods, intermediate inputs and services. A TFP shift variable in the model is specified as

$$ITFP_{ir} = 1 + IMS_{ir}$$

$$\times \left[\frac{NX_{ir}}{NX_{ir} + VA_{ir}} \times \left(\frac{\sum\limits_{j \in IM} \sum\limits_{s \in R} X_{jsr}}{\sum\limits_{j \in IM} \sum\limits_{s \in R} X0_{jsr}} \right)^{\sigma_{ir}} + \frac{NX_{ir}}{NX_{ir} + VA_{ir}} - 1 \right]$$

where $ITFP_{ir}$ is the total factor productivity shift variable; IMS_{ir} is the share of imported products embodied with advanced technology and used as intermediate products in total imports of i product at region r; NX_{ir} and VA_{ir} are intermediate inputs and primary inputs, respectively; $X0_{jsr}$ is trade flows in the base year; IM is the subset of i referring to those products embodied with advanced technology; and σ_{ir} is the elasticity. It is assumed that technology transfer flows in one direction only: from the more developed to the less developed regions. The improvements in income levels in the less developed countries, stemming from technological transfer, enhance their import demand for the exported products from the developed countries. This indirect effect of technology transfer helps to support export expansion in developed countries and might result in gross domestic product (GDP) and welfare gains in all countries throughout the world.

In equilibrium, the domestic commodity and factor prices will equalize the quantities supplied and demanded in all goods and factor markets within each region, and world prices will equalize the total supply of, and demand for, sectoral exports across the world economy.

4.4   Simulation Analysis

4.4.1   Simulation Design

In order to evaluate the economic impacts of global trade liberalization, a total of five simulations are carried out, the design of which is shown in table 4.6. In scenario A, the effects of global trade liberalization in commodity trade are investigated, whereas scenario B considers the effects of trade liberalization in goods as well as services. A comparison of the results

Table 4.6   **Simulation Design**

	Simulation Specification				
Scenario	Liberalization of Commodity Trade[a]	Liberalization of Services Trade[a]	Technology Spillover Effects	40% Cut in Shipping Costs[b]	Taiwan's WTO Accession[c]
A	✓				✓
B	✓	✓			✓
C	✓	✓	✓		✓
D	✓	✓	✓	✓	✓
E	✓	✓	✓		

[a]Liberalization of Commodity Trade and Liberalization of Services Trade refer to the reductions of barriers to commodity trade and services trade, respectively, in all regions, including Taiwan and mainland China, during the period 1995–2000.
[b]40% Cut in Shipping Costs refers to the reduction in trade transportation costs between Taiwan and Mainland China during the period 2001–2005.
[c]Taiwan's WTO Accession refers to the further liberalization of Taiwan and China during the period 2001–2005.

of scenario A with those of scenario B reveals the effects of trade liberalization in the services sectors.

It has been suggested that as a result of the technology spillover effects, access to imported capital goods, to intermediate durable goods, and to the services provided by the developed countries is crucial for increases in productivity in developing countries; see, for example, Robinson, Wang, and Martin (1999) and Markusen and Rutherford (1999). In addition to the liberalization of global trade in scenario B, scenario C introduces the technology spillover mechanism contained within our simulation. A comparison of the simulation results from scenarios B and C will reflect the technology spillover effect. Prior to the accession of both Taiwan and mainland China into the WTO, direct shipments across the Taiwan Strait continue to be prohibited for political reasons, but this transportation barrier may have to be eliminated once they both become full-fledged WTO members. If that is the case, then it is estimated that transportation costs can be reduced by 40 percent. Thus, in scenario D it is assumed that shipping costs for trade between Taiwan and mainland China will be cut by 40 percent.

In scenarios A through D, it was assumed that Taiwan and China would have joined the WTO by the end of 2000. However, both Taiwan and China had already carried out trade liberalization to some degree prior to their expected accession into the WTO, through bilateral negotiations with other countries. After their expected accession into the WTO, these liberalization actions would continue and expand further. Thus, the year 2000 was regarded as a crucial time point. In order to highlight the effects of both Taiwan's and China's accession into the WTO, in simulation E it was assumed that neither of them would carry out any further liberalization actions after 2000 and that there would be no shipping cost reductions. A comparison of the results of scenarios D and E will reveal the impacts of WTO accession by Taiwan and China on their domestic economies and on the world economy as a whole.

In all simulations, the degree of trade liberalization for all regions, with the exceptions of Taiwan and China, is based on the Uruguay Round agreement as summarized by Francois, McDonald, and Nordstrom (1995). Within the agricultural sector, six years after the Uruguay Round of talks, tariffs and tariff equivalents for nontariff barriers are reduced by 36 percent for developed countries, whereas production subsidies are calculated as aggregate measure of support (AMS) and reduced by 20 percent over six years. For the developing countries, the reductions in tariffs and nontariff barriers are moderately light, with a 24 percent reduction carried out over a ten-year period; the AMS reduction is 13.3 percent over ten years. Within the manufacturing sector most advanced countries have already exceeded the one-third reduction in tariff rates, a requirement of the Uruguay Round agreement by the year 2000.

In our simulations, the tariff rate reduction for the manufacturing sector is assumed to be 38 percent. For the service sector, we adopt the figure used by Robinson, Wang, and Martin (1999); thus, the tariff equivalent reduces to 50 percent by 2005. The degree of trade liberalization within the agricultural and manufacturing sectors in Taiwan is calculated based on the data provided by the Department of Customs Administration, Ministry of Finance, Taiwan. Table 4A.3 shows the degree of trade liberalization in the agricultural and manufacturing sectors with details provided of the reduction in both tariffs and tariff equivalents of nontariff barriers. The degree of trade liberalization for the service sector is calculated based on Hoekman's method, as described earlier, with the reduction in tariff equivalents for the service sector being shown in table 4A.4.

The degree of trade liberalization for China prior to 2000 is calculated by comparing the published tariff schedules for 1995 and 2000 based on data provided by the Industrial Development Bureau, Ministry of Economic Affairs, Taiwan. For the scope of trade liberalization beyond 2000, we collect information on reduction targets announced by Chinese leaders on various occasions and carry out calculations based on a comparison with the 2000 tariff schedule. The degree of tariff reductions for mainland China is shown in Table A-5.

4.4.2 Simulation Results

A comparison between the classifications of twenty seven subsectors and eight classes is provided in table 4.7.[7] Tables 4.8 to 4.13 summarize the re-

---

7. The classification of "heavy manufacturers" and "nonheavy manufacturers" is somewhat arbitrary here in order to provide readers with a more precise perception of Taiwan's structural change. We provide an additional three tables in the appendix showing the more detailed sectoral changes in Taiwan's output, exports, and imports (see tables 4A.6–4A.8).

Table 4.7   **Regional and Industrial Sector Classifications**

Regions

Taiwan	Korea	United States
China	ASEAN	Canada
Hong Kong	South Asia	European Union
Japan	Australia and New Zealand	Rest of world

Industrial Sectors

Agriculture and mining:	Agriculture
	Livestock
	Forestry
	Fishery
	Mining
Non-heavy manufacturing industries:	Processed foods
	Beverages and tobacco products
	Textiles
	Garments and apparel
	Leather and leather products
	Wood and wood products
	Paper and paper products
	Plastic and chemical products
	Other mineral products
	Other manufactured products
Heavy manufacturing industries:	Petroleum and petroleum products
	Steel
	Non-iron metals
	Metal products
	Motor vehicles
	Other transportation equipment
	Electronics, electrical and other machinery

Utility
Construction
Trade, transportation and communications
Business and financial services
Other services

sults of our simulations. To highlight the effects on the subsectors within the service sector, in tables 4.11 to 4.13, the subsectors are aggregated further into eight classes.

The Impact on World Trade

Table 4.8 shows the impact on world trade of trade liberalization in different simulations. The results of simulation A show that for sole trade liberalization in the agricultural and manufacturing sectors, the world trade volume increases by 3.96 percent, while the value of world trade increases by 4.12 percent, which includes a 0.15 percent element for price inflation stemming from liberalization. Simulation B shows that by adding service-sector liberalization, world trade volume increases by 10.99 percent, whereas the value of world trade increases by 11.33 percent. Comparing

Table 4.8   Impact on World Trade (%)

| | Simulation | | | | |
World Trade	A	B	C	D	C1
World trade volume	3.96	10.99	11.39	11.40	11.00
World price index	0.15	0.30	0.18	0.18	0.17
Value of world trade	4.12	11.33	11.59	11.60	11.19

simulations B and A, we find that trade liberalization in the service sector has a much greater impact on world trade than agricultural and manufacturing trade liberalization. These results partly reveal the truth that trade in services is currently much more restrictive than trade in the agricultural and manufacturing sectors. A 50 percent cut in service trade barriers means a significantly larger cut in barriers in absolute terms and reflects the larger impact on the world economy. These results also highlight the importance of trade negotiations within the service sector.

We incorporate the spillover effect and technology links into our operation in simulation C, where we find that world trade volume grows by 11.39 percent, with an increase in value of 11.59 percent. Compared with simulation B, these results show that spillover and technology links contribute to the volume of world trade by a mere 0.4 percent; however, the effect on world price levels is much more significant. A comparison of the global price inflation rates of 0.30 percent in simulation B and 0.18 percent in simulation C seems to imply that the expansion of production due to technology transfusion is helpful in smoothing out world price inflation.

In simulation D, a 40 percent cut in transportation costs is assumed in trade between Taiwan and China. Simulation D shows that the world trade volume expands by 11.40 percent and that the value of world trade increases by 11.60 percent. When we compare this with simulation C, the 40 percent cut in transportation costs between Taiwan and mainland China does not appear to contribute significantly to world trade; the increase in world trade volume is a mere 0.01 percent. We regard this result as reasonable; after all, trade between Taiwan and mainland China does not occupy a significant share in terms of the total volume of world trade.

In order to highlight the impacts of Taiwanese and Chinese accession into the WTO, we add simulation E for global trade liberalization without their accession. In simulation E, we assume that Taiwan and China are not members of the WTO; thus, they will not carry out further liberalization actions, and there will be no shipping cost savings on trade between Taiwan and China. This assumption results in an 11.00 percent increase in world trade volume and an 11.18 percent increase in the value of world trade. In a comparison with simulation D, the results show that the effects on the world economy from Taiwanese and Chinese accession into the WTO are a 0.40

percent increase in world trade volume and a 0.42 percent increase in the value of world trade.

These results indicate that the effects of liberalization on the world economy after 2000 are significantly lower than the effects prior to 2000. This result does not surprise us, however, because almost all of the necessary liberalization actions for most of the developed countries had already been completed by 2000, as discussed earlier. After 2000, only the new members, Taiwan and China, will continue to carry out their commitments. The fact that simulation D gave no significant differences, as compared to simulation E, was therefore foreseeable.

In all five simulations, world prices increased, a result that we regard as quite feasible. Although the immediate impact of trade liberalization is to bring down import prices, it is possible that the induced increase in demand from both lower prices and higher income will boost the price as long as the resources for production are fixed. In this case, one way to bring down the price is to improve technology. This becomes obvious when we compare simulations C and B; the technology transfusion from advanced countries to less developed countries helps to bring down the world price.

The Effects on Regional Economies

Tables 4.9 and 4.10 summarize the effects of trade liberalization on regional economies. Table 4.9 shows the effects of liberalization on regional GDP, and in percentage terms the country that stands to gain the most is China. This is understandable given that, in absolute terms, China is the country that is probably liberalizing the most. Southern Asian countries also show significant improvements in their GDP. Hong Kong's GDP increases by 5.11 percent in both C and D simulations, and Hong Kong is the second highest beneficiary in world trade liberalization. This is quite pos-

Table 4.9   Effects on GDP, by Country/Region (%)

Country/Region	Simulation				
	A	B	C	D	E
Taiwan	0.46	0.72	1.51	1.53	1.03
China	2.41	2.75	7.39	7.41	6.24
Hong Kong	0.06	2.01	5.11	5.11	5.06
Japan	0.24	0.45	1.60	1.60	1.59
Korea	0.49	1.03	2.84	2.83	2.82
ASEAN	0.51	1.87	2.77	2.77	2.75
South Asia	1.54	2.02	3.55	3.55	3.54
Australia and New Zealand	0.18	0.89	2.31	2.31	2.32
United States	0.02	0.26	0.60	0.60	0.60
Canada	0.00	0.47	0.81	0.80	0.81
European Union	0.03	0.61	1.53	1.53	1.53
Rest of world	0.24	0.80	2.12	2.12	2.12

sibly the result of the promotion of mainland China's position in world trade. As the most active commercial port, and the biggest window for world connections into China, the liberalization of China will surely benefit Hong Kong.

However, when we look at other simulation results, in simulation A, Hong Kong has only a 0.06 percent improvement in its GDP. This shows that the sole liberalization of the agricultural and manufacturing sectors is of little help to Hong Kong. Simulations B and C show that in the liberalization of the service sectors, the technology spillover effects play very important roles. This probably has something to do with Hong Kong's role as a commercial port and financial services center. In simulations C and D, South Korea, the Association of Southeast Asian Nations (ASEAN), Australia and New Zealand, and the Rest of world all have more than a 2 percent improvement in their GDP. Asian countries benefit the most from trade liberalization in the agricultural and manufacturing sectors (see simulation A), and with the exception of Hong Kong, they show much more significant improvements in GDP than all other regions. This reflects exactly the comparative advantage of Asian countries in the manufacturing sectors.

Of course, liberalization of the service sectors also plays a very important role. As our model design shows, the less developed countries benefit from the advanced countries via spillover effects and the technology transfusion from the importation of services, capital goods, and durable intermediate goods. Countries with the ability to import more, and those that currently have greater technology gaps, vis-à-vis their advanced trading partners, stand to benefit the most; Asian countries tend to be the countries with these characteristics. Compared to the benefits for China, the benefits for Taiwan from trade liberalization are much more moderate, and this is of course due to the fact that Taiwan is already much more liberalized than China. In terms of further trade liberalization, the improvements in efficiency of resource allocation that are available to Taiwan are therefore much lower than those available to China. The results of simulations D and E show that the effects on other regions from Taiwanese and Chinese accession into the WTO are rather insignificant; however, trade liberalization improves Taiwan's GDP by 0.5 percent and China's by 1.17 percent. This seems to suggest that, even though Taiwan and China's accession into the WTO is unlikely to contribute very much to the world economy, it is nevertheless an important event for their own benefit.

Table 4.10 shows the welfare improvements for regions under different simulations in terms of Hicksian Equivalent Variation. The greatest figures shown are for the EU and Japan, with the respective welfare improvements for the two regions in simulation D being US$126,493 million and US$86,909 million. The United States and the Rest of world follow. In terms of absolute values, the size of economy plays an important part, and

Table 4.10   **Effects on Welfare, by Country/Region (US$ millions)**

	Simulation				
Country/Region	A	B	C	D	E
Taiwan	3,869	4,767	6,358	6,893	5,653
China	6,031	7,207	40,632	41,161	36,471
Hong Kong	2,442	10,354	13,454	13,425	12,957
Japan	23,431	27,534	86,981	86,909	85,089
Korea	5,491	8,729	17,388	17,324	16,581
ASEAN	3,758	13,613	19,393	19,337	19,119
South Asia	2,978	4,045	11,261	11,253	11,321
Australia and New Zealand	2,455	4,052	10,245	10,233	10,160
United States	1,072	22,995	45,084	45,031	44,728
Canada	−663	950	2,264	2,258	2,249
European Union	273	50,668	126,506	126,429	125,594
Rest of world	6,411	24,545	80,987	80,923	81,197

Note: Hicksian equivalent variation is used to measure changes in welfare.

in simulation D, Taiwan, with a relatively smaller economy size, has only a US$6,893 million welfare improvement, whereas the figure for China is US$41,161 million.

Similar to the case of GDP improvement, simulations D and E in table 4.10 also show that Taiwan's and China's accession into the WTO has only a limited contribution to other regions in terms of welfare improvement; however, there are significant welfare improvements to be gained by Taiwan and China themselves through accession and continued liberalization after 2000. For other regions, although Taiwan's and China's accession again does not bring much in the way of benefits, Japan's improvement in welfare reaches US$1,820 million. In absolute terms, this is almost the sort of improvement that Taiwan has gained from its continued liberalization. Japan's benefit from Taiwan's and China's accession into the WTO and liberalization is both understandable and foreseeable from their geographical position and trading relationship.

The Effects on Taiwan's Structural Changes

In tables 4.11 to 4.13, we focus our attention on Taiwan's economy. Table 4.11 shows the changes in Taiwan's output by industry. As shown in simulation A, with agricultural and manufacturing liberalization there is expansion in only the nonheavy manufacturing industries, utilities, and construction, whereas all other sectors experience contraction. With the utilities and construction sectors both being nontradable sectors, these results seem to indicate that Taiwan's comparative advantage still relies heavily on the nonheavy manufacturing industries.

Simulation B adds the effects of service trade liberalization, with the re-

Table 4.11   Effects on Taiwan's Output, by Sector (%)

Sector	Simulation				
	A	B	C	D	E
Agriculture and mining	−0.05	−0.06	−0.33	−0.34	−0.27
Non-heavy manufacturing industries	1.03	1.09	0.90	1.00	0.81
Heavy manufacturing industries	−0.64	−0.61	−0.14	−0.23	−0.37
Utility	1.14	1.16	1.07	1.20	0.90
Construction	2.53	2.02	2.03	2.27	0.75
Trade, transportation and communications	−0.20	0.34	0.39	0.31	0.50
Business and financial services	−0.20	−0.84	−0.82	−0.84	−0.75
Other services	−0.59	−0.72	−0.269	−0.25	0.06
Total	0.36	0.39	0.46	0.47	0.20

sults showing that agricultural and mining, heavy manufacturing industries, business and financial services, and the other services sector are still not comparatively advantageous. The output of the nonheavy manufacturing industries increases by 1.09 percent; the trade, transportation, and communications sector expands by 0.34 percent, whereas transportation and communications is the only service sector that increases output due to liberalization. When we add in technology spillover effects and cuts in transportation costs, the pattern of Taiwan's output change remains the same. The utilities sector and the construction sector both expand, whereas the heavy manufacturing industries contract, as do the service subsectors such as business and financial services and other services. The nonheavy manufacturing industries sector and the trade, transportation, and communications sector are the two sectors that have comparative advantage.

In simulation D, the output from the nonheavy manufacturing industry increases by 1 percent, whereas the trade, transportation, and communications sector expands by 0.31 percent. When we compare simulations D and E, the continuation of liberalization in Taiwan and China beyond 2000 increases the output from Taiwan's nonheavy manufacturing industries by 0.18 percent; however, it does not help in terms of expansion of the trade, transportation, and communications sector. It seems that resources are largely shifted to the relatively non–trade-related sectors, such as utilities and construction.

Tables 4.12 and 4.13 report the changes in Taiwan's exports and imports. Simulation A in table 4.12 shows that liberalization of the agricultural and manufacturing sectors increases exports in the agriculture and mining sector by 0.02 percent, the nonheavy manufacturing industries increase by 4.92 percent, and the utilities sector increases by 0.53 percent. Exports in the heavy manufacturing industries and in all service subsectors decrease. This seems to be consistent with the results of output changes in table 4.11.

Table 4.12   **Effects on Taiwan's Exports, by Sector (%)**

Sector	Simulation				
	A	B	C	D	E
Agriculture and mining	0.02	0.02	–0.09	–0.08	–0.15
Non-heavy manufacturing industries	4.92	5.16	4.66	5.12	3.93
Heavy manufacturing industries	–0.37	–0.15	1.48	1.25	–0.27
Utility	0.53	0.37	0.73	0.85	0.71
Construction	–9.18	4.49	8.24	7.06	8.12
Trade, transportation and communications	–4.39	46.02	46.46	44.85	48.85
Business and financial services	–11.48	8.73	9.64	8.04	12.23
Other services	–10.61	25.29	27.66	26.02	30.01
Total	4.06	7.30	8.38	8.52	6.01

Table 4.13   **Effects on Taiwan's Imports, by Sector (%)**

Sector	Simulation				
	A	B	C	D	E
Agriculture and mining	1.14	1.13	0.96	1.01	0.76
Non-heavy manufacturing industries	3.57	3.68	3.71	4.01	2.47
Heavy manufacturing industries	3.02	3.00	3.11	3.28	1.30
Utility	0.53	0.37	0.73	0.85	0.71
Construction	4.01	13.12	12.03	12.62	12.51
Trade, transportation and communications	3.99	72.40	72.93	74.06	71.42
Business and financial services	6.43	29.71	29.71	30.66	28.00
Other services	2.47	15.98	15.93	16.38	15.66
Total	8.30	13.22	13.21	13.83	9.82

When we add trade liberalization in the service sectors, exports expand in all sectors, with the exception of the heavy manufacturing industries in simulation B. This expansion is most significant in the service sector. Exports in the trade, transportation, and communications sector and the other services sector expand by 46.23 percent and 25.29 percent, respectively.

Combined with the decrease in output (see table 4.11) and increase in imports (see table 4.13), export expansion in the service sector indicates the prosperity of service trade after liberalization in that area. Simulations C and D in table 4.12 demonstrate the same pattern of change as simulation B. When we compare simulations D and E, the continuation of trade liberalization in Taiwan and China after accession into the WTO does not benefit exports in the service sector; however, it does benefit the manufacturing sector. Exports in the nonheavy manufacturing industries and the heavy manufacturing industries expand by 1.19 and 1.52 percent, respectively.

Following trade liberalization, imports in all sectors increase in Taiwan. In terms of proportional change, imports in the trade, transportation, and communications sector increase the most, expanding by 74.06 percent. In tables 4.12 and 4.13, simulations B, C, and D, both exports and imports expand in all sectors, with the single exception of the agricultural and mining sector. The results obviously reveal the tremendous effect of liberalization on trade expansion. Trade becomes more vibrant, and this is shown not only in the expansion of total trade volume but also in the volume of intraindustry trade. In all five simulations, the nontradable sectors, utility and construction, seem to expand significantly. Although these are actually very small sectors in comparison to the other sectors included in our model, the expansion of these sectors might be attributed to the resource reallocation effect resulting from trade liberalization. Trade liberalization causes an increase in imports and thus squeezes the market for domestic products in import-competing sectors. Due to the shrinkage of the import-competing sectors, some resources are then released from these sectors and transferred to nontradable sectors as well as export-expanded sectors. In addition, the increase in GDP resulting from trade liberalization creates a rise in domestic demand for the services of the utility and construction industries.

4.5   Conclusions

This study employs Hoekman's method to estimate the barriers to trade in services in Taiwan, followed by the application of a multiregional computable general equilibrium model to analyze the impacts of trade liberalization in services as well as commodities. The technology spillovers from the developed countries to the developing countries are considered through the import of intermediate inputs, capital goods, and services, along with an investigation of the potential differences resulting from the accession of Taiwan and China into the WTO.

The tariff equivalents compiled show that Taiwan's barriers to service trade are lower than those in most developing countries and are close to the levels of Japan, Hong Kong, and Korea. Following the accession of both Taiwan and mainland China into the WTO, trade in commodities and services will be further liberalized. The tariff equivalent in trade, transportation, and communications services will be reduced much further, so that it will become even lower than the average rate in the developed countries.

Our simulation results indicate that the developed countries benefit more from global trade liberalization in services, whereas developing countries benefit from the liberalization in manufacturing goods. With Taiwan's accession into the WTO and global trade liberalization, its GDP will increase by 1.53 percent; outputs of nonheavy industries, utility, construction, and trade, transportation, and communications increase, whereas outputs of heavy industries and service subsectors such as business and financial services and other services decrease, indicating that Taiwan's comparative advantage still relies heavily on the nonheavy manufacturing industries. The decrease in outputs, combined with import and export expansion in the service sector, indicates the prosperity of trade in services in Taiwan after liberalization. Our results also show that Taiwan and China's WTO accession contributes very little to the world economy but that trade liberalization improves Taiwan and China's respective GDP by 0.5 percent and 1.17 percent. This seems to suggest that liberalization is very important for their own benefit.

There are two limitations to this study, the first of which is that the Hoekman method still leaves much to be desired. The need to develop a more reliable measure of barriers in trade in services seems clear, and Warren and Findlay (1999) have made some progress in this area. The other limitation is that our simulation model does not incorporate the linkage between trade liberalization and commercial presence, and the movement of natural persons. Instead, it is assumed that the shocks from trade liberalization in four modes of service trade can be determined by changes in tariffs equivalents. Further research aimed at developing much more sophisticated models therefore seems warranted.

Appendix

Table 4A.1   Tariff Equivalents for Two-Digit ISIC Services Sectors

ISIC	ISIC Sectors	Benchmark Tariff Equivalent	Taiwan (Pre-WTO Accession)	Taiwan (Post-WTO Accession)	China	Hong Kong	Japan	Korea	ASEAN 5	Singapore	Indonesia	Malaysia	The Philippines	Thailand
5	Construction	40.0	10.0	10.0	25.0	32.0	5.0	16.0	21.2	12.0	16.0	10.0	40.0	28.0
61	Motor vehicle repair	40.0	5.0	5.0	40.0	40.0	5.0	40.0	32.0	40.0	40.0	40.0	0.0	40.0
62	Wholesale	40.0	16.7	16.7	40.0	40.0	5.0	27.5	38.5	40.0	40.0	40.0	40.0	32.5
63	Retail	40.0	5.0	5.0	40.0	30.0	5.0	20.0	40.0	40.0	40.0	40.0	40.0	40.0
64	Hotels and restaurants	20.0	10.0	10.0	10.0	10.0	2.5	2.5	5.5	2.5	5.0	5.0	5.0	10.0
71	Land transport	50.0	35.8	35.8	47.5	50.0	37.5	46.3	42.3	50.0	50.0	50.0	21.3	40.0
72A	Maritime/waterway transport	50.0	50.0	50.0	43.2	39.6	35.4	38.0	39.7	39.1	43.8	37.5	39.1	39.1
72B	Maritime cabotage	200.0	200.0	200.0	200.0	200.0	200.0	200.0	200.0	200.0	200.0	200.0	200.0	200.0
73A	Supporting air transport	50.0	39.6	39.6	50.0	50.0	33.8	42.5	44.5	50.0	50.0	50.0	31.3	41.3
73B	Air transport proper	200.0	200.0	200.0	200.0	200.0	200.0	200.0	200.0	200.0	200.0	200.0	200.0	200.0
74	Auxiliary transport	50.0	42.4	25.0	27.5	50.0	32.5	41.3	43.0	50.0	50.0	50.0	20.0	45.0
75A	Postal (incl. courier)	200.0	150.0	150.0	200.0	100.0	200.0	200.0	140.0	75.0	200.0	200.0	25.0	200.0
75B	Basic telecommunications	200.0	200.0	75.0	200.0	200.0	75.0	200.0	191.4	200.0	200.0	157.1	200.0	200.0
75C	VA telecommunications	100.0	12.5	12.5	100.0	50.0	12.5	12.5	53.9	37.5	57.1	46.4	46.4	82.1
81+82	Financial services	50.0	45.2	35.3	38.3	27.9	25.0	35.4	25.3	22.5	27.5	27.1	13.8	35.4
81+82	Life insurance	200.0	200.0	200.0	200.0	200.0	200.0	200.0	200.0	200.0	200.0	200.0	200.0	200.0
83	Real estate	50.0	37.5	37.5	25.0	37.5	18.8	50.0	50.0	50.0	50.0	50.0	50.0	50.0
84	Rental	40.0	19.0	15.0	40.0	28.0	19.0	21.0	32.8	40.0	40.0	22.0	26.0	36.0
85	Computers	20.0	2.5	2.5	10.0	6.0	2.5	2.5	13.0	6.0	16.0	11.0	20.0	12.0
86A	R&D (social sciences)	20.0	2.5	2.5	20.0	20.0	14.2	17.5	17.2	14.2	16.7	15.0	20.0	20.0
86B	R&D (hard)	100.0	100.0	100.0	100.0	100.0	100.0	100.0	100.0	100.0	100.0	100.0	100.0	100.0
89	Business	40.0	17.4	16.7	29.3	30.9	17.0	18.7	29.0	21.7	34.1	21.3	38.7	29.1
92	Sewage	50.0	12.5	12.5	50.0	50.0	6.3	26.6	45.0	50.0	50.0	50.0	50.0	25.0
93	Education	50.0	30.0	30.0	46.3	50.0	22.5	50.0	47.8	50.0	50.0	50.0	50.0	38.8
94	Health/society	50.0	31.3	31.3	47.9	50.0	44.8	50.0	44.8	36.5	50.0	37.5	50.0	50.0
96	Recreation and culture	20.0	16.0	12.9	20.0	17.1	6.9	14.4	16.6	14.6	18.8	13.8	19.0	16.7

ISIC	ISIC Sectors	South Asia	India	Sri Lanka	Bangladesh	Pakistan	Australia and New Zealand	Australia	New Zealand	United States	Canada	European Union	Rest of World
5	Construction	37.0	34.0	40.0	40.0	34.0	8.5	12.0	5.0	5.0	6.0	10.0	30.4
61	Motor vehicle repair	40.0	40.0	40.0	40.0	40.0	40.0	40.0	40.0	5.0	5.0	5.0	33.3
62	Wholesale	40.0	40.0	40.0	40.0	40.0	5.0	5.0	5.0	5.0	10.0	12.5	35.0
63	Retail	40.0	40.0	40.0	40.0	40.0	15.0	7.5	22.5	5.0	10.0	10.0	32.4
64	Hotels and restaurants	9.4	10.0	10.0	10.0	7.5	2.5	2.5	2.5	2.5	5.0	5.0	5.7
71	Land transport	50.0	50.0	50.0	50.0	50.0	33.8	43.8	23.8	27.5	21.9	31.9	47.0
72A	Maritime/waterway transport	50.0	50.0	50.0	50.0	50.0	38.8	37.5	40.1	50.0	46.4	47.9	47.6
72B	Maritime cabotage	200.0	200.0	200.0	200.0	200.0	200.0	200.0	200.0	200.0	200.0	200.0	200.0
73A	Supporting air transport	50.0	50.0	50.0	50.0	50.0	40.7	32.5	48.8	41.3	35.0	32.5	45.9
73B	Air transport proper	200.0	200.0	200.0	200.0	200.0	200.0	200.0	200.0	200.0	200.0	200.0	200.0
74	Auxiliary transport	50.0	50.0	50.0	50.0	50.0	36.9	23.8	50.0	42.5	18.8	32.5	48.1
75A	Postal (incl. courier)	200.0	200.0	200.0	200.0	200.0	200.0	200.0	200.0	25.0	50.0	200.0	131.7
75B	Basic telecommunications	200.0	200.0	200.0	200.0	200.0	200.0	200.0	200.0	200.0	200.0	200.0	190.8
75C	VA telecommunications	88.4	64.3	100.0	100.0	89.3	18.8	25.0	12.5	12.5	12.5	25.0	74.0
81+82	Financial services	44.3	36.3	50.0	50.0	40.8	17.8	16.7	18.8	12.5	17.9	25.0	36.3
81+82	Life insurance	200.0	200.0	200.0	200.0	200.0	200.0	200.0	200.0	200.0	200.0	200.0	200.0
83	Real estate	50.0	50.0	50.0	50.0	50.0	12.6	18.8	6.3	6.3	12.5	18.8	46.6
84	Rental	40.0	40.0	40.0	40.0	40.0	8.5	12.0	5.0	19.0	5.0	17.0	35.3
85	Computers20.0	16.8	16.0	20.0	20.0	11.0	6.0	6.0	6.0	2.5	2.5	2.5	13.6
86A	R&D (social sciences)	18.5	18.3	20.0	20.0	15.8	17.1	14.2	20.0	20.0	14.2	15.0	17.7
86B	R&D (hard)	100.0	100.0	100.0	100.0	100.0	100.0	100.0	100.0	100.0	100.0	100.0	100.0
89	Business	38.6	38.9	40.0	40.0	35.4	20.3	12.2	28.3	10.6	17.2	13.3	32.6
92	Sewage	50.0	50.0	50.0	50.0	50.0	33.6	17.2	50.0	6.3	6.3	6.3	42.1
93	Education	50.0	50.0	50.0	50.0	50.0	23.8	23.8	23.8	33.8	50.0	20.0	43.9
94	Health/society	45.8	45.8	50.0	50.0	37.5	36.5	30.2	42.7	43.8	50.0	29.2	45.2
96	Recreation and culture	19.4	18.8	19.2	20.0	19.4	15.0	14.4	15.6	3.3	18.8	13.8	17.9

Sources: Details on Taiwan are compiled from this study; details on the remaining countries are obtained from Hoekman (1995).

Notes: The original GATS-specific breakdown of activities is contained in the United Nation's Central Product Classification (CPC), for industrial comparison, the code is converted to International Standard Industrial Classification (ISIC). The case of the construction sector in Taiwan is used to illustrate the derivation of the tariff equivalent (te) as follows: Since the coverage ratios (cr) for Taiwan's construction services, both before and after, are 0.5, and the benchmark tariff equivalent (te*) of the construction sector is 40%, following the tariff equivalent equation as shown in the text (i.e., te = (1 − cr) × te*), the tariff equivalent for the construction services in Taiwan is te = (1 − 0.5) × 40% = 20%.

Table 4A.2   Taiwan's Further Commitments, Post-WTO Accession

I. Horizontal Commitments
   1. Foreign businesses and individuals in Taiwan may invest directly in portfolios in companies whose shares are listed in the securities market of Taiwan, without previous ceiling limitations.
   2. Duration of business visitors extended from sixty days to ninety days.
   3. Intra-corporate transferees may stay for an initial period of two-years renewable each year, as compared to the previous period of stay of no more than three years.

II. Sector-Specific Commitments

Professional services:	Legal and Architectural services
Rental leasing:	Financial leasing of private cars
Telecommunications services:	Basic telecommunications services
	Voice telephone services
	Packet-switched data transmission services
	Circuit-switched data transmission services
	Telex, Telegraph and Facsimile services
	Private leased circuit services
Audiovisual services	Motion picture and video tape production and distribution services
	Motion picture projection services
	Sound recording
Insurance and its related services	Nonlife insurance services
	Reinsurance and retrocession
	Services auxiliary to insurance (including broking and agency services)
Banking and other financial services	Acceptance of deposits and other repayable funds from the public
	Lending of all types
	Financial leasing
	All payment and money transmission services
	Guarantees and commitments
	Trading for own account or for account of customers
	Participation in the issues of short-term bills
	Money broking, asset management
	Settlement and clearing services for financial assets
	Advisory and other auxiliary financial services
	Provision and transfer of financial information
	Other
Services auxiliary to all media of transport	Cargo-handling, storage and warehousing and freight transport agency services
	Other

Source: Draft Schedule for The Separate Customs Territory of Taiwan, Penghu, Kinmen, and Matsu.

Note: The listed sector-specific commitments are further liberalized either completely or partially in terms of four modes of supply.

Table 4A.3   Liberalization of Taiwan's Agricultural and Manufacturing Sectors (%)

Sector	2000	2005
Agriculture	–3.11	–50.11
Livestock	–0.60	–58.40
Forestry	–0.10	–33.53
Fishery	–3.16	–37.71
Mining	–3.69	–7.17
Processed foods	–2.72	–14.88
Beverages and tobacco products	–57.43	–75.97
Textiles	–13.24	0.74
Garments and apparel	0.00	–12.45
Leather and leather products	0.00	–28.33
Wood and wood products	–0.87	–46.91
Paper and paper products	–14.42	–99.66
Petroleum and petroleum products	0.00	–42.51
Plastics and chemicals products	–1.79	–29.24
Other mineral products	–5.97	–30.62
Steel	–1.98	–97.37
Non-iron metals	–0.73	–30.56
Metal products	–4.10	–41.04
Motor vehicles	–0.66	–40.13
Other transportation equipment	–3.65	–26.54
Electronics, electrical and other machinery	–16.80	–29.31
Other manufactured products	–3.20	–52.50

Table 4A.4   Reductions in Tariff Equivalents in Service Sectors (%)

	Tariff Equivalents			Reduction		
	Pre-2000 (1)	Post-2000 (2)	50% (3)	Pre-2000 (4)	Post-2000 (5)	Total (6)
Construction	10.00	10.00	0.00	–50.00	0.00	–50.00
Trade, transportation and communications	85.70	53.90	–37.11	–31.45	–18.55	–50.00
Business and financial services	24.30	21.40	–11.93	–44.03	–5.97	–50.00
Other services	26.20	25.60	–2.29	–48.85	–1.15	–50.00

Notes: Tariff equivalents for services sectors are calculated from the second section of this study. The 50% reduction in tariff equivalents is an arbitrary figure adopted in Robinson, Wang, and Martin (1999). (3) = [(2)/(1) – 1] * 100.

Table 4A.5   Tariff Reductions in Mainland China (%)

Sector	2000	2005
Agriculture	−29.84	−40.08
Livestock	−49.93	−66.69
Forestry	−64.38	−81.97
Fishery	−39.60	−53.94
Mining	−61.77	−58.50
Processed foods	−36.93	−44.50
Beverages and tobacco products	−45.26	−56.96
Textile	−56.89	−71.94
Garments and apparel	−56.19	−71.13
Leather and products	−62.02	−78.20
Wood and products	−62.56	−78.20
Paper and products	−46.81	−59.12
Petroleum and products	−26.30	−26.65
Plastic and chemical products	−53.63	−68.06
Other mineral products	−55.24	−69.99
Steel	−34.03	−39.89
Non-iron metals	−48.43	−61.32
Metal products	−59.97	−75.43
Motor vehicle	−63.07	−78.73
Other transportation equipment	−52.39	−66.49
Electronics, electrical machinery and other machinery	−48.54	−61.41
Other manufactures	−62.29	−77.91
Utility	−54.49	−68.30
Construction	−54.49	−68.30
Trade, transportation and communication	−54.49	−68.30
Business and financial services	−54.49	−68.30
Other services	−54.49	−68.30

Table 4A.6   Effects on Taiwan's Output, by Sector (%)

| | Sector | | | | |
Simulation	A	B	C	D	E
Agriculture and mining					
Agriculture	−8.716	−8.849	−14.593	−14.686	−10.977
Livestock	4.609	4.315	−6.822	−6.944	−6.312
Forestry	−1.248	−1.589	−1.698	−1.730	−1.607
Fishery	0.235	0.173	−0.502	−0.581	−0.425
Mining	−1.562	−1.862	−2.785	−2.715	−2.127
Non-heavy manufacturing industries					
Processed foods	8.161	7.766	−1.961	−2.085	−0.828
Beverages and tobacco products	−14.398	−14.515	−14.526	−14.598	−12.540
Textile	21.186	22.243	24.488	26.200	19.602
Garments and apparel	−1.031	1.826	3.838	3.018	4.754
Leather and products	7.395	9.674	14.156	17.529	12.279
Wood and products	−2.638	−2.783	−2.359	−2.792	−2.583
Paper and products	−0.158	−0.074	-0.055	0.158	0.617
Plastic and chemical products	5.155	5.368	5.746	6.335	4.503
Other mineral products	0.413	0.065	−0.105	0.131	−0.068
Other manufactures	−6.593	−7.147	−6.206	−6.510	−4.073
Heavy manufacturing industries					
Petroleum and products	0.866	0.914	0.621	0.827	0.913
Steel	−2.187	−2.961	−3.182	−3.409	−3.678
Non-iron metals	−0.156	−0.109	0.761	0.513	−0.631
Metal products	−16.037	−16.426	−15.391	−15.998	−6.582
Motor vehicle	12.436	12.812	18.422	18.070	9.813
Other transportation equipment	−4.595	−4.407	−1.869	−2.358	−3.113
Electronics, electrical and other machinery	5.257	6.011	6.977	7.152	4.447
Utility	1.143	1.160	1.065	1.197	0.900
Construction	2.526	2.022	2.033	2.273	0.750
Trade, transportation and communication	−0.204	0.335	0.389	0.309	0.502
Business and financial services	−0.199	−0.835	−0.817	−0.836	−0.753
Other services	−0.594	−0.720	−0.269	−0.251	0.061

Table 4A.7 **Effects on Taiwan's Exports, by Sector (%)**

Simulation	Sector A	B	C	D	E
Agriculture and mining					
Agriculture	-2.548	-2.180	-9.153	-8.199	-23.183
Livestock	15.349	15.592	-34.662	-34.840	-46.269
Forestry	-4.720	-5.814	-2.006	2.191	2.589
Fishery	-4.347	-3.502	3.459	2.876	2.162
Mining	3.124	4.328	0.875	8.136	7.867
Non-heavy manufacturing industries					
Processed foods	37.926	36.518	-4.970	-5.669	6.275
Beverages and tobacco products	30.535	37.121	37.776	35.865	29.659
Textile	26.927	27.847	30.485	32.858	24.272
Garments and apparel	-5.857	-1.175	1.884	0.245	4.900
Leather and products	12.583	15.372	20.783	25.757	17.865
Wood and products	-2.556	-2.868	-2.468	-2.918	-3.364
Paper and products	11.496	13.098	15.221	16.486	11.836
Plastic and chemical products	10.895	11.159	11.660	13.035	8.692
Other mineral products	6.336	7.135	8.320	10.285	7.792
Other manufactures	0.169	-1.639	-0.287	1.103	-4.012
Heavy manufacturing industries					
Petroleum and products	-4.595	-3.599	-4.971	-5.434	-4.327
Steel	6.342	5.155	3.016	3.700	0.673
Non-iron metals	5.048	5.256	6.524	6.357	2.433
Metal products	4.577	4.190	5.511	3.234	-14.294
Motor vehicle	29.132	30.054	39.217	38.963	21.904
Other transportation equipment	-4.137	-3.782	-0.852	-1.343	-2.495
Electronics, electrical and other machinery	8.542	9.482	10.542	10.912	6.325
Utility	0.532	0.369	0.726	0.853	0.709
Construction	-9.183	4.488	8.242	7.059	8.122
Trade, transportation and communication	-4.388	46.023	46.463	44.852	48.851
Business and financial services	-11.477	8.730	9.643	8.039	12.234
Other services	-10.611	25.285	27.656	26.017	30.010

Table 4A.8   Effects on Taiwan's Imports, by Sector (%)

Simulation	Sector A	B	C	D	E
Agriculture and mining					
Agriculture	25.449	25.082	19.499	19.884	13.641
Livestock	0.942	1.083	17.927	18.894	22.953
Forestry	0.214	0.129	0.244	−0.245	−0.758
Fishery	20.686	20.720	14.061	15.012	11.621
Mining	1.586	1.561	0.941	1.577	1.368
Non-heavy manufacturing industries					
Processed foods	1.545	2.101	6.298	7.109	7.256
Beverages and tobacco products	100.613	100.763	102.631	104.138	82.315
Textile	14.126	16.029	15.930	17.425	14.199
Garments and apparel	17.354	12.394	10.898	14.210	7.308
Leather and products	18.604	19.632	20.928	25.623	17.374
Wood and products	3.925	3.455	2.910	3.373	0.739
Paper and products	9.395	9.549	9.207	9.810	4.336
Plastic and chemical products	7.860	8.345	8.016	8.846	5.568
Other mineral products	8.204	8.773	8.309	9.355	4.183
Other manufactures	11.990	12.723	11.848	12.557	1.819
Heavy manufacturing industries					
Petroleum and products	11.570	12.264	11.819	12.481	6.347
Steel	0.707	1.141	1.662	1.692	0.020
Non-iron metals	12.211	12.437	12.437	13.262	4.760
Metal products	27.431	27.539	28.133	28.941	8.554
Motor vehicle	12.750	12.700	12.890	13.536	8.314
Other transportation equipment	3.031	2.906	3.082	3.319	1.523
Electronics, electrical and other machinery	10.141	10.018	10.182	11.054	4.697
Utility	0.532	0.369	0.726	0.853	0.709
Construction	4.006	13.115	12.030	12.620	12.512
Trade, transportation and communication	3.991	72.395	72.927	74.060	71.419
Business and financial services	6.434	29.707	29.713	30.661	28.001
Other services	2.473	15.981	15.934	16.382	15.660

References

Brown, D. K., A. V. Deardorff, A. K. Fox, and R. M. Stern. 1995. The liberalization of services trade: Potential impacts in the aftermath of the Uruguay Round. In *The Uruguay Round and the developing countries*, ed. W. Martin and L. A. Winters, 292–315. Cambridge: Cambridge University Press.

Francois, J. F., B. McDonald, and H. Nordstrom. 1995. Assessing the Uruguay Round. In *The Uruguay Round and the developing economies*, ed. W. Martin and L. A. Winters, 117–214. Cambridge: Cambridge University Press.

Hertel, T. W. ed. 1997. *Global trade analysis: Modeling and applications*. Cambridge: Cambridge University Press.

Hertel, T. W. 2000. Potential gains from reducing trade barriers in manufacturing, services, and agriculture. *Federal Reserve Bank of St. Louis Review* 82 (4): 77–100.

Hoekman, B. 1995. Assessing the General Agreement on Trade in Services. In *The*

Uruguay Round and the developing economies, Discussion Papers no. 307, ed. W. Martin and L. A. Winters, 327–64. Washington, D.C.: World Bank.

———. 1996. Assessing the General Agreement on Trade in Services. In *The Uruguay Round and the developing countries,* ed. W. Martin and L. A. Winters, 88–124. Cambridge: Cambridge University Press.

Markusen, J. R., and T. R. Rutherford. 1999. Foreign direct investment in services and the domestic market for expertise. Paper presented at the *Second Annual Conference on Global Economic Analysis.* 20–22 June, Copenhagen, Denmark.

McDougall, R. A., A. Elbehri, and T. P. Truong. 1998. *Global trade assistance and protection: The GTAP 4 database.* Purdue University. Center for Global Trade Analysis.

Robinson, S., Z. Wang, and W. Martin. 1999. Capturing the implications of services trade liberalization. Paper presented at *the Second Annual Conference on Global Economic Analysis.* 20–22 June, Ebberup, Denmark.

Warren, T., and C. Findlay. 1999. Measuring impediments to trade in services. Paper presented at *the Second Annual Conference on Global Economic Analysis.* 20–22 June, Ebberup, Denmark.

World Trade Organization (WTO). 1999. *An introduction to the GATS.* Geneva: WTO.

Comment   Ponciano S. Intal, Jr.

As Warren and Findlay, Hoekman and Braga, and Philippa Dee have all noted, the literature on the modeling efforts on services trade liberalization is still miniscule. Thus, the Chou et al. paper is a welcome addition, especially because it tackles economywide effects of services trade liberalization in a multiregional model. Drawing from the Philippine experience, the discussion, analysis, and policy decisions on services trade deregulation and liberalization have been primarily at the industry level (e.g., banking and telecommunications), with barely any quantitative analysis on the effects on the rest of the economy. This is in sharp contrast to the policy debate on agriculture and manufacturing protection and liberalization, which benefited from the availability of estimates of the structure of tariffs and industry protection and economywide analyses. In the service sectors, it has been very difficult to estimate the "tariff equivalents" of service-sector regulation and protection. As a result, it was not possible to undertake economywide quantitative analysis of the effects of service-sector deregulation and liberalization.

My comments on the paper focus on the following points: the need to refine or modify the Hoekman-type estimates; the technology spillover effect; the structural effects of trade liberalization on Taiwan's economy; and Taiwan's and China's nonaccession to the World Trade Organization (WTO).

Ponciano S. Intal, Jr. is professor of economics and executive director of the Angelo King Institute for Economic and Business Studies, De La Salle University, Manila, the Philippines.

Need to Refine or Modify the Hoekman-Type Estimates

The authors used the Hoekman method to estimate the tariff equivalent of service-sector regulation and protection for Taiwan and used the Hoekman estimates for the other countries in their multicountry, multi-industry computable general equilibrium (CGE) analysis. Aware of the weaknesses of the Hoekman method, the authors nevertheless chose to use it because Hoekman's estimates are the only consistent set of estimates for the whole world that is apparently available.

Be that as it may, it still merits consideration to modify at least some of the more indefensible and unrealistic Hoekman-type estimates. The credibility of the CGE analysis hangs on the reasonableness of the estimates of the tariff equivalents of service-sector regulation and protection. Thus, the more robust the estimates of tariff equivalents, the better.

Some examples of the puzzling Hoekman estimates can be cited here. It is very hard to believe that the tariff equivalent for the trade, transport, and telecommunications sector in the Philippines is lower than in Singapore. This is especially so if there is a presumption that the lower the tariff equivalent, the more contestable and efficient the industry (which is apparently the implicit assumption in the paper). A statement that the Philippine trade, transport, and telecommunications sector, given its lower tariff equivalent, is more efficient than Singapore's would be totally laughable. The sector is one of the sources of Singapore's international competitiveness; in contrast, the transport and telecommunication industries have been one of the major bottlenecks to Philippine development. It is also worth noting that there are many foreign retailers in Singapore, whereas the Philippine retail industry was closed to foreigners from the late 1950s until 2000.

Similarly, I find it very surprising that Japan's trade, transport, and telecommunications sector has a significantly lower average tariff equivalent than that of the United States, especially given the usual American complaint about the Japanese distribution sector and the apparently more open and dynamic American sector compared to Japan's.

The Hoekman-type results show particularly high tariff equivalent in the transportation, storage, and communication sector in virtually every country in the world, including the United States, Canada, and the European Union. Most of the estimates are beyond 100 percent. Developed countries have very low tariff rates on goods: for example, virtually zero for Singapore and an average of 6.5–6.6 percent for Japan and the United States. Given the importance of transport, storage, and communication in much of the goods sectors in these countries, the high tariff equivalent implies that the effective rates of protection in the goods sectors could be close to zero or even negative. This seems unrealistic. Again, this rather awkward result suggests the need to refine the Hoekman-type estimates, especially for the transport, storage, and communication sector.

In this regard, the recent work of T. Warren and C. Findlay (1999) and their colleagues in Australia and the rest of East Asia on measuring the tariff equivalents of the barriers to trade in services is worth exploring as a possible alternative to the Hoekman-type estimates. At least, it would be useful to calibrate the model by drawing from the results from Warren, Findlay, and their colleagues in order to determine whether or not the model's results are robust.

On the Technology Spillover Effect

I find the incorporation of the technology spillover effect interesting and potentially very insightful. However, I find the simulation results puzzling. By assumption, the technology spillover effect is only from the developed countries to the developing countries. However, a comparison between the results of scenario B and scenario C in table 4.9 (on the effects on gross domestic product [GDP] by region) indicates that, with the exception of China, the GDP growth rate rose more proportionately in the developed countries (i.e., Japan, the European Union, Canada, Australia-New Zealand, and the United States) than in the developing countries (i.e., the Association of Southeast Asian Nations [ASEAN]). This is inconsistent with what could be expected from a technology spillover effect that is presumably meant to benefit more the developing countries by assumption. It is important for the authors to explain the counterintuitive results of the impact of the technology spillover effect on GDP growth.

Structural Effects of Trade Liberalization on Taiwan's Economy

The results of the simulations on the economic structure of Taiwan are interesting but also puzzling. For example, in simulation A, the results indicate that goods liberalization under the Uruguay Round will largely benefit nonheavy manufacturing, utilities, and construction and will hurt heavy manufacturing and all the service subsectors (see table 4.11). As the authors pointed out, this result in simulation A in table 4.11 seems to indicate that Taiwan's comparative advantage still lies in nonheavy manufacturing. This seems somewhat at odds with the popular impression that the country has become a world leader in semiconductors and other electronics industries (included in heavy manufacturing in the authors' classification) and is losing competitiveness in the more labor-intensive industries like garments (included in nonheavy manufacturing in the authors' classification). Given the rising real wages in Taiwan and the secular appreciation of the New Taiwan dollar, Taiwan has been in the process of economic restructuring away from relatively unskilled labor–intensive industries like garments and plastics (under nonheavy manufacturing) toward skilled labor– and capital-intensive industries like electronics and electrical machinery. It is likely that there is an aggregation problem here where both losing and gaining industries are included in the same classification under heavy manufacturing or nonheavy manufacturing. It may be better to give information on the output effects of the twenty-seven industry subsectors rather than the more aggregated sectors in table 4.11 in order to help the readers gain better insight on the impact of trade liberalization on the Taiwanese economy.

Another likely explanation for the positive (negative) impact of the goods liberalization under the Uruguay Round on Taiwan's nonheavy (heavy) manufacturing output can be gleaned from the schedule of Taiwan's liberalization in tandem with the Uruguay Round. Specifically, Taiwan's schedule suggests that the nonheavy manufacturing industries (e.g., leather goods, garments and textiles, processed food) have the lowest tariff reductions. Thus, the positive output effect of Taiwan's proposed goods liberalization on the nonheavy manufacturing sector stems from the shift in *relative* industrial protection toward the nonheavy manufacturing sector.

However, it is puzzling that the outputs of the trade, transportation, and communication sector and the other service sectors declined *despite the comparatively high rates of protection.* Two possible explanations can be pointed to here. The first one is that the goods-sector liberalization resulted in significant real wage increase, thereby making the service sectors (being presumably labor intensive) less competitive, resulting in lower exports, greater imports, and less domestic output. The second possible explanation is that the heavy manufacturing sector, which registered a reduction in output, is particularly intensive in the use of services, thereby dragging down the output performance of the service sectors. I find the second possible explanation less convincing.

Clearly, it is best for the authors to provide an economic underpinning, including what is happening to the relative factor prices in conjunction with the factor intensities of the various sectors, to the somewhat surprising results in scenario A.

A comparison of the total output effects under scenarios A and B in table 4.11 shows a marginal increase in the growth rate from 0.35 percent to 0.39 percent. That is, the table indicates that services trade liberalization has miniscule impact on the overall output of the economy. This is partly because services trade liberalization results in the restructuring of the service sector itself, with lower outputs in services except in trade, transport, and communication and a corresponding increase in both exports and imports of services. The only noticeable impacts of the services trade liberalization on the structure of the economy are the reversal in the output growth of trade, transport, and communication sector; the marked deceleration in the growth rate of the construction industry; and a marginal acceleration in the growth rate of nonheavy manufacturing. These results are interesting and deserve elaboration. What is behind these structural effects? What is happening with factor prices? What is happening on the exchange rate? Is there a real currency appreciation that has dampened the output effect of the services trade liberalization? How do these results on Taiwan of services trade liberalization—that is, miniscule overall output effect but significant industry restructuring effect within the service sector—compare with similar policy experiments in other countries?

Taiwan's and China's Nonaccession to the World Trade Organization

The authors present an interesting simulation on the effect of Taiwan's and China's not being admitted into the WTO. The authors found that the accession of Taiwan and China into the WTO would not have a significant effect on the world economy and the economies in the region, but the concomitant trade liberalization in the two countries will nonetheless have substantial benefits to Taiwan and China. For Taiwan, the accession will provide significant boost to trade in manufactures and a decline in exports of services. The sectoral and trade effects in Taiwan are not very surprising because much of tariff reductions in goods are backloaded after the year 2000, whereas much of the service-sector liberalization occurred before 2000. The minimal impact of China's accession into the WTO on the countries in the region is somewhat surprising, given the growing importance of the China market, the large tariff reductions planned, and the competition that China presents to other developing countries (e.g., ASEAN and South Asia). It seems reasonable to expect that with the accession of China into the WTO and the further opening up of the Chinese economy, a further round of industrial restructuring could occur in East Asia and the Pacific in the next decade similar to the so-called "flying geese" that occurred in the late 1980s and early 1990s in the region. The authors might like to elaborate further on their results with respect to the nonaccession of China and Taiwan into the WTO.

Reference

Warren, Tony, and Christopher Findlay. 1999. How significant arc the barriers? Mearuring imediments to trade in services. Paper presented at the *Services 2000: New Directions in Services Trade Liberalization Conference.* 1–2 June, the University Club, Washington, D.C.

Comment   June-Dong Kim

This paper simulates the effects of services trade liberalization using a multiregional computable general equilibrium model and also investigates the effects of the accession of Taiwan and Mainland China to the World Trade Organization (WTO).

June-Dong Kim is a research fellow at the Korea Institute for International Economic Policy.

This study improves upon previous works by incorporating technical change in international transport services due to the accession of Taiwan and mainland China into the WTO. It also endogenizes the balance of trade by setting up a global bank in order to capture the efficiency gains from the trade liberalization, particularly in services trade.

Even though the authors mention limitations to their study, let me iterate some comments on their measures of tariff equivalents in service sectors. The authors adopt Hoekman's method for calculating tariff equivalents in service sectors, which is based on the sectoral coverage ratios of specific commitments by each of the member countries. More specifically, Hoekman's calculation of a country-specific tariff equivalent in each service sector is defined as one minus the coverage ratio multiplied by the benchmark tariff equivalent, which takes a value of 200 percent for the most restrictive sectors such as transportation and basic telecommunications, and a value of 20 to 50 percent for the rest.

The resulting measure of tariff equivalent may be useful for making comparisons of relative restrictiveness in services trade across countries. However, one needs to be cautious when using this measure in simulating the effects of services trade liberalization.

First, the simulation results using this measure may be sensitive to the absolute magnitude of the benchmark tariff equivalent set for each sector. The benchmark tariff equivalent can be interpreted as "prohibitive" rates, where services trade is completely restricted. Setting a benchmark value of 200 percent for certain sectors and a value of 20 to 50 for the rest can be indicative of the relative restrictiveness of each service sector across countries. However, applying this measure in simulating percentage changes of some macro variables may lead to overestimation or underestimation, depending on the absolute magnitude of the benchmark tariff equivalent that is arbitrarily set for each sector.

Furthermore, multiplying the benchmark tariff equivalent by country- or sector-specific coverage ratios is problematic in that it double-counts the country- or sector-specific trade barriers. The relative restrictiveness of services trade is already incorporated in the country- or sector-specific coverage ratios. Hoekman's method of setting the benchmark tariff equivalent depending on the relative degree of restrictiveness reiterated the measurement of the trade restrictiveness of each sector and country, magnifying the difference in tariff equivalents across sectors.

Therefore, simulations using this measure tend to overestimate the effects in those sectors where trade is relatively more restrictive, and, thus, a relatively high benchmark tariff equivalent is assigned. The simulation results, that the effects of trade liberalization in transport and communications are a lot greater than those of the other sectors, may be attributed to these problems.

To avoid this kind of problem, one may assign an equal benchmark tariff equivalent across sectors (for example, 100 percent) and multiply them by the country-specific coverage ratios. The resulting measure of country- and sector-specific tariff equivalents may yield more reasonable simulation results.

In addition, as the authors mention, the simulation model does not take into account commercial presence and movement of natural persons. Because Hoekman's tariff equivalents cover all of the four modes of supply, there is an inconsistency between the coverage of tariff equivalents and the simulation model. In order to be consistent, it is necessary to construct tariff equivalents incorporating only cross-border supply and consumption abroad.

Concerning the computable general equilibrium (CGE) simulation using the GTAP model, simulation results of the CGE model tend to be sensitive to the elasticity parameters used in the simulation. If the elasticity parameters of the service sectors are set relatively higher than those of agricultural or manufacturing sectors, it might lead to the simulation results that trade liberalization in the service sectors has much greater impact on world trade than trade liberalization in agriculture or manufacturing. Reporting the elasticity parameters used in the simulation will be informative in this regard.

It might also be helpful if the authors provide some intuitive explanation on what brings about the rise in inflation rates after trade liberalization, even in simulations C and D. In addition, it may be interesting to further investigate why the utilities and construction sectors in Taiwan show a great increase in output due to trade liberalization.

5

The Growth and Potential of Taiwan's Foreign Trade in Services

Li-min Hsueh, An-loh Lin, and Su-wan Wang

5.1 Introduction

Taiwan's economy has performed remarkably well since 1951, growing at an annual average rate of 8.41 percent during the half-century period from 1951 to 1999, and foreign trade has undoubtedly played an important role in this remarkable growth rate. Taiwan's exports and imports of goods and services in real terms was only 16.9 percent of its gross domestic product (GDP) in 1951, but this share had increased steadily to 68.9 percent in 1981 and reached a staggering 95.7 percent in 1999, an extremely high percentage in comparison with many other countries.

Over the past two decades, as Taiwan has continued to maintain its high economic growth rate, there has also been rapid growth in Taiwan's foreign trade in services, with the domestic-service sector as a percentage of real GDP rising from 49.3 percent in 1971 to 56.8 percent in 1991, and reaching a peak of 62.9 percent in 1999. Thus, the percentage of trade in services in real GDP, as given in the national income accounts, has increased from 11.3 percent in 1971 to 14.2 percent in 1991 and to 15.1 percent in 1999. This share of the services trade is expected to increase further in the future as Taiwan gains accession into the World Trade Organization (WTO) and its economy becomes fully liberalized and internationalized.

Still, the services trade has played a minor role in Taiwan's total trade and it has always been in deficit. The current status of Taiwan's services trade is closely related to Taiwan's developmental stage, which went through the peak of industrialization of the economy and reached the stage of expand-

Li-min Hsueh and Su-wan Wang are research fellows at the Chung-hua Institution for Economic Research, and An-loh Lin is a professor in the department of economics at Shih Hsin University.

ing the economy's services sector in the mid-1980s. Government policies play an important role in the development of Taiwan's service sector. Before the mid-1980s, the government's industrial policies and its tax and financial incentive schemes were mainly directed toward manufacturing. Until the late 1980s, the scope of operation for several major service industries, such as banks, other financial institutions, highway passenger transportation, and air transportation, was restricted and new entries were only occasionally granted. Although Taiwan's service sector has grown rapidly in recent decades, its competitiveness is still weak.

To examine the past and future development of Taiwan's trade in services, this paper plans to (a) present a historic account of Taiwan's services trade and examine some of the factors responsible for its growth; (b) analyze the sources of growth of Taiwan's trade in producer services within the input-output framework; and (c) discuss the potential for future growth of Taiwan's services trade in the light of world trade developments.

The remainder of this paper is structured as follows: Section 5.2 examines the past development of the Taiwanese services trade, based on data from the national income accounts and some of their determinants. Section 5.3 analyzes Taiwan's producer import services and the sources of its growth based on data from the official input-output tables. Section 5.4 examines the trend in the services trade and discusses Taiwan's comparative advantage, based on data from its balance-of-payments statistics, with those of ten other developed and newly industrialized countries. It also provides a further examination of the trade in several specific services, including transportation, communications, and business services, and their potential for future development when Taiwan enters the WTO. Conclusions are drawn in Section 5.5.

5.2 Taiwan's Trade in Services

In this section, imports and exports of services are defined as covering all imports and exports except for those of goods valued on a free on board (f.o.b.) basis. They include such items as transport, warehousing, communication, insurance, financial services, business services, overseas travel, royalties and license fees, and other service items relating to imports or exports. The data available for this paper come from three sources: the *National Income Statistics*, input-output (I-O) tables, and the international balance-of-payments (BOP) statistics. The *National Income Statistics* (Directorate-General of Budget, Accounting, and Statistics, various issues) provide time-series data on the aggregate exports and imports of goods and services and on some crude components. The I-O tables (Directorate-General of Budget, Accounting, and Statistics, various issues) provide data on import services by the producing industry but only in intermittent years. The BOP statistics (Central Bank, various issues) contain annual information on a number of service items relating to imports or exports. Data from these three sources are not exactly in agreement with each other in terms of definition and coverage, and thus are not strictly comparable. Except for the aggregate data on the imports and exports of goods and services from the *National Income Statistics*, the data are available only in current dollars and in units of local or U.S. currency, and these create problems of comparability over time if prices change. Despite all these deficiencies, the data from these three sources are analyzed in turn in the present and following sections.

5.2.1   Import and Export Trends

Table 5.1 lists separate annual data on Taiwan's imports and exports of goods and services, import and export service shares, and balance of trade in goods and services for the period 1951–99 and the average annual growth rates of these variables over each decade and over the whole period. Data are available on aggregate imports and exports for the components of goods and services in current and constant prices from the national income accounts (NIA), but merchandise transport (including storage and communication) and insurance charges, considered items of services, are included in the goods component and thus excluded from the services component. We have deducted these transport and insurance costs from the goods component and added them to the services component by using information on foreign transactions (available from the NIA for the years covering 1981–99) as well as information available from the BOP for the years prior to 1981. The resultant adjusted imports and exports and their services shares are listed in the first six columns. The next two columns are separate net exports for goods and services expressed in US$ millions. The last two columns are shares of services based on the original, unadjusted data. These are given for a comparison of the movements of the service shares with or without inclusion of the merchandise transport and insurance in the service components. In addition, average annual growth rates for each item are given at the foot of the table for each decade from 1951 to 1999 and for the whole period.

Several observations can be made from table 5.1. First, Taiwan's services imports started with a much smaller base and grew faster on average than its goods imports. The average annual growth rate is 13.3 percent over the entire period 1951–99 for the former, as compared with 11.8 percent for the latter. Both growth rates have come down considerably in reflection of the slowing rate of overall economic growth over the last decade, with a 6.2 percent average growth rate obtained for the imports of services as compared with a higher growth rate of 8.5 percent reached for the imports of goods. The share of service imports rose from 12.4 percent in 1951 to 31.6 percent in 1985, declined to 27.3 percent in 1986, and then by 1999 fell back to 20.8 percent—which remained 5 percentage points higher in comparison with the United States' share. The faster increase in the imports of goods relative

Table 5.1 Taiwan's Imports and Exports of Goods and Services (1951–99)

Year	Goods and Services Imports in 1991[a] (constant NT$ millions)			Goods and Services Exports in 1991[a] (constant NT$ millions)			Balance of Trade (current US$ millions)		Share of Services[b] (%)	
	Goods	Services	Share of Services (%)	Goods	Services	Share of Services (%)	Goods	Services	Imports	Exports
1951	15,736	2,232	12.42	10,026	1,133	10.15	-46	-10	7.81	7.38
1952	21,298	2,806	11.64	10,469	721	6.44	-64	-14	6.99	3.55
1953	22,729	3,530	13.44	12,702	761	5.65	-60	-23	8.89	2.73
1954	22,396	4,773	17.57	9,419	601	6.00	-100	-35	13.23	3.09
1955	19,845	3,419	14.70	11,902	1,054	8.14	-60	-23	10.21	5.29
1956	23,245	4,173	15.22	13,070	1,460	10.05	-75	-20	10.76	7.27
1957	23,490	3,908	14.26	14,600	2,279	13.50	-70	-12	9.75	10.82
1958	26,493	4,245	13.81	14,611	2,527	14.74	-100	-15	9.28	12.11
1959	31,121	4,923	13.66	18,735	1,973	9.53	-94	-24	9.11	6.73
1960	33,757	5,887	14.85	20,644	2,882	12.25	-105	-24	10.37	9.54
1961	40,132	5,904	12.83	25,133	4,391	14.87	-115	-9	8.24	12.24
1962	41,164	6,107	12.92	25,997	4,094	13.61	-90	-13	8.33	10.93
1963	45,317	7,038	13.44	33,960	5,480	13.89	-18	-7	8.50	11.09
1964	54,225	8,680	13.80	44,844	5,666	11.22	38	-19	8.82	9.65
1965	67,211	10,828	13.87	55,389	7,993	12.61	-65	-19	8.82	10.82
1966	69,972	10,738	13.30	63,534	12,396	16.33	-10	19	8.27	14.74
1967	88,149	16,419	15.70	72,199	15,151	17.35	-72	-3	8.85	13.41
1968	109,229	29,832	21.45	90,102	22,338	19.87	-72	-49	14.46	17.63
1969	134,996	28,692	17.53	115,141	23,980	17.24	-10	-25	11.00	15.62
1970	164,712	37,127	18.39	154,168	24,290	13.61	93	-96	11.82	12.23
1971	200,428	45,923	18.64	210,313	29,310	12.23	293	-131	12.51	11.05
1972	245,895	55,392	18.38	286,432	36,588	11.33	649	-156	13.81	10.47
1973	305,273	68,814	18.40	349,909	56,580	13.92	742	-173	12.08	13.36
1974	353,956	73,135	17.12	320,718	57,925	15.30	-823	-295	11.62	14.53
1975	326,715	77,464	19.17	321,068	58,843	15.49	-250	-257	14.99	14.58
1976	411,584	91,538	18.19	447,404	72,246	13.90	689	-288	14.28	13.19
1977	441,799	85,648	16.24	504,988	81,398	13.88	1,194	-132	12.30	13.17

Year										
1978	483,831	122,079	20.15	624,830	97,815	13.54	2,227	−500	15.72	12.94
1979	561,014	161,570	22.36	654,235	113,934	14.83	1,358	−984	17.73	14.09
1980	623,643	139,107	18.24	717,567	111,664	13.47	335	−827	13.68	12.59
1981	617,287	160,787	20.66	779,503	128,769	14.18	1,940	−954	15.73	12.28
1982	576,894	186,414	24.42	792,760	133,670	14.43	3,759	−1,250	19.47	12.30
1983	618,763	227,238	26.86	943,917	141,902	13.07	6,412	−1,891	21.50	10.81
1984	683,937	282,249	29.21	1,127,040	154,530	12.06	9,347	−2,727	24.25	9.84
1985	637,020	293,767	31.56	1,152,439	157,734	12.04	11,313	−2,758	26.54	9.00
1986	810,064	304,148	27.30	1,478,598	205,047	12.18	17,204	−2,216	21.46	9.46
1987	1,072,698	347,950	24.49	1,808,334	198,989	9.91	21,437	−3,687	19.32	6.98
1988	1,289,124	414,951	24.35	1,872,919	238,627	11.30	18,006	−4,814	19.54	7.83
1989	1,427,570	465,981	24.61	1,951,900	266,926	12.03	17,365	−6,141	19.80	8.70
1990	1,509,600	497,147	24.77	1,972,139	260,800	11.68	15,288	−7,318	20.45	8.11
1991	1,760,350	549,647	23.79	2,224,816	296,983	11.78	15,982	−7,852	19.79	8.22
1992	1,941,608	660,367	25.38	2,365,432	324,658	12.07	15,360	−11,574	21.43	10.95
1993	2,082,436	730,009	25.96	2,475,356	412,750	14.29	14,413	−10,767	22.17	13.54
1994	2,201,652	702,792	24.20	2,641,044	394,468	13.00	14,986	−10,903	20.52	12.30
1995	2,422,122	766,483	24.04	2,978,275	431,958	12.67	16,895	−12,494	20.61	12.03
1996	2,603,033	777,924	23.01	3,170,401	469,688	12.90	20,662	−11,225	19.80	12.27
1997	3,015,937	829,608	21.57	3,456,228	514,236	12.95	17,202	−11,000	18.42	12.32
1998	3,179,270	910,025	22.25	3,493,596	572,549	14.08	13,231	−10,291	19.28	13.46
1999	3,379,959	888,615	20.82	3,975,850	480,099	10.77	18,066	−12,862	17.95	10.15
Annual Average Growth Rate (%)										
1951–61	9.81	10.22	0.32	9.63	14.51	3.89	−9.50	0.54	0.53	5.20
1961–71	17.45	22.77	3.81	23.67	20.90	−1.94	13.54	−30.12	4.27	−1.02
1971–81	11.91	13.35	1.04	14.00	15.95	1.49	20.80	−21.92	2.31	1.07
1981–91	11.05	13.08	1.42	11.06	8.72	−1.84	23.48	−23.47	2.33	−3.93
1991–99	8.50	6.19	−1.66	7.53	6.19	−1.11	1.54	−6.36	−1.22	2.66
1951–99	11.84	13.28	1.08	13.27	13.43	0.12	13.25	−16.09	1.75	0.67

Source: Constructed from *National Income Statistics*, Directorate-General of Budget, Accounting, and Statistics, Republic of China (various issues).

[a]Merchandise shipping and insurance are included in services.

[b]Merchandise shipping and insurance are excluded in services.

to that of services after 1985 resulted from a combination of factors, such as a strong recovery of the Taiwanese economy beginning in 1986, a substantial fall in the world prices of commodities, huge appreciation of the New Taiwan (NT) dollar, reductions of import duties, lessening of trade restrictions, upgrading of the industrial structure, and so on. These factors helped increase imports of consumption goods, materials, parts, and equipment more than imports of services.

Second, Taiwan's exports of services have been growing at 13.4 percent per year during the entire period, a rate close to the 13.3 percent for its exports of goods. Similar to the growth patterns of imports, the growth rates of exports have also slowed down over the last decade, having a 6.2 percent average growth rate for the exports of services as compared with an 8.5 percent growth rate for the exports of goods. The services share of exports fluctuated over the entire period. It was 10.2 percent in 1951 and attained 19.9 percent in 1968. The ratio has declined since then, however, and was only 10.8 percent in 1999 as compared with a much higher ratio of more than 28 percent for the United States.

Third, Taiwan is a small open economy and is dependent on foreign trade. Its trade position rather depends on the underlying comparative advantages combined with government policies, market structure, and changes in technologies. Net exports of goods have been in surplus since 1971, with the exception of the oil-crisis years 1974, 1975, and 1980, but the services-trade balance has been constantly in deficit ever since 1951, apart from a small positive figure in 1966. The average annual rate of change for the period 1951–99 was 13.3 percent for the former and –16.1 percent for the latter. The situation does seem to improve after 1991, with the average rate of increase in services-trade deficits declining to 6.4 percent for the last decade (from more than 20 percent for the previous two decades).

Fourth, when merchandise transport and insurance are excluded, the share of services in imports or exports has risen faster as opposed to the share when these items are included, as shown in the last two columns of table 5.1. That is, the imports and exports of other services have increased much faster than the freight and insurance cost of the merchandise imports and exports. During the period 1951–99 the share of these other services had increased by 1.8 percent per year over total imports and 0.7 percent per year over total exports, as compared with the corresponding shares of 1.1 percent and 0.1 percent with the inclusion of the merchandise freight and insurance cost. As table 5.2 shows, both insurance and freight, each as a percentage of the cost plus insurance and freight (c.i.f.) to Taiwan, have been falling since 1981. The share of insurance had declined from 0.48 percent in 1981 to 0.31 percent in 1999, and the share of freight from 5.38 percent to 3.05 percent, over the same period. The declines are due partly to the falling shipping and insurance costs and partly to the locational shifts in imports and exports as trades between Taiwan and other Asian countries

Table 5.2   **Merchandise Imports: Cost (f.o.b.), Insurance, and Freight**

Year	CIF (US$ millions)	FOB (%)	Insurance (%) Total	Insurance (%) FF	Insurance (%) DF	Freight (%) Total	Freight (%) FF	Freight (%) DF
1981	21,773	94.14	0.48	0.39	0.10	5.38	4.42	0.96
1982	19,197	93.85	0.46	0.31	0.15	5.69	4.38	1.31
1983	19,935	93.17	0.47	0.30	0.16	6.37	4.75	1.62
1984	22,283	93.45	0.44	0.28	0.15	6.12	4.45	1.66
1985	20,498	93.17	0.43	0.24	0.19	6.40	4.41	1.99
1986	24,299	92.57	0.45	0.28	0.17	6.98	5.15	1.83
1987	34,476	93.59	0.45	0.31	0.14	5.96	4.34	1.62
1988	44,921	94.02	0.44	0.31	0.13	5.54	4.04	1.51
1989	51,875	94.00	0.40	0.28	0.11	5.60	4.26	1.34
1990	54,389	94.56	0.39	0.27	0.12	5.04	3.59	1.46
1991	62,693	95.01	0.39	0.27	0.12	4.60	3.30	1.30
1992	71,839	94.97	0.37	0.25	0.12	4.66	3.34	1.32
1993	77,009	95.13	0.36	0.26	0.10	4.52	3.65	0.87
1994	84,716	95.37	0.34	0.25	0.09	4.28	3.50	0.78
1995	102,203	95.68	0.35	0.27	0.09	3.97	3.25	0.71
1996	101,963	96.00	0.34	0.23	0.10	3.66	2.94	0.72
1997	112,024	96.13	0.33	0.24	0.09	3.53	2.84	0.70
1998	103,412	96.31	0.31	0.22	0.09	3.38	2.71	0.67
1999	109,481	96.64	0.31	0.22	0.10	3.05	2.47	0.58
			Annual Average					
1981–85	20,738	93.55	0.45	0.30	0.15	6.00	4.48	1.52
1986–90	41,992	93.85	0.42	0.29	0.13	5.73	4.20	1.53
1991–95	79,692	95.28	0.36	0.26	0.10	4.36	3.40	0.96
1996–99	106,720	96.27	0.32	0.23	0.10	3.41	2.74	0.67

Source: National Income Statistics, Directorate-General of Budget, Accounting, and Statistics, Republic of China.
Note: FF = foreign firms; DF = domestic firms.

have increased substantially. Table 5.2 also indicates that the declines of both insurance and freight as a percentage of c.i.f. were greater for foreign firms than for domestic firms over the period.

5.2.2   Determinants of Trade in Services

Next, we examine the factors affecting the annual growth of Taiwan's trade in services for the period 1951–99, as reported above. In this regard the economic theory of demand for a commodity can be applied to explain the demand for imports of services as well as the demand for exports of services. Income or output and relative price are thus two typical determinants to be considered, and some other variables, such as trade in goods, industrial structure, and government policy, are also relevant. Our procedure is (a) to discuss the explanatory variables employed, (b) to conduct a unit root test, (c) to perform a regression analysis and cointegration test, and (d) to estimate the corresponding error-correction model. We deal with imports of services first and then with exports of services.

Imports of Services

To estimate the demand for imports of services, the explained variable is real imports of services (IMPS) as listed in table 5.1. These figures are computed from *National Income Statistics* to include the imports of merchandise transport and insurance services. The explanatory variables include per capita real national income, relative price of imports to domestic services, real imports of goods, GDP share of the service sector, and a liberalization-policy variable. The selection of each of these variables is discussed below in turn.

1. *Per capita real national income (NIP)*. Imported services are purchased for intermediate or final use. The former refers to producer services and is affected by output, whereas the latter refers to final consumer services and is affected by income. An example is overseas trips. Higher income will induce more personal overseas trips and higher output will encourage more business trips abroad, and thus will increase imported services. Since income and output are closely related, we use income as an explanatory variable. Because NIP is a better measure of economic welfare and empirically performs better than total real national income, it is chosen as the income variable in our imports of services equation, with its data taken from *National Income Statistics*. NIP is expected to have a positive effect on IMPS.

2. *Relative price of import to domestic services (RPIMPS)*. Many import and domestic services are substitutable and thus import services will be affected by their relative prices. We measure this variable by the ratio of the implicit deflator of IMPS to the implicit GDP deflator for the service sector. The two deflators are calculated from *National Income Statistics*. Because the price of import services will reflect changes in the foreign exchange rate, an appreciation of local currency will lower the price of imported services and thus will increase their demand. The sign of RPIMPS should be negative.

3. *Imports of goods (IMPG)*. There are reasons that imports of goods will have a positive effect on imports of services. First, transport and insurance accompany shipments of goods; thus, more imports of goods increases the demand for transport and insurance services. Second, international businesses require communications, traveling, and financial services; thus, more imports of goods generates more demand for these services. Third, imports of equipment, materials, or finished goods may require importers to pay royalties, consultation fees, or assembly costs to foreign firms and thus may increase imported services. Fourth, trade in goods, a dominant component of foreign trade, can reflect the current status of an economy

and constitutes a good proxy for the output variable, thus supplementing the use of the income variable given above. Although IMPS is closely related to IMPG, their relationship is found to be nonlinear. We thus add a quadratic term as an additional variable. IMPG, with data taken from *National Income Statistics* as listed in table 5.1, is expected to affect IMPS positively, but the effect of its square term is somewhat uncertain.

4. *Share of the service sector (SS)*. This variable is used as a measure of the maturity of an economy in pursuit of economic development, which typically evolved from an agrarian economy to a manufacturing economy and then to a service economy. Taiwan has become a service economy since the mid-1980s, with SS now more than 60 percent. Expansion of the service sector should have a net negative effect on imports of services because more and more imported services would be replaced by domestic services. Whether shifts in Taiwan's industrial structure toward a service economy would have such an effect can be tested by the presence of the share variable. Data on SS in real terms, taken from *National Income Statistics*, are used.

5. *Policy of liberalization (DETR)*. The government began to relax foreign exchange control and trade restrictions on goods and services in the mid-1980s and has continued this liberalization policy since then. Therefore, more and more financial and other services can be provided by foreign firms to domestic users, and this should increase imports of services. In order to measure the effect of the liberalization policy, we employ a proxy variable that equals zeros for the period 1951–85 and equals the effective tariff rate for the period 1986–99. The effective tariff rate, which is the ratio of import tariff revenues to custom imports of goods as computed from *Tax Revenues Statistics* (Ministry of Finance, various issues), has declined from 7.79 percent in 1986 to 2.82 percent in 1999. The gradual declines of the effective tariff rate appear to mirror the increasing efforts of the government in its implementation of the liberalization policy. The sign of DETR should be negative because a smaller tariff rate means greater liberalization and thus more imports of services.

The explained variable and the explanatory variables, except SS and DETR, are expressed in log form. All of these variables are found to have a unit root according to the augmented Dickey-Fuller unit root tests as shown at the bottom of table 5.3. Two regressions are run and found cointegrated of order 1 based on the Phillips-Perron tests. The first regression contains NIP, RPIMPS, IMPG and its square, and SS, whereas the second includes the same set of explanatory variables plus the policy variable, DETR.

To summarize in equation form, the second equation is given by

$$(1) \quad \ln(\text{IMPS}_t) = i_0 + i_1\ln(\text{NIP}_t) + i_2\ln(\text{RPIMPS}_t) + i_3\ln(\text{IMPG}_t) + i_4[\ln(\text{IMPG}_t)]^2 + i_5(\text{SS}_t) + i_6(\text{DETR}_t) + u_t$$

Table 5.3 Regression Results on Imports and Exports of Services

Imports of Services

Explanatory Variables	ln(IMPS)$_t$ Equation (1)	ln(IMPS)$_t$ Equation (2)	Δln(IMPS)$_t$ Equation (1')	Δln(IMPS)$_t$ Equation (2')
ln(NIP)$_t$, Δln(NIP)$_t$	1.653 (5.74) [0.000]	2.032 (6.06) [0.000]	0.429 (0.62) [0.541]	0.360 (0.50) [0.617]
ln(RPIMPS)$_t$, Δln(RPIMPS)$_t$	−0.793 (−3.79) [0.000]	−0.837 (−4.12) [0.000]	−0.390 (−1.42) [0.162]	−0.479 (−1.75) [0.088]
ln(IMPG)$_t$, Δln(IMPG)$_t$	2.492 (4.68) [0.000]	2.178 (4.06) [0.000]	2.144 (2.19) [0.034]	1.745 (1.80) [0.080]
[ln(IMPG)]$_t^2$, Δ[ln(IMPG)]$_t^2$	−0.088 (−3.46) [0.001]	−0.081 (−3.25) [0.002]	−0.075 (−1.79) [0.080]	−0.057 (−1.37) [0.180]
SS$_t$, ΔSS$_t$	−0.015 (−1.04) [0.302]	−0.026 (−1.72) [0.093]	−0.019 (−1.03) [0.310]	−0.022 (−1.21) [0.233]
DETR$_t$, ΔDETR$_t$		−0.029 (−2.03) [0.049]		−0.012 (−0.81) [0.425]
û$_{t-1}$			−0.695 (−4.69) [0.000]	−0.763 (−4.88) [0.000]
C	−23.630 (−4.78) [0.000]	−24.587 (−5.14) [0.000]	0.077 (1.88) [0.067]	0.077 (1.85) [0.071]
Adjusted R^2	0.9959	0.9962	0.4419	0.4645
SE	0.1274	0.1230	0.1118	0.1095
DW	1.404	1.543	1.915	1.863
N	49	49	48	48
CT	−5.16	−5.51		

Exports of Services

Explanatory Variables	ln(EXPS)$_t$ Equation (3)	Δln(EXPS)$_t$ Equation (3')
ln(USGDP)$_t$, Δln(USGDP)$_t$	0.475 (3.37) [0.002]	−0.083 (−0.47) [0.637]
ln(RPEXPS)$_t$, Δln(RPEXPS)$_t$	−1.635 (−2.66) [0.011]	−1.468 (−1.98) [0.055]
ln(EXPG)$_t$, Δln(EXPG)$_t$	3.003 (8.74) [0.000]	3.126 (3.49) [0.001]
[ln(EXPG)]$_t^2$, Δ[ln(EXPG)]$_t^2$	−0.097 (−6.59) [0.000]	−0.107 (−2.79) [0.008]
SS$_t$, ΔSS$_t$	0.054 (4.22) [0.000]	0.046 (1.68) [0.102]
û$_{t-1}$		−0.635 (−4.05) [0.000]
C	−16.558 (−7.66) [0.000]	0.044 (0.93) [0.360]
Adjusted R^2	0.9941	0.4844
SE	0.1647	0.1405
DW	1.441	1.300
N	49	48
CT	−5.62	

Augmented Dickey-Fuller Tests of Unit Roots with Constant and Trend[a]

Imports	ln(IMPS)	ln(NIP)	ln(RPIMPS)	ln(IMPG)	$[\ln(\text{IMPG})]^2$	SS	DETR
t test	−1.11	−2.34	−1.92	−1.28	−1.97	−0.75	−2.44
Exports	ln(EXPS)	ln(USGDP)	ln(RPEXPS)	ln(EXPG)	$[\ln(\text{EXPG})]^2$	SS	
t test	−0.60	−2.58	−1.88	−0.54	−1.41	0.02	

Notes: Equations are estimated by ordinary least squares. Figures in parentheses are *t*-statistics and those in brackets are *p*-values. Variables: NIP = per capita real national income; US-GDP = U.S. per capita real GDP; IMPS (IMPG) = imports of services (goods); EXPS (EXPG) = exports of services (goods); RPIMPS = relative price of import to domestic services; RPEXPS = relative price of export to import services; SS = share of the service sector; DETR = dummy times effective tariff rate; C = intercept; ln = natural log; SE = standard error of the estimate; DW = Durbin-Watson statistic; *N* = sample size; and CT = cointegration test (Phillips-Perron) with the critical value = −4.70. Explanatory variables with Δ are for equations (1′), (2′), and (3′).

[a] The critical value for rejecting the null hypothesis of having a unit root is −3.13.

where IMPS is real imports of services; NIP is per capita real national income at factor cost; RPIMPS is relative price of import to domestic services; IMPG is real imports of goods; SS is real share of the service sector in GDP; DETR is a proxy variable for import liberalization, assuming zeros for the period 1951–85 and equal to the effective tariff rate for the period 1986–99; u is an error term; t is the year; and $i_1 > 0$, $i_2 < 0$, $i_3 > 0$, $i_4 = ?(\leq 0$ or $\geq 0)$, $i_5 < 0$, $i_6 < 0$. The first equation is the same as equation (1) without the policy variable DETR. These two equations are estimated by ordinary least squares (OLS) with the results reported in table 5.3.

Column (1) of table 5.3 shows that the sign of the income variable is positive and that of the price variable is negative, as expected, and both are statistically very significant. The estimated income elasticity is 1.65 and the price elasticity is –0.80. Import of goods has a positive sign but its squared term a negative sign, implying that more imports of goods induces more imports of services but less proportionately. Share of the service sector has a negative effect on imports of services but the effect is not statistically very significant. Even so, the expansion of the service sector seems to have potential to replace part of the services imported. Addition of the policy variable appears to make the income and price effects larger and more significant. As indicated in column (2), the estimates of income and price elasticities become 2.03 and –0.84, respectively, which clearly indicates that the imports of services are both income elastic and price inelastic. The effect of imports of goods has remained about the same and that of the service share becomes greater and more significant. The significant, negative sign of the policy variable, measured by the declining effective tariff rate, implies that liberalization has benefited imports of services.[1]

---

1. Our regression results indicate that using per capita real national income performs better than using total real national income in terms of R^2, t-statistic, and Durbin-Watson statistic. The two regressions are given below for a comparison:

	C	ln(Y)	ln(RP)	ln(IG)	[ln(IG)]²	SS	DETR	R^2/SE	DW
Equation (2)	–24.6	2.032	–0.837	2.178	–0.081	–0.026	–0.029	0.9962/ 0.1230	1.543
t-statistic	–5.76	–6.06	–4.12	–4.06	–3.25	–1.72	–2.03		
Equation (2′)	–16.5	1.237	–1.124	1.544	–0.049	–0.023	–0.018	0.9959/ 0.1283	1.426
t-statistic	–4.32	5.52	–4.95	2.70	–1.93	–1.45	–1.31		

where Y = per capita real national income in equation (2) but = total real national income in equation (2′); RP = relative price; IG = imports of goods; SS = real share of service sector; DETR = proxy for import liberalization policy; R^2/SE = adjusted R^2/standard error of the estimate; and DW = Durbin-Watson statistic. It can be seen that using per capita real national income as the income variable has lower serial correlation than using total real national income and hence has a higher Durbin-Watson statistic and higher t-statistics for all the estimated coefficients except for the relative price. The overall fit is also better for the former than for the latter.

The above two regressions, both of which are cointegrated of order 1, show the long-run relationship between the imports of services and the four or five explanatory variables examined. To see their short-run relationship, two corresponding error-correction models are estimated and the results are presented in columns (1') and (2') of table 5.3. The explained variable is now in change in ln(IMPS) or the exponential growth rate in the imports of services, and its explanatory variables are also in change in each variable and include an error-correction term, which is the estimated OLS residual from column (1) or (2) lagged one period. The results indicate that the signs of the explanatory variables remain the same but their significances are considerably reduced. The sign of the error-correction term is negative and statistically very significant, with the estimated adjustment coefficient equal to −0.695 and −0.763, respectively. Thus the short-run behavior of services imports appears to follow its long-run relationship with per capita income, relative service price, goods import, service-sector share, and government policy as its determinants.

Exports of Services

To estimate the demand for export services, the explained variable is real exports of services (EXPS) as listed in table 5.1, whose figures are computed from *National Income Statistics* to include the exports of merchandise transport and insurance services. The explanatory variables include per capita real U.S. GDP, relative price of export to import services, real exports of goods, and GDP share of the service sector, as discussed below.

1. *Per capita real GDP of the United States (USGDP).* This variable is used as a measure of per capita real income for the rest of the world because the United States has been the most important trade partner of Taiwan and has also played a very influential role in the world economy. The variable is the product of U.S. per capita nominal GDP and the real exchange rate between Taiwan and the United States. The latter is obtained by multiplying the nominal exchange rate (total of NT$ per US$) by the ratio of U.S. and Taiwanese GDP deflators. Since both per capita or total real U.S. GDP perform equally well empirically, either can be employed. However, to be in line with the import equation, per capita real U.S. GDP is used as the income variable in our export equation. The data are taken from the United States' *Survey of Current Business* (Department of Commerce, various issues), *Statistical Abstract of the United States* (Department of Commerce, various issues), and Taiwan's *Financial Statistics* (Central Bank, various issues) and *National Income Statistics* (Directorate-General of Budget, Accounting, and Statistics, various issues). The variable is expected to have a positive effect on Taiwan's exports of services.

2. *Relative price of services between Taiwan and the rest of the world (RPEXPS).* We use Taiwan's export price of services as a measure of the

Taiwanese service price and Taiwan's import price of services as a measure of the world service price. Both prices are implicit deflators taken from *National Income Statistics*. The price ratio has taken into account changes in the foreign exchange rate. Thus an appreciation of local currency will increase the price ratio by either lowering the import price in NT$ or raising the export price in US$. The effect of the relative price on Taiwan's exports of services is expected to be negative.

3. *Exports of goods (EXPG)*. For the inclusion of exports of goods, reasons can be given similar to those for imports of services. First, services of transport and insurance accompany shipments of goods; thus, more exports of goods means greater demand for transport and insurance services, which can benefit domestic firms. Second, communications, traveling, and financial services go with international businesses; thus, more exports of goods generates more demand for these services. Third, exports of equipment, materials, or finished goods may require foreigners to pay royalties, consultation fees, or assembly costs to domestic firms, and thus may increase export services. The relationship between services export and goods export is found to be nonlinear as in the case of services imports. We thus add a quadratic term as an additional variable. EXPG, with data taken from *National Income Statistics* and listed in table 5.1, is expected to affect EXPS positively whereas the effect of its square term is more likely to be negative.

4. *Share of the service sector (SS)*. This variable measures the developmental stage of an economy. As the domestic service sector expands, the capability of the domestic firms to serve exporters or importers should also expand. Expansion of the service sector, measured in real terms, is thus expected to have a positive effect on exports of services.

The explained variable and the explanatory variables (except SS) are expressed in log form. All of these variables are found to have a unit root according to the augmented Dickey-Fuller unit root tests as shown at the bottom of table 5.3. A regression containing ln(USGDP), ln(RPEXPS), ln(EXPG), [ln(EXPG)]^2, and SS as the explanatory variables is run and found to be cointegrated of order 1 based on the Phillips-Perron tests. The equation is given by

$$(2) \quad \ln(\text{EXPS}_t) = e_0 + e_1 \ln(\text{USGDP}_t) + e_2 \ln(\text{RPEXPS}_t)$$
$$+ e_3 \ln(\text{EXPG}_t) + e_4 [\ln(\text{EXPG}_t)]^2 + e_5(\text{SS}_t) + v_t,$$

where EXPS is real exports of services; USGDP is per capita real GDP of the United States; RPEXPS is relative price of export to import (world) services; EXPG is real exports of goods; SS is real share of service sector in GDP; v is an error term; t is the year; and $e_1 > 0$, $e_2 < 0$, $e_3 > 0$, $e_4 = ?(\leq 0$ or $\geq 0)$, $e_5 > 0$. This equation is estimated by OLS and the results are given in column (3) of table 5.3.

The result shows that the sign of the income variable is positive and that of the price variable is negative, as expected, and both are statistically very significant. The estimated income elasticity is 0.48 and the price elasticity is –1.64 in contrast to 2.03 and –0.84 for the case of service imports. Thus Taiwan's exports of services are income inelastic but price elastic, whereas its imports of services are income elastic but price inelastic. This finding may explain in part why Taiwan's trade balance in services has been in deficit for the last fifty years. Signs of the exports of goods and its square are significantly positive and negative, respectively, as their counterpart on the imports of services, with the estimated values of the square term roughly the same. Thus more exports of goods induces more exports of services but less proportionately. Share of the service sector has a significant positive effect on exports of services, whereas it has a negative effect on imports of services. Growth of the service sector will thus help reduce the deficits of trade in services.[2]

The regression, which is cointegrated of order 1, shows the long-run relationship between the exports of services and the four explanatory variables examined. To illustrate their short-run adjustment relationship we estimate the corresponding error-correction model; the results are presented in column (3') of table 5.3. The explained variable is now in change in ln(EXPS) or the exponential growth rate in the exports of services, and its explanatory variables are also in change in each variable in equation (2) and include an error-correction term, which is the estimated OLS residuals from equation (2) lagged one period. The results indicate that signs of the explanatory variables remain the same and significant except for Δ ln(USGDP), which becomes negative but statistically insignificant. The short-run effect of income is thus almost nil but the effects of other variables remain intact. Sign of the error-correction term is negative and statistically significant

2. The two regressions based on per capita or total real GDP of the United States are given below for a comparison:

	C	ln(Y)	ln(RP)	ln(XG)	[ln(XG)]²	SS	R²/SE	DW
Equation (3)	–16.6	0.475	–1.635	3.003	–0.097	0.054	0.9941/ 0.1647	1.441
t-statistic	–7.66	3.37	–2.66	–8.74	–6.59	4.22		
Equation (3')	–18.1	0.462	–1.690	2.896	–0.093	0.052	0.9942/ 0.1624	1.434
t-statistic	–8.55	3.58	–2.78	8.33	–6.38	4.09		

where Y = per capita real USGDP in equation (3) but = total real USGDP in equation (3'); RP = relative price; XG = exports of goods; SS = real share of service sector; R^2/SE = adjusted R^2/standard error of the estimate; and DW = Durbin-Watson statistic. It can be seen that using per capita real USGDP as the income variable has a slightly higher Durbin-Watson statistic but slightly lower adjusted R^2 than using total real USGDP. The estimates of coefficients are about the same in terms of size and significance. Therefore, both perform about equally well and either one can be employed as the income variable in our export equation.

with the estimated adjustment coefficient equal to –0.635. The short-run behavior of services import appears to follow its long-run relationship with per capita world income, relative service price, goods export, and service-sector share as its determinants.

We conclude this section by pointing out that domestic or world income, relative price of services, trade in goods, and a country's service sector are four important determinants of the trade in services found in our regression analysis. Besides, government liberalization policies can play an important role in importing services.

5.3   Taiwan's Imports of Producer Services from Input-Output Tables

Input-output tables provide another source of data on Taiwan's imports and exports of services for analysis, but they are available only on an intermittent basis. In this section, we shall first summarize the information on Taiwan's imports and exports of goods and services in a sequence of five-year intervals covering the years 1966, 1971, 1976, 1981, 1986, 1991, and 1996.[3] We then focus on the imports of producer services in the framework of I-O tables[4] and decompose changes in the total imports of producer services between 1986 and 1991 and between 1991 and 1996 to examine the sources of the changes in terms of changes in I-O coefficients, in final demands, and in economic growth.

5.3.1   Trade Statistics from the Input-Output Tables

Table 5.4 provides a summary of statistics on the import and export of goods and services tabulated from the officially available I-O tables. The goods and services components of the imports are each listed by intermediate and final use, and net exports are given for goods and services. The amount, share, and growth are shown separately for the listed items; the amount is recorded in US$ millions converted from the local currency by use of the prevailing exchange rate. The share is nominal, but calculation of growth is based on local currency in real terms. Although the figures are given only for 1966, 1971, 1976, 1981, 1986, 1991, and 1996—and thus would be affected by the economic conditions of those particular years—the data over a thirty-year time span should display a general tendency of the growth of Taiwan's trade in goods and services.

Several observations can be made from the table. First, total imports of services has increased faster than total imports of goods. The annual rate of growth was 16.6 percent for the former and 11.8 percent for the latter over the entire period 1966–96. This was also the case for the imports of services

---

3. Input-output data are compiled based on censuses conducted independently every five years.

4. Producer services are defined as services that are used for production by the producing sectors of the economy.

Table 5.4   Amount, Share, and Growth of Trade in Goods and Services

	Imports of Goods and Services							Exports of Goods and Services			Net Exports		
	Goods				Services								
Year	Total	Sum	Intermediate Use	Final Use	Sum	Intermediate Use	Final Use	Total	Goods	Services	Total	Goods	Services
Amount (US$ millions)													
1966	734	693	453	240	41	24	17	707	598	110	-27	-96	69
1971	2,211	2,092	1,448	644	119	47	72	2,285	2,056	229	73	-37	110
1976	8,434	7,789	5,862	1,927	645	435	210	9,298	8,181	1,116	863	392	471
1981	24,460	21,773	17,041	4,732	2,687	2,046	641	24,821	21,580	3,241	362	-193	554
1986	28,316	24,663	20,061	4,602	3,653	2,664	989	43,196	38,238	4,959	14,880	13,575	1,306
1991	74,384	62,608	46,916	15,691	11,776	7,925	3,851	84,147	72,101	12,046	9,763	9,493	269
1996	103,075	84,868	72,210	12,658	18,207	12,478	5,730	131,127	108,236	22,891	28,051	23,368	4,684
Share (%)													
1966	100	94.4	61.7	32.7	5.6	3.2	2.3	100	84.5	15.5	100	356.0	-256.0
1971	100	94.6	65.5	29.1	5.4	2.1	3.2	100	90.0	10.0	100	-50.2	150.2
1976	100	92.4	69.5	22.9	7.6	5.2	2.5	100	88.0	12.0	100	45.4	54.6
1981	100	89.0	69.7	19.3	11.0	8.4	2.6	100	86.9	13.1	100	-53.3	153.3
1986	100	87.1	70.8	16.3	12.9	9.4	3.5	100	88.5	11.5	100	91.2	8.8
1991	100	84.2	63.1	21.1	15.8	10.7	5.2	100	85.7	14.3	100	97.2	2.8
1996	100	82.3	70.1	12.3	17.7	12.1	5.6	100	82.5	17.5	100	83.3	16.7
Growth (%)													
1966-71	23.4	23.1	24.5	20.2	25.0	15.7	34.7	24.5	25.8	17.1	-11.9	-9.6	10.6
1971-76	15.0	13.7	15.6	8.8	29.1	43.6	14.0	18.5	17.5	26.6	36.3	92.2	23.3
1976-81	9.4	8.1	9.0	5.4	21.3	24.3	14.0	10.5	10.2	13.4	20.1	38.0	-5.7
1981-86	6.7	6.4	7.2	3.2	5.6	4.7	8.3	13.1	13.8	9.0	39.1	41.8	22.1
1986-91	15.4	15.8	13.9	22.8	15.2	13.4	19.7	8.5	8.3	8.4	-14.4	-12.7	-34.9
1991-96	4.8	4.7	7.4	-5.6	5.7	6.1	4.9	7.6	6.9	10.1	24.1	18.6	72.2
1966-96	12.3	11.8	12.8	8.7	16.6	17.3	15.6	13.6	13.6	13.9	16.0	13.1	9.9

Source: Input-Output Tables, Directorate-General of Budget, Accounting, and Statistics, Republic of China (various issues).

Notes: Shares are computed as a percentage of total imports or exports or net exports. Exports of services also include domestic services relating to exports. Computation of growth rates is based on 1991 constant-price data. In computing a growth rate for net exports, if the sign of the base is negative, it is changed to positive. For example, if a net export changes from −5 to 5, it is treated as 10 divided by 5.

in intermediate or final use. Over the same period, the imports of producer and final services had grown 17.3 percent and 15.6 percent per year, respectively, as compared with the annual growth rates of 12.8 percent and 8.7 percent for the imports of intermediate and final goods. The growth in the imports of services was faster because it started from a base much lower than the imports of goods.

Second, the share of total imported services rose from 5.6 percent in 1966 to 17.7 percent in 1996. In particular, the imports of producer services as a percentage of the total imports of goods and services increased from 3.2 percent in 1966 to 12.1 percent in 1996, and the import share for final services rose at a slower rate, from 2.3 percent to 5.6 percent. Thus the percentage of imported producer services in total imported services increased from 58 percent in 1966 to 68 percent in 1996 as the share of imported consumption services decreased.

Third, the declining share of merchandise imports from 94.4 percent to 82.3 percent between 1966 and 1996 resulted mainly from the decline in final use, which fell from 32.7 percent in 1966 to 12.3 percent in 1996. The share of imported goods for production actually rose from 61.7 percent in 1966 to 69.5 percent in 1976, and then remained steady at 70.1 percent in 1996. If the goods and services components for intermediate use are combined, the share was 64.9 percent in 1966 and 82.2 percent in 1996. All these figures reflect the fact that the Taiwanese economy has increasingly depended on imports of intermediate inputs for production while the imports of goods for final use have gradually been substituted by domestic final goods.

Fourth, the exports of goods increased by 13.6 percent per year during 1966–96, a rate slightly below the 13.9 percent annual rate of increase for the exports of services. It is noted that domestic services that come as a result of the exports of goods, such as transport services to ports and export-related financial services, are treated as the exports of services in the I-O tables, contrary to the convention of national income accounting, which treats export-related inland services as part of the f.o.b. cost of the goods shipped abroad. This is why the net exports of services, as shown in the final column of table 5.4, are positive and not negative as seen in table 5.1.

The figures in tables 5.1 and 5.4 are therefore not strictly comparable due to the above-mentioned definitional problem, and also due to the differences in the units (i.e., real NT$ or nominal US$) used in the two tables. The imports and exports of services in table 5.1 are in local currency expressed in constant prices, and their shares are in real terms. These items in table 5.4 are calculated in current prices and in US$. However, the shares in both tables moved in tandem. For example, in table 5.4 the share of import services in 1966 was 5.6 percent (in nominal terms), but in table 5.1 this was 13.3 percent (in real terms), and these shares were 17.7 percent and 23.0 percent respectively, in 1996. In spite of the differences, they had all moved upward and maintained roughly the same level of disparity. However, the share of export services in table 5.4 had moved up from 15.5 percent in 1966 to 17.5 percent in 1996, while there was a fall in table 5.1 from 16.3 percent to 12.9 percent between 1966 and 1996, partly due to the definitional problem just mentioned.

Finally, we note that Taiwan has been well known for its development of hi-tech industries and its strong showing in the exports of hi-tech products, but its exports of hi-tech information services are still very limited. Statistics from the I-O tables show that the exports of hi-tech products, such as computer products, computer peripheral equipment, data-storage media, and computer components, were US$7,364 million in 1991 and US$15,965 million in 1996, and the corresponding imports were US$7,250 million in 1991 and US$1,474 million in 1996, resulting in a trade surplus of US$114 million in 1991 and US$14,491 million in 1996. The exports of hi-tech information services, including software designs, on the other hand, were US$259 million in 1991 and US$393 million in 1996; the corresponding imports were US$458 million in 1991 and US$712 million in 1996, yielding a trade deficit of US$199 million in 1991 and US$319 million in 1996. This is again an indication of weak showing for Taiwan's trade in services.

5.3.2   Decomposition of Growth in Total Imported Producer Services

The total amount of services imported by all industries for intermediate use, which is defined as producer service imports, was NT$54,867 million in 1986, NT$213,177 million in 1991, and NT$292,996 million in 1996, all in 1991 constant dollars. This increased on average by 31.2 percent per year over the period 1986–91 and 6.6 percent per year over the period 1991–96. In order to examine the causes of changes in these two periods, the amount of increase during each period is decomposed into several sources based on I-O tables and on an equation derived from Han (1995).

The Model

The decomposition for a vector of imported producer goods and services is given by

(3) $\Delta S = (M_t - M^*)R_0 Y_{D0}$ (import substitution effect)

$+ (M^* - M_0)R_0 Y_{D0}$ (changes in import intensity)

$+ M_0(R_t - R_0)Y_{D0}$ (changes in intermediate input)

$+ (M_t - M_0)(R_t - R_0)Y_{D0}$ (interaction effect)

$+ M_0 R_0 (y_t^c - \lambda^c y_0^c)$ (effect of changes in consumption)

$+ M_0 R_0 (y_t^g - \lambda^g y_0^g)$ (effect of changes in government expenditure)

$$+ \mathbf{M_0}R_0 (\mathbf{y_t^i} - \lambda^i y_0^i) \qquad \text{(effect of changes in investment)}$$

$$+ \mathbf{M_0}R_0 (\mathbf{y_t^{iv}} - \lambda^{iv} y_0^{iv}) \qquad \text{(effect of changes in stock change)}$$

$$+ \mathbf{M_0}R_0 (\mathbf{y_t^e} - \lambda^e y_0^e) \qquad \text{(effect of changes in export)}$$

$$+ \mathbf{M_0}R_0\mathbf{Y_{D0}} (\tilde{\lambda} - \tilde{\theta}) \qquad \text{(effect of changes in final demand component structure)}$$

$$+ \mathbf{M_0}R_0 (\theta - 1)\mathbf{Y_{D0}} \qquad \text{(economic growth)}$$

$$+ (\mathbf{M_t}R_t - \mathbf{M_0}R_0)(\theta - 1)\mathbf{Y_{D0}} \qquad \text{(growth multiplied by technical change effect)}$$

$$+ (\mathbf{M_t}R_t - \mathbf{M_0}R_0)(\mathbf{Y_{Dt}} - \theta\mathbf{Y_{D0}}) \qquad \text{(effect of interaction between technical change and changes in final demand structure)}$$

where S is a subtotal imported goods and services demand vector by categories (m × 1) measured in value; M is an import coefficient matrix by goods and services and by sector (m × n) with the coefficients measured in terms of value required per unit output; $\mathbf{M}^* = \mathbf{M_0}(\widehat{\mathbf{uM_t}})(\widehat{\mathbf{uM_0}})^{-1}$ with $\mathbf{u}$ being a unit row vector (1 × m) and (^) denoting the diagonal matrix of the vector in the parentheses; $\mathbf{Y_D}$ is a final domestic demand vector (n × 1) measured in value terms and $R = (\mathbf{I} - \mathbf{D})^{-1}$ with $\mathbf{D}$ being a domestic technical coefficient matrix (n × n) measuring the input requirements per unit output in value terms; and $\mathbf{I}$ is an identity matrix (n × n). $\theta = TDP_t / TDP_0$ is the expansion rate of the total domestic product (TDP) between any two years. Moreover,

$$\lambda^c = \frac{\mu\mathbf{y_t^c}}{\mu\mathbf{y_0^c}}, \quad \lambda^g = \frac{\mu\mathbf{y_t^g}}{\mu\mathbf{y_0^g}}, \quad \lambda^i = \frac{\mu\mathbf{y_t^i}}{\mu\mathbf{y_0^i}}, \quad \lambda^{iv} = \frac{\mu\mathbf{y_t^{iv}}}{\mu\mathbf{y_0^{iv}}}, \quad \lambda^e = \frac{\mu\mathbf{y_t^e}}{\mu\mathbf{y_0^e}}$$

where $\mathbf{y^c}, \mathbf{y^g}, \mathbf{y^i}, \mathbf{y^{iv}}$, and $\mathbf{y^e}$ represent the vectors of consumption, government expenditure, investment, inventory change, and exports, respectively, all in (n × 1), and μ is a unit row vector (1 × n). So λ^c is the ratio of total consumption in period t over that in period 0, and likewise of $\lambda^g, \lambda^i, \lambda^{iv}$, and λ^e. Moreover, $\tilde{\lambda} = (\lambda^c, \lambda^g, \lambda^i, \lambda^{iv}, \lambda^e)'(5 \times 1)$, and $\tilde{\theta} = (\theta, \theta, \theta, \theta, \theta)'(5 \times 1)$. Thus the first four terms in the right-hand side of equation (3) comprise the technological change effect, the next six terms comprise the effect from the final demand change, the eleventh term is the effect of economic growth and the last two terms capture interaction effect.

Equation (3) was applied to Taiwan I-O transactions tables of domestic goods and services and of import goods and services (c.i.f.) for 1986, 1991, and 1996, at 1991 constant prices, to compare the sources of demand fluctuation for total imported producer services over two periods, 1986 to 1991 and 1991 to 1996. All the I-O tables have thirty-nine–sector classification whereas imports are classified as goods; transportation, warehousing, and communications; financial and insurance services; food, beverage, and hotel services; business services; and other producer services.

Decomposition over 1986–91

Over the period 1986 to 1991 Taiwan's TDP grew by 45.0 percent. Had there not been any structural changes, then Taiwan's demand for imported producer services should also have grown by 45.0 percent. That is to say, *ceteris paribus*, economic growth in terms of final domestic product would cause imported producer services to increase by NT$56,798 million (table 5.5). In reality, however, imported producer services increased by NT$158,310 million, far higher than the increased value caused by TDP. Hence, economic growth accounts for only 35.9 percent of the total growth of imported producer services. The main reason Taiwan's imported producer services grew faster than TDP was the technical change in Taiwan's economy, which raised import intensity and increased by NT$71,388 million in imported services, accounting for 45.1 percent of the total growth. However, this was offset somewhat by the substitution-effect changes in producer services of NT$12,125 million.

In addition to the technical change factor, the increase in imported producer services was also partly due to changes in the structure of final demand. All other things being equal, a change in the structure of final demand would cause demand for imported producer services to increase by NT$11,474 million, which accounts for 7.2 percent of the total increase in imported producer services. The main source for this change comes from the structural changes in the component elements of final demand. The interaction term for technical change and final demand would also cause an increase in imported producer services of 20.4 percent.

Viewed in terms of demand for individual services, economic growth played a more important role in increases in the import of transportation, warehousing, and communications, and of finance and insurance services by inducing more imports of goods, thus confirming the finding in section 5.2 that merchandise imports contribute significantly to increases in the import of services. If we look at subitems of these changes, we find that a change in import intensity is even more important than the factor of economic growth in contributing to the import of transportation, warehousing, and communications. Moreover, the effect of technical change on other producer services and business services is far more significant than the effect of economic growth, although it might be partly induced by economic growth itself.

The increase in imported producer services caused by the change in the structure of final demand had the greatest impact (41.3 percent) on the import of finance and insurance. Among the subitems in the changes in final

Table 5.5   Decomposition of Changes in Taiwan's Producer Services Import Between 1986 and 1991 (NT$ millions at 1991 constant prices)

Item	Goods Amount	Goods %	Transportation, Warehousing, and Communications Amount	%	Finance and Insurance Amount	%	Food, Beverage, and Hotel Amount	%	Business Services Amount	%	Other Services Amount	%	Producer Services Subtotal Amount	%	Total Amount
Technical change	125,101	22.7	6,502	31.2	−189	−3.4	2,771	36.8	16,633	41.7	32,039	37.9	57,756	36.5	182,857
Substitution effect	6,091	1.1	−4,524	−21.7	−3,498	−62.8	−124	−1.6	8,388	21.0	12,367	−14.6	−12,125	−7.7	−6,034
Changes in import intensity	205,143	37.2	12,173	58.5	2,403	43.2	2,961	39.3	9,329	23.4	44,522	52.7	71,388	45.1	276,531
Changes in intermediate input	−70,359	−12.8	−1,027	−4.9	925	16.6	−49	−0.7	−237	−0.6	−358	−0.4	−746	−0.5	−71,105
Interaction term	−15,774	−2.9	−120	−0.6	−19	−0.3	−17	−0.2	−847	−2.1	242	0.3	−761	−0.5	−16,535
Changes in final demand structure	44,732	8.1	1,182	5.7	2,296	41.3	175	2.3	1,371	3.4	6,450	7.6	11,474	7.2	56,206
Changes in consumption	−16,161	−2.9	−569	−2.7	1,773	31.9	−16	−0.2	−126	−0.3	−115	−0.1	947	0.6	−15,214
Changes in government expenditure	271	0.0	3	0.0	−6	−0.1	4	0.1	13	0.0	4	0.0	18	0.0	289
Changes in capital formation	3,603	0.7	77	0.4	8	0.1	−3	0.0	75	0.2	−16	0.0	141	0.1	3,744
Changes in stock investment	17,384	3.2	−41	−0.2	−166	−3.0	−229	−3.0	−1,730	−4.3	22	0.0	−2,144	−1.4	15,240
Changes in export	30,962	5.6	807	3.9	12	0.2	−47	−0.6	212	0.5	−142	−0.2	842	0.5	31,804
Changes in final demand component structure	8,673	1.6	905	4.3	675	12.1	466	6.2	2,927	7.3	6,697	7.9	11,670	7.4	20,343
Economic growth	320,042	58.1	9,315	44.8	4,021	72.2	3,034	40.3	12,849	32.2	27,579	32.6	56,798	35.9	376,840
Interaction of technical change and change in final demand	61,362	11.1	3,814	18.3	−562	−10.1	1,548	20.6	9,005	22.6	18,477	21.9	32,282	20.4	93,644
Growth in multiple technical change	56,316	10.2	2,927	14.1	−85	−1.5	1,247	16.6	7,488	18.8	14,423	17.1	26,000	16.4	82,316
Interaction term	5,046	0.9	887	4.3	−477	−8.6	301	4.0	1,517	3.8	4,054	4.8	6,282	4.0	11,328
Total	551,237	100.0	20,813	100.0	5,566	100.0	7,528	100.0	39,858	100.0	84,545	100.0	158,310	100.0	709,547

Source: Input-Output Tables, Directorate-General of Budget, Accounting, and Statistics, Republic of China (1986, 1991).

demand structure, changes in consumption (which accounted for 31.9 percent) was the most important factor.

In summary, for transportation, warehousing, and communications services; food, beverage, and hotel services; business services; and other producer services, the main sources of the increase in imports over the period 1986 to 1991 was economic growth and technical change. However, the increase in imported financial and insurance services was brought about by economic growth and changes in final demand.

Decomposition over 1991–99

Over the period 1991 to 1996, TDP in Taiwan grew by 37.9 percent. Other things being equal, imported producer services demand should have increased by NT$108,281 million (table 5.6). However, the demand for imported producer services increased by only NT$79,819 million, less than the rate of TDP growth. This was mainly because structural changes in final demand caused it to fall by NT$31,218 million, which accounts for a negative 39.0 percent of the total increase in the import of producer services. Changes in the structure of the final demand component and changes in government expenditure were the main causes.

In addition, technical change led to import density change, which also caused imported producer services demand to fall by NT$23,432 million, a negative 29.4 percent of the total increase in the import of producer services. However, substitutability between imported producer services caused demand to increase by NT$23,234 million (29.1 percent). Furthermore, changes in the intermediate input coefficient and the interaction term have both negatively affected the demand for imported producer services. On balance, the effect of technical change reduced demand for imported producer services.

As far as individual sectors are concerned, economic growth increased demand for imported producer services in all sectors. The impact was particularly great in the other producer-services sector in terms of value and in the finance and insurance sector in terms of contributing to the rate of growth. However, as changes in technology created a fall in transportation, warehousing, and communications services; financial and insurance services; and food, beverage, and hotel services, there was a significant increase in the substitution effect for business services and other producer services. Change in import intensity had a large negative impact on import of the financial and insurance services sector and on other producer services. If one looks at the overall effect of technical change, one can see that the effects were negative for the transportation, warehousing, and communications sector; for the financial and insurance services sector; and for the food, beverage, and hotel services sector.

The main sources of import change for the transportation, warehousing, and communications sector; financial and insurance services sector; and

Table 5.6   Decomposition of Changes in Taiwan's Producer Services Import Between 1991 and 1996 (NT$ millions at 1991 constant prices)

Item	Goods Amount	%	Transportation, Warehousing, and Communications Amount	%	Finance and Insurance Amount	%	Food, Beverage, and Hotel Amount	%	Business Services Amount	%	Other Services Amount	%	Producer Services Subtotal Amount	%	Total Amount
Technological change	-108,709	-27.9	-4,552	-35.0	-6,354	-364.3	-3,468	-162.7	9,509	23.3	483	1.9	-4,382	-5.5	-113,091
Substitution effect	-7,700	-2.0	-6,022	-46.3	-3,955	-229.1	-4,475	-209.9	4,864	11.9	32,862	128.7	23,234	29.1	15,534
Changes in import intensity	-21,201	-5.4	1,768	13.6	-1,714	-98.3	1,241	58.2	6,564	16.1	-31,291	-122.5	-23,432	-29.4	-44,633
Changes in intermediate input	-77,118	-19.8	-346	-2.7	-1,001	-57.4	-235	-11.0	-2,063	-5.0	-328	-1.3	-3,973	-5.0	-81,091
Interaction term	-2,690	-0.7	48	0.4	356	20.4	1	0.0	144	0.4	-760	-3.0	-211	-0.3	-2,901
Changes in final demand structure	120,613	30.9	3,870	29.7	2,283	130.9	739	34.7	5,791	14.2	43,811	-171.6	-31,128	-39.0	89,485
Changes in consumption	-7,909	-2.0	-267	-2.1	1,157	66.3	-13	-0.6	353	0.9	-124	-0.5	1,106	1.4	-6,803
Changes in government expenditure	-338	-0.1	-243	-1.9	123	7.1	-173	-8.1	-41	-0.1	-6,154	-24.1	-6,488	-8.1	-6,826
Changes in capital formation	4,319	1.1	-70	-0.5	18	1.0	51	2.4	-233	-0.6	-11	0.0	-245	-0.3	4,074
Changes in stock investment	1,338	0.3	43	0.3	-3	0.2	18	0.8	68	0.2	-17	-0.1	109	0.1	1,447
Changes in export	76,396	19.6	-973	-7.5	-272	15.6	-82	-3.8	-1,233	-3.0	1	0.0	-2,559	-3.2	73,837
Changes in final demand component structure	46,807	12.0	5,380	41.3	1,260	72.2	938	44.0	6,877	16.8	-37,506	-146.9	-23,051	-28.9	23,756
Economic growth	478,153	122.5	15,965	122.6	5,538	317.5	5,500	258.0	26,034	63.7	55,244	216.3	108,281	135.7	586,434
Interaction of technical change and change in final demand	-99,779	-25.6	-2,263	-17.4	-3,211	-184.1	-639	-30.0	-461	-1.1	13,622	53.3	7,048	8.8	-92,731
Growth in multiple technological change	-41,187	-10.6	-1,725	-13.2	-2,407	-138.0	-1,314	-61.6	3,603	8.8	183	0.7	-1,660	-2.1	-42,847
Interaction term	-58,592	-15.0	-538	-4.1	-804	-46.1	675	31.7	-4,064	-9.9	13,439	52.6	8,708	10.9	-49,884
Total	390,278	100.0	13,020	100.0	-1,744	-100.0	2,132	100.0	40,873	100.0	25,538	100.0	79,819	100.0	470,097

Source: Input-Output Tables, Directorate-General of Budget, Accounting, and Statistics, Republic of China (1991, 1996).

food, beverage, and hotel services sector over the period 1991 to 1996 were economic growth and structural change in final demand. For business services, the main sources of change were economic growth and technical change. For other producer services, the main sources of change were economic growth and the interaction effect between technical change and final demand. The main sources of import-service change thus differed for the various producer-services sectors.[5]

In conclusion, regardless of whether one looks at the period from 1986 to 1991 or from 1991 to 1996, with the exception of the financial and insurance services sector, demand has been increasing in every producer-services sector. While major changes have all been the result of economic growth, the secondary sources of change have been more diverse. In the financial and insurance services sector, in both stages, change in final demand has been the secondary source, whereas in the business-services sector it has been technical change. As far as the transportation, warehousing, and communications sector is concerned, over the period 1986 to 1991 the source of secondary changes was technical change, whereas in the period 1991 to 1996 it was change in final demand. There was a similar change in the producer-services sector where, during the period 1986 to 1991, the secondary source of changes was technical change, but during the period 1991 to 1996 it was final demand.

5.4 Comparative Advantage and Potential Development of Taiwan's Services Trade

The last two sections analyze the growth and determinants of Taiwan's trade in services based on national income accounts for the period 1951–99 and the sources of growth in producer services based on I-O tables for the years 1986, 1991, and 1996. To complete our analysis this section aims at examining features of service statistics as published in the official BOP, comparing the comparative advantage of service trade among several countries including Taiwan, South Korea, Singapore, and eight advanced countries based on their BOP data, and discussing the potential development of Taiwan's service exports in several specific areas.

---

5. We have so far left out discussion of the decomposition for imported producer goods due to our focus on imported producer services. For a comparison we also list decomposition for changes in imported producer goods that occurred between 1986 and 1991 in table 5.5 and between 1991 and 1996 in table 5.6. It is seen that the main sources of the changes were economic growth and technical change (particularly change in import intensity) during the first period, and economic growth and structural change in final demand (particularly change in export) for the second period. A major difference between imports of producer goods and services is that structural change in final demand was a positive contributor (30.9 percent) to increases in imported producer goods, but was a negative contributor (–39.0 percent) to increases in imported producer services. Other than this the patterns of changes in the imported producer goods and services are quite similar.

5.4.1   Service Statistics as from BOP

The BOP provides trade statistics based on actual international payments, so its coverage is not as comprehensive as that of the NIA-based statistics. Therefore, any services paid in kind or not yet paid must be estimated and included in the NIA trade statistics. Items such as transportation and insurance carried out by domestic firms for merchandise imports do not appear in the BOP but are counted as both imports and exports of services in the NIA trade statistics. However, the BOP is still the main source of services-trade data for the NIA statistics. Moreover, the BOP provides services-trade statistics by detailed category and in U.S. dollars, and which can be used for analysis of services trade for different categories and for computation of revealed comparative advantage (RCA) for comparison.

Although BOP figures differ from NIA constant-price figures as presented in table 5.1 due to coverage and changes in prices and in the exchange rate, they tend to move in tandem. As a comparison table 5.7 lists growth rates and shares of services trade as calculated from the BOP and NIA figures, respectively, for the period 1964–99. The growth rates are for five-year average annual rates while the shares refer to annual shares of service exports (imports) in their total exports (imports) of goods and services. It is seen from table 5.7 that the two sets of growth rates or shares moved together but that those based on the BOP fluctuated more widely than those based on the NIA, particularly during 1974–79, when large price changes

Table 5.7   Growth and Share of Taiwan's Total Service Trade as Computed from BOP and NIA, 1964–99 (%)

Selected Year	BOP Growth Rate EX	IM	BOP Share EX	IM	NIA Growth Rate EX	IM	NIA Share EX	IM
1964			10.5	14.6			11.2	13.8
1969	32.3	27.0	16.1	17.4	33.5	27.0	17.3	17.5
1974	26.6	23.6	10.8	13.3	19.3	20.6	15.3	17.1
1979	20.8	22.3	9.9	15.7	14.5	17.2	14.8	22.4
1984	7.3	14.6	7.5	20.1	6.3	11.8	12.1	29.2
1989	23.7	20.5	9.8	21.3	11.6	10.6	12.0	24.6
1994	13.1	9.2	12.5	20.7	8.1	8.6	13.0	24.2
1999	2.1	3.1	10.8	18.8	4.0	4.8	10.8	20.8

Source: Balance of Payments, Central Bank (various issues); *National Income Statistics,* Directorate-General of Budget, Accounting, and Statistics, Republic of China (various issues).

Notes: BOP = balance of payments; NIA = national income accounts; EX = exports of services; IM = imports of services. Growth rate is five-year average rate; share is service trade as a percentage of total exports or imports of goods and services. Blank cells represent starting year.

occurred worldwide, and during 1984–94, when there was sharp appreciation of the New Taiwan dollar against the U.S. dollar.

To examine detailed categories of services trade as provided by the BOP, data on exports and imports of services are listed in table 5.8 for transportation, travel, and other services for the period 1984–99. Several observations can be made from table 5.8:

1. *Passenger services.* Exports grew substantially between 1984 and 1994 before beginning to decline from 1995 onward. However, imports maintained a growth trend throughout this period, and thus the deficit in passenger services has enlarged in recent years. In 1999, the value of exports was US$486 million while imports were almost three times as large, amounting to US$1,324 million.

2. *Freight services.* The level of exports and imports is more balanced. Both imports and exports grew very rapidly between 1984 and 1994, with the growth in exports being faster than that of imports. The value of exports outweighed that of imports for several years in the mid-1990s. However, there has been a slight decline in recent years in the level of both imports and exports.

3. *Other transport services.* These include seaport and airport services, cargo handling, maintenance and repair of aircraft, and others. The trade balance has been in deficit, with import values being around four to six times the level of export values.

4. *Business and personal travel.* Both business and personal travel in Taiwan grew rapidly between 1984 and 1994 but declined slightly after 1996 due to the Asian financial crisis. The trade deficit has been large for both business and personal travel throughout 1984 and 1999. In 1999 the import of business travel was about twice that of the export, at US$2,222 million and US$1,238 million, respectively, whereas the import of personal travel was more than twice that of the export, at US$5,176 million and US$2,324 million, respectively.

5. *Other services.* These include communications, financial, computer and information, and other business services, which are the fastest growing areas of the service trade.[6] As indicated in table 5.8, other services in Taiwan grew rapidly for both imports and exports between 1984 and 1994, with the import amount being greater than the export. Export growth between 1989 and 1994 was much faster than that of imports, so the gap between imports and exports narrowed in the mid-1990s. However, exports started to decline in 1998 and 1999, while imports continued to grow (although at a slower pace) after 1994. In this area, then, the trade deficit has

---

6. As proposed by the International Monetary Fund in 1994, Taiwan adopted the new method in 1996 to provide more detailed information on trade in other services and traced data back to 1984.

Table 5.8   Service Trade of Taiwan (1984–99) (US$ millions)

Item	1984	1989	1994	1995	1996	1997	1998	1999	Annual Growth Rate (%)		
									1984–89	1989–94	1994–99
Total service, export	2,472	7,147	13,205	15,016	16,260	17,144	16,768	14,642	23.66	13.06	2.09
Total service, import	5,327	13,555	21,070	24,053	24,381	24,888	24,169	24,552	20.54	9.22	3.11
Transportation and Travel											
Transportation service, export	640	2,227	3,732	4,548	4,237	3,777	3,656	3,597	28.32	10.88	-0.73
Transportation service, import	1,822	3,782	5,308	6,333	6,235	6,488	5,774	5,818	15.73	7.01	1.85
Passenger, export	192	286	590	698	583	540	428	486	8.30	15.58	-3.80
Passenger, import	288	561	969	1,030	954	1,090	1,064	1,324	14.27	11.55	6.44
Freight, export	329	1,705	2,783	3,418	3,257	2,853	2,917	2,771	38.97	10.30	-0.09
Freight, import	992	2,200	2,964	3,328	2,998	3,187	2,805	2,854	17.27	6.14	-0.75
Other, export	119	236	359	432	397	384	311	340	14.68	8.75	-1.08
Other, import	542	1,021	1,375	1,975	2,283	2,211	1,905	1,640	13.50	6.13	3.59
Travel, export	1,067	2,699	3,210	3,287	3,636	3,403	3,372	3,562	20.39	3.53	2.10
Travel, import	2,012	4,922	7,618	8,457	8,152	8,198	7,331	7,398	19.59	9.13	-0.58
Business travel, export	150	552	892	919	1,090	1,120	1,175	1,238	105.67	10.07	6.78
Business travel, import	773	1,082	2,126	2,196	2,206	2,423	2,302	2,222	6.96	14.46	0.89
Personal travel, export	917	2,147	2,318	2,368	2,546	2,283	2,197	2,324	18.55	1.54	0.05
Personal travel, import	1,239	3,840	5,492	6,261	5,946	5,775	5,029	5,176	25.39	7.42	-1.18

Other Services

Other services, export	765	2,221	6,263	7,181	8,387	9,964	9,740	7,483	23.76	23.04	3.62
Other services, import	1,493	4,851	8,144	9,263	9,994	10,202	11,064	11,336	26.58	10.92	6.84
Communications, export	45	200	480	563	603	613	629	412	34.76	19.14	-3.01
Communications, import	17	132	436	493	563	530	519	437	50.67	26.99	0.05
Construction, export	15	11	69	111	136	127	160	167	6.01	44.37	19.34
Construction, import	16	—	284	275	255	235	342	525	—	—	13.08
Insurance, export	44	140	369	418	324	471	699	376	26.05	21.39	0.38
Insurance, import	136	199	482	508	563	460	526	519	7.91	19.35	1.49
Financial, export					741	722	712	680	—	—	-1.7[b]
Financial, import[a]	21	676	11	7	1,101	877	900	807	100.24	-56.12	136.10
Computer and information, export					8	28	23	40	—	—	37.97[b]
Computer and information, import[a]	14	21	43	45	48	75	98	118	8.45	15.41	22.37
Royalties and License fees, export	2	85	210	241	256	237	317	245	111.68	19.83	3.13
Royalties and License fees, import	121	476	803	937	1,234	1,148	1,419	1,637	31.51	11.03	15.31
Other business services, export	598	1,701	5,045	5,759	6,208	7,620	7,069	5,413	23.25	24.29	1.42
Other business services, import	686	2,528	5,333	5,775	5,418	5,902	6,172	6,074	29.80	16.10	2.64

Source: Balance of Payments Statistics, Central Bank, Republic of China.

Note: Dashes indicate that information is not available.

[a]Figures before 1996 for financial and computer and information trade are imports net of exports.

[b]The growth rate is between 1996 and 1999.

grown in recent years. Among the individual items of other services, communications, insurance, and other business services have a relatively balanced trade flow in Taiwan, while in construction, finance, computers and information, and particularly royalties and license fees, the trade flow has clearly favored imports rather than exports.

In summary, according to the BOP statistics, Taiwan's total exports of services grew at an annual average rate of 23.7 percent during 1984–89, 13.1 percent during 1989–94, and 2.1 percent during 1994–99, as compared with the annual average rates of 20.5 percent, 9.2 percent, and 3.1 percent, respectively, in total imports of services for the three periods. The growth of service exports was faster than that of service imports before 1994 mainly because the service exports started from a much smaller base. As table 5.8 indicates, the imports actually exceeded the exports every year by more than 44 percent and incurred a large deficit throughout the period. In fact, the deficit existed almost every year for all categories of services. One eye-catching services category in this table is royalties and license fees, whose import has increased rapidly because of the development of hi-tech industries by Taiwan (with the amount reaching US$1,637 million in 1999) but whose export has remained stagnant in recent years (with the amount totaling only US$245 million for the same year). The weak position of Taiwan's trade in services seems to attest to its lack of comparative advantage in exports of services, to which we now turn.

5.4.2   Comparative Advantage of Services Trade among Countries

Balassa (1965) proposed an RCA index to measure the relative advantage of a trade product for a country to export to the rest of the world. This index can be applied to services trade for Taiwan and a number of other countries for comparison. For an ith country and jth service, the index is given by

(4) $$RCA_{ij} = \frac{E_{ij}/TE_i}{WE_j/WTE},$$

where E_{ij} is the export value of service j of the ith country, TE_i is the total export of goods and services of the ith country, WE_j is the world total export of service j, and WTE is the world total export of goods and services.

Table 5.9 lists yearly RCA indexes of total services, transportation, and other services for eleven countries from 1992 to 1998. It shows that the RCA index of total services has values less than 1 for Taiwan, Japan, Korea, Singapore, Canada, and Germany and has values greater than 1 for the United States, the United Kingdom, the Netherlands, France, and Italy, and this pattern was maintained throughout the 1990s. This shows that developed countries, except Japan, Canada, and Germany, have a clear advantage over Asian newly industrialized economies (NIEs) in services trade. The difference in the RCA indexes between countries probably can be partially ex-

Table 5.9   Index of Revealed Comparative Advantage of Service Trade

Country	1992	1993	1994	1995	1996	1997	1998
Total Service Export							
Taiwan	0.54	0.64	0.61	0.61	0.62	0.62	0.65
United States	1.36	1.35	1.38	1.39	1.40	1.38	1.39
Japan	0.62	0.62	0.64	0.67	0.73	0.73	0.71
The Netherlands	1.06	1.10	1.14	1.10	1.12	1.19	1.17
United Kingdom	1.19	1.17	1.19	1.23	1.21	1.26	1.33
Korea	0.59	0.65	0.74	0.79	0.77	0.80	0.78
Canada	0.64	0.61	0.62	0.61	0.63	0.62	0.62
France	1.38	1.43	1.21	1.18	1.15	1.12	1.09
Germany	0.66	0.69	0.65	0.69	0.71	0.70	0.66
Italy	1.19	1.12	1.08	1.06	1.04	1.10	1.08
Singapore	0.94	0.91	0.94	1.03	0.97	0.98	0.71
Transportation Export							
Taiwan	0.60	0.65	0.73	0.78	0.72	0.62	0.64
United States	1.25	1.21	1.20	1.23	1.22	1.16	1.09
Japan	0.98	0.94	0.94	0.99	1.03	1.03	1.09
The Netherlands	1.87	1.89	1.87	1.87	2.01	2.17	2.05
United Kingdom	1.12	1.09	1.10	1.09	1.10	1.11	1.15
Korea	1.02	1.16	1.36	1.36	1.28	1.51	1.46
Canada	0.56	0.53	0.53	0.53	0.54	0.55	0.53
France	1.15	1.24	1.16	1.22	1.23	1.20	1.18
Germany	0.68	0.72	0.69	0.70	0.73	0.73	0.73
Italy	0.88	0.86	0.81	0.79	0.76	0.79	0.77
Singapore	0.67	0.66	0.69	0.75	0.74	0.75	0.77
Other Service Export							
Taiwan	0.67	0.87	0.72	0.71	0.77	0.83	0.88
United States	1.13	1.11	1.25	1.24	1.26	1.27	1.36
Japan	0.82	0.85	0.91	0.97	1.05	1.02	0.97
The Netherlands	1.09	1.21	1.30	1.19	1.19	1.24	1.24
United Kingdom	1.47	1.40	1.49	1.49	1.49	1.59	1.68
Korea	0.48	0.51	0.60	0.66	0.71	0.68	0.56
Canada	0.72	0.71	0.72	0.70	0.73	0.70	0.69
France	1.78	1.82	1.30	1.21	1.13	1.05	1.01
Germany	0.63	0.68	0.67	0.78	0.83	0.80	0.77
Italy	1.23	1.05	0.93	0.92	0.91	0.98	0.97
Singapore	1.12	1.10	1.22	1.42	1.33	1.40	0.82

Source: Balance of Payments Year Book, IMF (1999).

plained by the stage of development in their service industries as measured by the share of service production in GDP. Take average figures of three years (1993, 1995, and 1997) for demonstration. A pair of numbers (the first for RCA index and the second for service-sector share in GDP) for each of these eleven countries is as follows: United States (1.37, 72), France (1.25, 70), United Kingdom (1.22, 66), the Netherlands (1.13, 70), Italy (1.09, 66), Singapore (0.97, 64), Korea (0.75, 50), Germany (0.69, 62), Japan (0.68, 59),

Table 5.10   Balance of Trade in Services (US$ millions)

Country	1992	1993	1994	1995	1996	1997	1998
Taiwan	−9,128	−7,896	−7,865	−9,037	−8,121	−7,744	−7,401
United States	57,300	60,770	65,740	74,340	85,050	89,980	80,700
Japan	−43,960	−43,080	−48,060	−57,350	−62,240	−54,150	−49,420
The Netherlands	265	676	1,411	2,262	3,835	5,756	5,201
United Kingdom	9,910	9,890	10,000	14,050	13,960	20,340	20,290
Korea	−2,884	−2,126	−1,801	−2,979	−6,179	−3,200	628
Canada	−10,812	−10,418	−8,146	−7,345	−6,433	−6,563	−4,755
France	19,378	16,861	19,512	17,973	16,254	17,493	18,705
Germany	−27,721	−31,538	−39,021	−45,362	−44,018	−41,294	−42,978
Italy	−4,438	50	1,448	6,569	8,055	7,764	4,170
Singapore	6,551	7,413	10,014	12,056	10,255	11,095	330

Source: *Balance of Payments Yearbook,* IMF (1999).

Taiwan (0.62, 60), and Canada (0.61, 64).[7] These figures clearly show that countries with RCA values greater than 1 tend to have higher service-sector shares than do those with RCA values greater than 1 tend to have higher service-sector shares than do those with RCA values smaller than 1. Among them, the United States, France, and the Netherlands have service-sector shares higher than 70 percent in GDP, whereas Asian countries, Germany, and Canada have service-sector shares lower than 65 percent. The two variables are highly correlated, with their coefficient of correlation equal to 0.75.

Moreover, data (table 5.10) also indicate that countries with high RCA values (such as the United States, France, the United Kingdom, the Netherlands, and Italy) tend to have surpluses in services trade, whereas those with low RCA values (such as Korea, Germany, Japan, Taiwan, and Canada) tend to have deficits in services trade. Singapore is a country in the margin, having an RCA value near 1 in total services but a surplus in services trade. This is probably because Singapore has an RCA value much greater than 1 in other services. The RCA index and the balance of services trade across the eleven countries are actually highly correlated, with the correlation coefficient equal to 0.79, again in three-year averages for 1993, 1995, and 1997. The correlation coefficient between balance of trade in services and service-sector share is also as high as 0.59. The high correlation between RCA index and GDP share of service industries, and between RCA index and balance of trade in services, is consistent with our finding in section 5.2 that service-sector share has a positive effect on services export but a negative effect on services import.

As shown in table 5.9, the RCA index of transportation or other services follows a pattern similar to that of total services, with a few exceptions. The

7. RCA index numbers are taken from table 5.9 and shares in percent are from various issues of the *World Development Report* by World Bank.

United States, the Netherlands, the United Kingdom, and France had maintained high RCA status throughout the period 1992–98 in both transportation and other services, but the index for Italy was less than 1 in both categories except for the years 1992 and 1993 in other services. Japan's index exceeded 1 in recent years in both categories, whereas Korea saw values much higher than 1 throughout the period in the category of transportation. Another exception is for Singapore, whose index had greatly exceeded 1 in the category of other services for the entire period other than 1998, a year when the Asian financial crisis broke out. The remaining countries are all having the value of the index less than 1 throughout the period.

Similar to Canada, the RCA index of Taiwan ranks low among the countries listed in table 5.9 either in total services, in transportation, or in other services. The index had never exceeded 0.65 in total services, 0.78 in transportation, and 0.88 in other services during the period from 1991 to 1998. Taiwan therefore had unfavorable exports of services and still has a large deficit in services trade. However, its RCA index appears to show signs of slight improvement in recent years, as can be seen from table 5.9.

5.4.3   Potential Development of Taiwan's Export Services Trade

To conclude this section, we discuss the potential development of some important export services in the areas of transportation, communications, and business services. Related government policies are also addressed.

Transportation

In a study of the flow of trade among developed and developing countries, Langhammer (1989) argued that developing countries had better positions in developing factor-enriched services, such as ocean transportation and travel, than in developing knowledge-based services. Taiwan's freight trade development seems to be a case in point. Its freight exports caught up with imports in the early 1990s because Taiwan's abundant capital and fast-growing foreign trade throughout the 1980s provided excellent opportunities for ocean freight services to develop. However, further development of freight services was restrained because a large proportion of Taiwan's export-oriented, labor-intensive manufacturing firms had moved their operations to mainland China, and direct shipping between the mainland China and Taiwan are still not permitted by the government.

In air transportation, during the last decade, a new Taiwanese airline, EVA, joined China Airlines in the operation of international flights. However, due to Taiwan's very limited official diplomatic relationships around the world, negotiations aimed at expanding flight routes have been so difficult as to limit the growth of these two airlines.

With Taiwan geographically placed at the center of the East Asian countries, the distance to other major East Asian cities can be minimized via Taipei. To make use of this locational advantage, since 1995 Taiwan's government has been pursuing plans to develop Taiwan into a regional sea and air transportation center. These plans are part of a bigger project, which aims to transform Taiwan into an Asia Pacific Regional Operations Center (APROC). For sea transportation the goal is to make Taiwan an East Asian hub for container transshipment, and efforts have been made to improve the efficiency of container handling, custom processing, and other details in pursuit of this plan. Some existing port services were privatized and liberalized to allow foreigners to participate. Most important, to overcome the ban of nondirect shipping to mainland China, an offshore shipping center has been set up in Kaohsiung harbor to handle goods transshipped to and from mainland China. According to the government's review (Council for Economic Planning and Development [CEPD] 1999), the volume of container handling at Kaohsiung harbor increased by 27 percent between 1994 and 1998.

As for the air transportation center, due to the government's effort in improving facilities and deregulation, an air freight transit hub is already taking shape. United Parcel Service (UPS) established its regional headquarters at CKS International Airport in 1996 and Federal Express (FedEx) has also expanded its operations at the airport.

Communications

As in many other countries, the communications industry in Taiwan has long been a monopolistic public enterprise and the liberalization process has just begun in recent years. The most significant change was the opening of the market to private cellular phone services in 1997, when eight new licenses were issued, including two licenses covering the whole of Taiwan and 6 licenses for regional coverage. In the last two years the market penetration rate of cellular phones grew enormously, from less than 20 percent to more than 50 percent. Foreign technology and foreign capital participated actively in this market opening and almost every new entrant to the market came in the form of a joint venture of a local enterprise and a world-famous communications giant, such as AT&T, GTE, Sprint, or others. Foreign firms will be allowed to invest up to 20 percent when Taiwan joins the WTO. The fixed net service (local phone services) was also opened to private entry (with three licenses being issued in March 2000) and to the participation of foreigners similar to that of the cellular phone services. At this stage Taiwan is still a net importer in the area of communications.

Business Services

Business services in Taiwan flourished with the rapid growth of manufacturing industries, and the growth of foreign direct investment (FDI) in Taiwan's manufacturing industries was also accompanied by foreign business services. The most significant services are advertising and accounting. In 1997, for example, among the 36 largest advertising companies in Taiwan, 10 companies were solely owned by foreigners and 12 were joint ventures between local and foreign providers. Foreign advertising services came into Taiwan's market initially with multinational firms that produced and marketed consumer products in Taiwan, and gradually began providing services to local consumer-product producers. They provided services first by cooperating with local firms and then either through the formation of joint ventures with local firms or by buying up local firms. Foreign advertising companies brought in new technologies and different business practices and have totally revolutionized this industry in the last decade.

In recent years, Taiwan's advertising companies have also begun to export their services, accompanying Taiwan's FDI, to other developing countries, with mainland China being the most prominent destination.

The accounting industry in Taiwan is also very internationalized even though some areas of accounting services are restricted to performance by only locally certified public accountants. World leading accounting firms, such as PriceWaterhouseCoopers, Arthur Andersen, and KPMG, have all established their operations in Taiwan mostly by acquiring existing Taiwanese accounting firms. Local independent accounting firms also seek to be members of an international accountants' group or confederation in order to meet the increasing needs of local firms for globalization. Taiwan's accounting firms are also beginning to provide services to Taiwanese businesses that have invested in mainland China.

In the area of business consulting services, engineering consulting services, technical services, and design services, local firms have been dominated by small and medium-sized enterprises and foreign presence has not been as prominent as in the advertising and accounting service industries. Foreign service providers operate mainly through the presence of natural persons for a short period of time rather than by establishing a permanent branch. This is probably because the scale of economy in these various areas is smaller and a large foreign firm may not be as competitive as local small firms. They may also cooperate with local firms on a case-by-case basis (Hsueh, Tu, and Wang 1995; Hsueh and Tu 1999).

In the legal services industry, foreign legal service providers can form a cooperative relationship with licensed lawyers in Taiwan except in the form of a joint liability and profit-sharing arrangement. The government will grant recognition of "attorney of foreign legal affairs" (AFLA) after Taiwan enters the WTO. Most importantly, AFLA will be allowed to establish partnerships with or employ locally licensed lawyers three years after Taiwan gains accession into the WTO. Thus we can expect that trade in legal services in Taiwan will increase in the near future.

5.5 Conclusion

In this paper we employ trade data from the NIA, the I-O tables, and the BOP to examine Taiwan's trade in services and its future development. Our analysis is very much limited by the availability and accuracy of the trade data on services. Unlike the trade in goods, which is registered through customs ports and produces abundant information, data on the services trade come mainly from records on foreign exchange transactions provided by the central bank in the BOP and are not comprehensive. The statistics on services trade in the NIA, which are based on the BOP and other information, are comprehensive but far too aggregate; they lack information on detailed components. In particular, information on the services trade is not available in constant prices except for the aggregate imports and exports of services exclusive of the merchandise transport and insurance. This makes our comparison over time difficult. Thus, provision of more information on the services trade should be improved and built on a common framework to facilitate comparisons over time and among countries.

Both NIA and BOP statistics demonstrate that Taiwan's trade in services has grown faster than its trade in goods and that service imports have also grown faster than service exports in the past forty years. In fact, Taiwan's trade in services has constantly been in deficit. This may be partly explained by the regression result of this research, which shows that Taiwan's exports of services are income inelastic but price elastic, whereas its imports of services are income elastic but price inelastic. Our regression result also indicates that expansion of domestic service-sector share contributes positively to balance of trade in services both by reducing imports of services and by increasing exports of services. The relatively weak position in Taiwan's services trade can also be seen from the comparison of revealed comparative advantage indexes or services-sector share between developed countries and Asian NIEs. This suggests that Taiwan needs to improve its position of comparative advantages in trade in services by strengthening its service sector.

As shown in the I-O tables, services used by the producing sector accounted for more than two thirds of the imported services in 1996. A complex decomposition analysis reveals that in the 1986–91 period, import coefficient changes and economic growth contributed most to the growth in imported producer services. However, in the 1991–96 period, import coefficient changes and change in final demand structure both contributed negatively to its growth, with the source of growth mainly coming from economic growth.

Our analysis also shows that in the individual services sector, no matter whether in transportation, communications, financial services, or business services, the trade balance has also been in deficit. As part of the overall commitment to WTO accession, commitment to the market opening of various service sectors will increase the import of services in the near future. However, with the increase in Taiwan's FDI toward other countries and the continued growth of service-sector GDP share, Taiwan's service sector will have good potential for export growth. Hence Taiwan's trade in services is likely to achieve substantial growth in the future.

References

Balassa, Bela. 1965. Trade liberalization and "revealed" comparative advantage. *Manchester School of Economic and Social Studies* 33:99–123.

Central Bank. Various issues. *Balance of payments.* Taipei: Central Bank, Republic of China.

———. Various issues. *Financial statistics.* Taipei: Central Bank, Republic of China.

Council for Economic Planning and Development (CEPD). 1999. *The performance and prospect of the Asian Pacific regional operations center project* (in Chinese). Taipei: CEPD.

Department of Commerce. Various issues. *Statistical abstract of the United States.* Washington, D.C.: Department of Commerce.

———. Various issues. *Survey of current business.* Washington, D.C.: Department of Commerce.

Directorate-General of Budget, Accounting, and Statistics. Various issues. *Input-output tables.* Taipei: Directorate-General of Budget, Accounting, and Statistics.

———. Various issues. *National income statistics.* Taipei: Directorate-General of Budget, Accounting, and Statistics.

Han, X. 1995. Structure change and labor requirement of the Japanese economy. *Economic System Research* 7 (1): 47–65.

Hsueh, L., and Y. Tu. 1999. *The competitive advantage of the business service industry in Taiwan* (in Chinese). Chung-Hua Institution for Economic Research's Economic Papers no. 187. July. Taipei: Chung-Hua Institution for Economic Research.

Hsueh, L., Y. Tu, and S. Wang. 1995. *The development and prospect of producer service industries in Taiwan: An international comparison* (in Chinese). Chung-Hua Institution for Economic Research Contemporary Economic Issues Series, no. 2. Taipei: Chung-Hua Institution for Economic Research.

International Monetary Fund (IMF). 1999. *Balance of payments yearbook.* Washington, D.C.: IMF.

Langhammer, R. J. 1989. North-south trade in services: Some empirical evidence. In *Services in world economic growth symposium*, ed. H. Giersch, 248–71. Tübingen, Germany: Mohr.

World Bank. Various issues. *World development report.* Washington, D.C.: World Bank.

Comment   Philippa Dee

The paper uses three different data sets to explore the growth potential of Taiwan's trade in services. These are as follows:

- Time-series national accounts data to econometrically estimate the determinants of Taiwan's service imports and exports over time
- Input-output data for three separate years to decompose the growth of Taiwan's total imports of producer services into the components aris-

Philippa Dee is assistant commissioner at the Productivity Commission, Australia.

ing from (a) changes in the intensity of imported service use in each producing industry and (b) changes in industrial structure
- Balance-of-payments data for a range of years to calculate indexes of revealed comparative advantage for Taiwan's services, to compare with similar indexes for other economies.

Together, the separate pieces of analysis paint an interesting picture of the role of services in economic development. I have some specific comments on particular aspects of the analysis, and some general comments about the broad themes of the paper.

Any researcher who has worked in the services area is aware of the difficulty of obtaining good primary data. Accordingly, I would like to know more about the data sources used, and the properties of some of the data.

First, it would help to know whether all three data sets were derived independently, or whether the national accounts and input-output data ultimately come from balance of payments sources. Such clarification would help to establish the information-richness of the data.

Second, it would help to know whether each of the three input-output tables were based on independent surveys of firms' costs and sales structures, or were imputed from surveys for other years. The reader can have greater confidence in the growth decomposition if it uses input-output data based on separate surveys.

Third, it would help to know about the time-series properties of the national accounts data used for the econometric estimation. If the data are stationary, then the estimation techniques and methods of statistical inference used by the authors are appropriate. If, however, the data are not stationary (in the sense of containing unit roots), as is likely, then it is not clear that the authors have used the appropriate techniques. If the residuals from their estimating equations are themselves not stationary, the regressions they present are spurious. If the relationships they present are genuine cointegrating relationships between nonstationary variables, the authors do not appear to have used the appropriate tests of statistical significance.

Further elaboration on a few of the econometric results would be useful. The authors note that Taiwan's exports of services seem to depend negatively on Taiwan's per capital income. The authors offer no interpretation, but the result could indicate that Taiwan's service exporters seek out foreign markets more vigorously when the growth in domestic demand slows. As the authors note, however, a more conventional economic specification would include foreign rather than domestic income as a determinant of Taiwan's service exports.

The finding that higher tariffs in manufacturing help raise exports of services is also puzzling. It is at odds with conventional economic analysis, in which tariffs in manufactures act as a tax on the exports of other sectors, either because other sectors use manufactures directly as inputs, or because import protection for manufacturing leads to a real appreciation of the exchange rate.

In the input-output analysis, it is not always clear whether the service input requirements being reported are direct requirements, or total (direct and indirect) requirements. Although most of the analysis seems to report direct requirements, an interesting piece of additional analysis would be to compute, for each sector of the economy, the direct and indirect service input requirements per unit of final demand. Since one particular category of final demand is exports, the computation would reveal how important an *indirect* role services have played in Taiwan's recent and phenomenal non-service export performance.

The paper has a really interesting underlying theme about the complementarity between production services and other sectors of the economy. The authors highlight that merchandise exports need transport services to be shipped around the world. The complementarities extend beyond that, however. Every business needs banking services. Every business needs telecommunications services. The facilitation (or margins, in input-output terminology) role of services extends well beyond transport.

The complementarities show up in the econometric results, where services imports increase as goods imports increase, where services exports increase as goods exports increase, and where services exports increase as services output increases.

The complementarities also show up in the increasing service import intensity of other (nonservice) activities. It is a very interesting finding that this is not due to changes in the sectoral composition of activity, but instead due to economic growth and technical change.

Here, however, it is important to understand why the technical change in favor of imported service inputs appears to be reversed over 1991–96. Is this a temporary phenomenon, a statistical artifact (perhaps due to some subtle classification change), or a permanent feature?

It would also be useful to know whether the longer term technical change in favor of imported services is because Taiwanese businesses are contracting out activities that they previously undertook in-house.

The complementarities show up finally in the stories told about the prospects for Taiwan's services industries in the future. By focusing on measures of revealed comparative advantage based on measures of cross-border trade, the authors (I suspect) paint too bleak a picture of Taiwan's future service export prospects. This is because the indexes of revealed comparative advantage are based on data that ignore services delivered via commercial presence. In the same way that U.S. advertising firms came to Taiwan to support U.S. manufacturing multinationals, perhaps Taiwanese service firms are moving offshore to support Taiwan's outbound manufacturing foreign direct investment (FDI). Any evidence on this, either anecdotal or via survey, would be extremely useful. There are theoretical antecedents for this view. Conventional trade theory has already noted possible complementarities between trade in goods and trade in capital, when the goods trade is motivated by things other than differences in factor intensities (i.e., other than comparative advantage, narrowly defined). I think similar reasoning can support complementarities between trade in manufactures and trade (via FDI or cross-border) in services.

Comment   Mario B. Lamberte

The paper first discussed Taiwan's economic growth rate over the period 1951–99, which is impressive by international standards, and the rapid structural change in the country's economy, particularly the shift from an agrarian economy to a services-dominated economy. As the authors have noted, Taiwan's external trade in services has increased rapidly as its entire services sector has expanded. The paper shows that, in general, Taiwan is quite strong when it comes to exports of goods but remains weak when it comes to exports of services. Thus, its trade in services has been in deficit during the period of analysis. Incidentally, the Filipino contract workers in Taiwan in a way have contributed to that deficit. My comments on the paper are organized around five main points, as follows.

Policy Environment for the Services Sector

It would greatly help the readers appreciate the results of the descriptive and econometric analyses if the authors discussed at the outset the changes in the policy environment for the services sector. Although this is somewhat alluded to in section 5.2 of the paper, a much more extended discussion of the policy environment is needed.

Descriptive Analysis

Regarding the analysis on import and export of services, it might be worthwhile to add information on the sources of import of services and destination of export of services of Taiwan. My interest lies on the extent of trade in services between Taiwan and other Asian countries. This may also help the authors make some qualifications to their observation that Taiwan is quite weak in its export of services. I suspect that its export of services to Asia is relatively large.

Determinants of Trade in Services

The authors have examined the determinants of Taiwan's trade in services. They concluded that "trade in goods, per capita income, and the

Mario B. Lamberte is president of the Philippine Institute for Development Studies (PIDS).

relative price of services are three important determinants of the services trade." The authors have included as explanatory variables nominal exchange rate and effective tariff rate, which is defined as the ratio of total import tariff revenues to total custom imports, to reflect changes in government policies. These may not capture entirely the government's policies for the services sector during the period of analysis. I think the authors must find a way of incorporating in their model nontariff barriers, which could have affected the dependent variables during the period of analysis. This is why a discussion about the policy framework at the very start of the paper would be greatly useful.

I am not convinced that the effective tariff rate, as defined in the paper, is a good candidate for an explanatory variable in the exports of services equations. Usually, countries do not impose tariffs on their exports of goods and services. What is more worrisome is that the result shows that the effective tariff rate has a significant, positive effect on exports of services. It means that raising the tariff on trade in services can boost Taiwan's exports of services!

There may also be a simultaneity problem in the model. For instance, growth in imports of services, which is an independent variable in the model, could have been affected by the relative price of imports and the nominal exchange rate. The same can be said of the growth in export services.

The use of Taiwan's per capita income as proxy for per capita income for the rest of the world is, I think, not appropriate. I suggest that the authors use the weighted average per capita income of Taiwan's major trading partners, which can be easily obtained from publications of multilateral agencies.

Intensity of Import Services

The data in table 5.7 about the intensity of import services of Taiwan are interesting. I expected the trends in import intensities of the thirty nine sectors to behave in a more stable manner, going up or down. However, I notice significant fluctuations in import intensities between 1986 and 1996. For instance, import intensities of trade, real estate, travel services, and business significantly dropped in 1991. There must be some explanation for these fluctuations.

One thing that seems surprising is the sharp rise in the import intensity of electricity between 1991 and 1996. How can such a phenomenal rise be explained?

Relative Comparative Advantage

That Asian economies are shown to have weak comparative advantage in trade in services today is not surprising at all. However, the authors should also emphasize the fact that the computed revealed comparative advantages (RCAs) of some Asian economies, including that of Taiwan, have been increasing during the period of analysis. This can be observed in certain sectors such as transportation. This may have occurred when Asian economies gradually liberalized their services sectors. As Asian economies deepen and broaden the liberalization of their services sector, further improvement of their RCAs can be expected. This is perhaps what the authors expected of Taiwan upon accession to the World Trade Organization.

6

Liberalization of Trade in Services and Productivity Growth in Korea

Jong-Il Kim and June-Dong Kim

6.1 Introduction

Korea's economic development over the past twenty five years has been based on industrialization, with priority being given to the manufacturing sectors at the expense of services. However, since the financial crisis of late 1997, the importance of the service sector has been increasingly recognized, and comprehensive reforms in the service sector were recommended in order to restore the crisis-ridden economy to its previous growth path (McKinsey 1998).

The liberalization of services can bring potential gains in productivity in service sectors that are subject to technology transfers and economies of scale. These are similar to the productivity effects of foreign direct investment (FDI) in the manufacturing sector, because a significant portion of service supplies occurs through FDI. Various studies show positive evidence of the productivity spillovers of FDI (Caves 1974; Globerman 1979; Blomstrom and Persson 1983; Borensztein, de Gregorio, and Lee 1998). Foreign investment may also raise productivity by enhancing competition. Based on an analysis of approximately 670 U.K. companies, Nickell (1996) showed that competition, as measured by increased numbers of competitors or by lower levels of rents, is associated with a significantly higher rate of total factor productivity growth. Using firm-level panel data of U.S. automobile component manufacturers, Chung, Mitchell, and Yeung (1994) found that productivity gains among the host country suppliers largely stem from the increase in competition created by FDI.

Anecdotal evidence shows that foreign-invested firms may raise produc-

Jong-Il Kim is an associate professor of economics at Dongguk University, and June-Dong Kim is a research fellow at the Korea Institute for International Economic Policy.

tivity by spinning out skilled workers, providing technical guidance to sub-contractors, bringing in new capital goods and technology, introducing advanced management know-how, conducting in-house research and development, and enhancing competition (Kim and Hwang 2000, 272). Most of these channels of raising productivity through FDI apply to both the manufacturing and service sectors.

Moreover, the liberalization of trade in services may result in improved productivity in other sectors, including manufacturing, due to the resulting access to a broader variety, better quality, and lower cost of inputs. Using a model of increasing returns due to specialization, Rivera-Batiz and Rivera-Batiz (1992) argued that FDI in the business service sector stimulates specialization and raises the productivity of the industry that uses them. Markusen (1989) also demonstrated that allowing trade in producer services is superior to allowing trade in final goods only, due to the complementarity between domestic and foreign producer services.

This paper investigates the changes in productivity growth rates of Korean service and manufacturing subsectors in relation to the liberalization of trade in services. Since Korea underwent accelerated liberalization of the service sector in the 1990s, we try to examine whether the service subsectors that were liberalized and the manufacturing subsectors that use liberalized services as inputs experienced productivity gains in this period.

This paper is organized as follows: Section 6.2 reviews the evolution of liberalization in services in Korea as well as the recent trends of trade in services. Section 6.3 illustrates the case of distribution services, which were liberalized almost completely in the 1990s. Changes in productivity in the service and manufacturing subsectors are explored in section 6.4 with a tabulation of the trends of labor and total factor productivity. We then investigate whether liberalized service subsectors posted relatively higher productivity growth and contributed to productivity gains in the manufacturing subsectors. Concluding remarks and policy implications are provided in section 6.5.

6.2   Evolution of Services Liberalization and Recent Trends of Trade in Services

6.2.1   Evolution of Services Liberalization

Unlike the manufacturing sector, in which FDI had been liberalized since the early 1980s, much of the services liberalization has only taken place since the mid-1990s. Table 6.1 shows that the Korean government has liberalized 154 business categories (at the Korean standard industrial classification [KSIC] five-digit level) in the service sector, completely or partially, since 1993. Many of these service subsectors were liberalized as a result of

Table 6.1   Korea's FDI Liberalization, 1993–2000 (as of May 2000)

Classification	Total	Liberalized								Remaining Restricted
		1993	1994	1995	1996	1997	1998	1999	2000	
Manufacturing	585	2	1	0	6	1	2	2	0	0
Services	495	9	23	42	39	16	20	3	2	2 (22)
Others	68	5	6	2	4	10	0	0	1	2 (2)
Total	1,148[a]	16	30	44	49	27	22	5	3	4 (24)

Source: Ministry of Finance and Economy, *Five-Year Foreign Investment Liberalization Plan* (various years), and Ministry of Commerce, Industry, and Energy, *Consolidated Public Notice for Foreign Investment* (May 2000).

Notes: The business categories are at the Korean Standard Industrial Classification (KSIC) five-digit level. "Others" denotes agriculture, fisheries and mining. "Liberalized" includes both complete and partial liberalization. The number of partially restricted business categories is in parentheses.

[a]Business categories that include government services and nonprofit organizations, where FDI is prohibited by domestic law, are not counted.

the Uruguay Round negotiations and Korea's accession to the Organization for Economic Cooperation and Development (OECD) in 1996. Additional liberalization took place after Korea suffered from economic crisis in 1997. In 1998, as a way of attracting more foreign investment and enhancing efficiency, the Korean government accelerated the liberalization of the service sector beyond the level of its OECD and World Trade Organization (WTO) commitments.

Comparison of service subsectors in which FDI was restricted as of January 1990 (table 6A.1) with those as of November 1997 (table 6A.2) shows that distribution services, business services, entertainment and recreational services, and other personal services have been liberalized since 1990. Also, transportation services, financial services, and telecommunication services were partially liberalized during this period.

More drastic liberalization has been implemented since the financial crisis of late 1997. Twenty-two business categories, most of which are in the service sector, including real estate rental and sales, land development, waterworks, and investment companies, fully opened in 1998. By May 1999, three more service business categories, the publishing of books, outer maritime transportation, and the operation of casinos, fully opened. Furthermore, existing ceilings on foreign equity ratios were raised in 1999 in six business categories: newspaper publishing, cable broadcasting, wire telegraph and telephone, and wireless telegraph and telephone.

As a result, only twenty four business categories in the service sector remained to be completely liberalized as of May 2000. Among them, radio and television broadcasting are the two categories in which FDI is wholly restricted. Foreign direct investment in twenty two business categories, including the publishing of newspapers, coastal water transport, air trans-

Table 6.2   Service Business Categories in which FDI is Restricted, Korea
             (as of May 2000)

Wholly Restricted	Partially Restricted
Radio broadcasting Television broadcasting	Wholesale of meats Publishing (newspapers, periodicals) Processing of nuclear fuel Electric power generation Coastal water transport (passenger, freight) Air transport (scheduled, non-scheduled) Telecommunications (leased line, wired, mobile, cellular, resellers, other) Domestic banking (special banking) Investment trust companies Program supplying Cable broadcasting, satellite broadcasting News agency activities Radioactive waste disposal

Source: Ministry of Commerce, Industry, and Energy, *Consolidated Public Notice for Foreign Investment* (May 2000).

port, telecommunications, investment trust companies, and electric power generation, are partially restricted (table 6.2).[1]

6.2.2   Recent Trends of Trade in Services

The service sector is gaining importance in the Korean economy, with its share of GDP and employment having increased from 43.9 percent and 39.5 percent in 1980 to 52.7 percent and 59.8 percent in 1998, respectively. However, the share of the service sector in the domestic economy is lower than that of the United States, Singapore, and Japan, where its portion of the GDP in 1996 was 74.1 percent, 70.9 percent, and 64.4 percent, respectively.

Table 6.3 shows Korea's trade in services by mode of supply in the 1990s. The sum of exports and imports, of cross-border supply, which is measured by commercial services in balance of payments (BOP), except for tourism, increased from about $16 billion in 1991 to $39.6 billion in 1998. Trade in services by the three modes of supply (cross-border supply, consumption abroad, and movement of natural persons), except commercial presence, increased from $22.8 billion in 1991 to $49 billion in 1998. In 1998, the total amount of Korea's trade in services, except commercial presence, was almost 20 percent of the amount of trade in goods. The share in the world's total trade in services, except commercial presence, also rose from 1.2 percent in 1991 to 1.8 percent in 1998.

Table 6.4 reveals that a significant increase in trade in services has occurred through commercial presence since the 1980s. Foreign direct investment inflows in services increased from $1.6 billion in 1982–90 to $6.3 billion in

---

1. Even though FDI in legal services is not restricted, foreign lawyers are not allowed to practice unless they acquire a domestic license.

Table 6.3   Trade in Services by Modes of Supply, Korea: 1991, 1995, 1998
(US$ millions)

	1991 Exports	1991 Imports	1995 Exports	1995 Imports	1998 Exports	1998 Imports
Cross-border supply[a]	7,158	8,953	17,677	19,465	18,647	21,053
Transportation	3,873	4,897	9,272	9,645	10,204	8,983
Communications	353	204	561	642	656	1,133
Consumption abroad[b]	2,856	3,214	5,150	6,341	5,933	2,898
Commercial presence	n.a.	n.a.	n.a.	n.a.	n.a.	n.a.
Movement of natural persons[c]	604	54	774	132	446	42
Total	10,618	12,221	23,601	25,938	25,026	23,993
	(1.2)	(1.3)	(1.8)	(2.0)	(1.8)	(1.7)

Source: International Monetary Fund, Balance of Payments Statistics Yearbook (1999).
Note: Percentage shares in the world's trade in services are in parentheses.
[a]BOP commercial services minus travel.
[b]BOP travel.
[c]BOP compensation of employees.

Table 6.4   FDI Inflows in Service Subsectors, Korea: 1962–99 (%)

Subsector	1962–81	1982–90	1991–95	1996–97	1998–99
Total FDI in services (US$ millions)	412.2	1,600.2	2,078.7	2,213.1	6,330.9
Electricity and gas	0	0	26.1	0	378.7
Construction	10.4	40.1	21.4	79.8	9.6
Wholesale and retail	0	20.1	103.4	586.6	956.7
Trading	0.4	55.5	394.7	306.5	336.1
Restaurants	0	4.2	60.2	7.1	9.4
Hotels	206.0	956.9	362.3	211.4	64.5
Transportation	28.7	9.6	9.9	150.2	9.4
Financial	109.7	384.9	710.3	480.8	2,292.9
Insurance	3.0	77.3	158.0	23.2	407.9
Real Estate	0	0	1.8	0.1	33.0
Others	53.9	51.4	230.5	367.4	1,832.5
Total FDI into Korea (US$ millions)	1,477.8	4,385.1	5,057.2	5,394.2	15,489.7

Source: Ministry of Commerce, Industry, and Energy, Trends in Foreign Direct Investment (January 31, 2000).
Note: Based on actual investment.

1998–99. Hotels were the largest recipients through the 1980s. In the 1990s, FDI increased remarkably in distribution services (wholesale and retail), transportation services, financial services, and other services, which are mainly composed of telecommunication and business services. Foreign direct investment in distribution services increased from $20.1 million in 1982–90 to $586.6 million in 1996–97. Foreign direct investment in transportation ser-

Table 6.5   Comparison of Productivity in Distribution Services, Korea and Japan: 1994 (US$ thousands)

	Wholesale		Retail	
	Korea	Japan	Korea	Japan
Sales per establishment	693	11,724	117.8	935.2
Sales per employee	170	1,099	57.8	190.0

Source: National Statistical Office, R.O.K., *Annual Report on the Survey of Wholesale and Retail Trade as of 1994,* and Ministry of Industry and Trade, Japan, *Annual Statistical Report of Commerce in 1994.*
Note: Applied exchange rates are US 1$ = 716.4 Korean won; US 1$ = 102.18 Japanese yen.

vices also increased, from $9.9 million in 1991–95 to $150.2 million in 1996–97. Foreign direct investment in financial services and other services experienced a sharp increase after the financial crisis. Foreign direct investment in financial services increased from $480.8 million in 1996–97 to $2.3 billion in 1998–99. The increase in FDI in other services was almost sixfold during the same period, from $367.4 million in 1996–97 to $1.8 billion in 1998–99.

6.3   The Experience of Liberalization in Distribution Services

In this section, we focus on the distribution sector, which experienced significant liberalization during the 1990s, to illustrate how liberalization affects the productivity of a specific sector.

Distribution services had been one of the least developed sectors in Korea up to the mid-1990s, along with financial services. Mom-and-pop stores having fewer than five employees accounted for approximately 80 percent of Korea's $116 billion retail market in 1996. The productivity of Korea's wholesale and retail service sector, in terms of sales per establishment or sales per employee, was far below that of Japan in 1994 (table 6.5).

We may attribute the low productivity of Korean distribution services to the regulations on zoning and land development and to the restrictions on FDI. The regulations on zoning and land development reduced the availability of land, limiting the scale of operation, and the restrictions on FDI prevented exposure to foreign best practices.[2]

However, a remarkable transformation has taken place in Korea's distribution industry since the government lifted some of the restrictions that kept foreign service suppliers out of the country before 1996 (table 6.6).[3] In

2. In terms of deregulation of zoning, the semiagricultural and forest areas were redefined to allow retail stores occupying less than 30,000 square meters to be built in 1993. In 1996, large discount retailers under 10,000 square meters were allowed to do business in the green areas, where development is regulated by the law. The objective was to promote discount stores (Mckinsey 1998).
3. In most of the service subsectors, the Korean government implemented domestic deregulation and external liberalization almost simultaneously. It used external commitment to liberalization in reducing any opposition or resistance to domestic deregulation or implemented domestic

Table 6.6   **Liberalization of Distribution Services, Korea: 1989–2000**

Year	Liberalization Measures
1989	Allow FDI in wholesale of medicine
	Expand permissible imports by branches of foreign companies
1991	Allow FDI in retailing, up to 10 stores of 1,000 m^2 or less for each foreign-invested company
1993	Expand store and space-related limits to 20 stores of 3,000 m^2 or less for each company
1996	Eliminate requirements on the number of stores and space (allowed establishment of hyper-markets)
	Liberalize 5 business categories, including commodity chains, and the retailing of meat
1997	Liberalize 10 business categories, including general trading and the retailing of grain
1998	Abolish economic needs tests on department stores and shopping centers
	Liberalize operation of gas stations
2000	Allow FDI in the wholesaling of meat

Source: Ministry of Commerce, Industry, and Energy, Department of Distribution.

Table 6.7   **Trends in the Establishment of Hyper-markets in Korea: 1997–2000**

		Number of Stores			
Name	Year of Entry[a]	1997	1998	1999	2000
Carrefour	1996	3	6	11	20
Wal-Mart	1996	4	4	5	10
Costco	1998(1994)	2	3	3	5
Promodes	1999	—	—	2	5
Tesco	1999(1997)	1	1	2	7
Total for foreign companies[b]	—	10	14	23	47
		(15%)	(16%)	(20%)	(29%)
Total for Korean companies[b]	—	55	74	92	117
		(85%)	(84%)	(80%)	(71%)
Total	—	65	88	115	164

Source: Korean Association of Retailers, *Management Revolution in 21st-Century Asian Retailing* (December 27, 1999).
[a]Entry year of the acquired local company in parentheses.
[b]Shares in total number of stores in parentheses.

particular, store and space-related limits on retailing were eliminated for both domestic and foreign retail firms. As a result, a number of large-sized discount stores or hypermarkets have been established by both domestic and foreign firms since 1996. The total number of hypermarket stores reached 164 in 2000, and almost 30 percent of them have been established by foreign firms (table 6.7).

deregulation to help domestic firms establish market position before foreign penetration. Hence, it is difficult to differentiate the impact of domestic deregulation from external liberalization.

The increasing number of hypermarkets is changing the manufacturer-dominated structure of the Korean retail industry, which had deterred productivity improvements and price competition. The increased buying power of the hypermarkets puts price determining in the hands of retailers rather than manufacturers, leading to price competition. Foreign retail firms also transferred advanced techniques in merchandising and inventory management as well as new technologies such as point of sales (POS) systems.

Table 6.8 presents the change in number of establishments per 1,000 res-

Table 6.8   **Trends in the Sizes of Establishments in Distribution Services, Korea and Japan: 1982–98**

Year	Korea		Japan	
	Wholesale	Retail	Wholesale	Retail
Number of Establishments per 1000 Residents				
1982	1.2	13.8	3.3	14.5
1985	—	—	3.1	13.5
1986	1.7	15.5	—	—
1988	1.9	16.0	—	—
1990	2.1	16.6	3.8	12.8
1992	2.6	16.9	—	—
1994	2.7	17.0	—	—
1996	3.2	16.9	—	—
1998	3.1	15.6	—	—
Workers per Establishment				
1982	3.8	1.7	9.3	3.7
1985	—	—	9.4	3.9
1986	5.0	1.9	—	—
1988	5.4	1.9	—	—
1990	5.5	1.9	—	—
1992	4.7	1.9	—	—
1994	5.1	2.0	—	—
1996	4.3	2.1	—	—
1998	4.2	2.0	—	—
Floor Space per Establishment (m²)				
1982	—	—	—	55.4
1985	—	—	—	58.0
1986	—	—	—	—
1988	—	—	—	—
1990	—	—	—	—
1992	75.7	35.6	—	—
1994	92.7	38.7	—	—
1996	129.4	45.8	—	—
1998	136.4	52.8	—	—

Sources: The data on Korea are constructed from various issues of the *Annual Report on the Survey of Wholesale and Retail Trade,* published by the Korean National Statistical Office. The data on Japan are from Ito and Maruyama (1991) and Anwar and Taku (1993).

Note: Dashes indicate that data are not available.

idents, workers per establishment, and floor space per establishment since 1982. The Korean distribution sector has experienced growth in terms of number of establishments as well as the size of establishments. Particularly, the number of wholesale establishments has grown quickly from 1.2 per 1,000 residents in 1982 to 3.1 in 1998. The number of retail stores reached 16.6 per 1,000 residents in 1990, far surpassing Japan. Although the Japanese distribution sector is accused of inefficiency due to the presence of many small establishments, the Korean distribution sector may be regarded as worse, with much smaller establishments in terms of size.[4] However, in the mid-1990s, the number of establishments in retailing began to decline, while the size continued to grow. During this period, the domestic retailing sector began to be exposed to foreign competition as foreign firms started to enter the market, as shown in table 6.7.

Figure 6.1 decomposes the growth of sales into the growth of the number of establishments and the growth of sales per establishment. The total amount of sales has grown steadily except for the period 1996–98, when Korea fell into a severe recession due to the financial crisis. In wholesale services, the opening of new establishments contributed to the growth of sales. However, in retail services, the growth of sales came largely from the growth of sales per establishment. Particularly, in contrast to the wholesale sector, opening of new retail stores slowed down in the 1990s, and the number of establishments even declined from 1996 to 1998.

Figures 6.2 and 6.3 show sales per employee and sales per establishment. The sales per employee and sales per establishment, which are widely used as measures of productivity and efficiency of the distribution system, show that the productivity of the Korean distribution sector has continually increased over time. Sales per both worker and establishment increased notably in 1996, which may be a result of the service liberalization and resulting FDI inflow. However, we have to wait to see whether this trend will continue after the economy recovers from the deep recession of 1998.

Figure 6.4 breaks down the sales per employee of retail stores according to their size. It shows that sales per employee of large retail stores, with five or more employees, recorded a noticeable increase in 1998, whereas sales per employee of small retail stores, with fewer than five employees, have been stagnant since 1996. This may be because liberalization of the retail sector brought about enhanced competition in the large-sized retail stores through the establishment of hypermarkets by foreign retailers. The role of liberalization in enhancing competition may be ascertained by the lower price margins of the supermarkets and department stores, from 17.8 percent and 24.2 percent in 1995 to 13.6 percent and 21.7 percent in 1998, re-

---

4. For a discussion of the efficiency of the Japanese distribution system, see Ito and Maruyama (1991) and Anwar and Taku (1993).

Wholesale

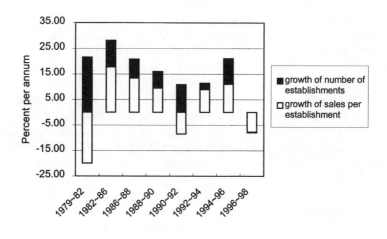

Retail

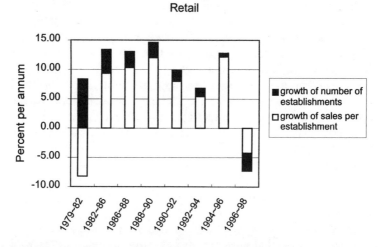

Fig. 6.1 Decomposition of sales growth in distribution services, Korea (1979–98)

Source: National Statistical Office, Republic of Korea, "Annual Report on the Survey of Wholesale and Retail Trade," various years.

Note: The amount of sales is deflated using the producer and consumer price index for wholesale and retail, respectively.

spectively (table 6.9). This reveals that supermarkets and department stores face direct challenges from foreign competitors.

In sum, a rough observation of the measures of efficiency points to enhanced productivity of the Korean distribution services with liberalization in the 1990s, although we cannot provide definite evidence due to the limited data. Particularly, the inflow of FDI with the opening of hypermarkets

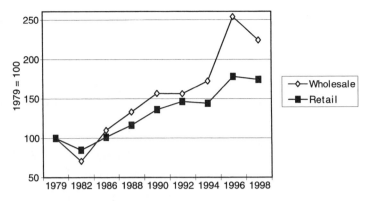

Fig. 6.2   Sales per employee in distribution services, Korea (1979–98)

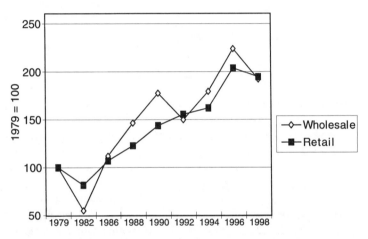

Fig. 6.3   Sales per establishment in distribution services, Korea (1979–98)
Source: See figure 6.1 note
Note: See figure 6.1 note

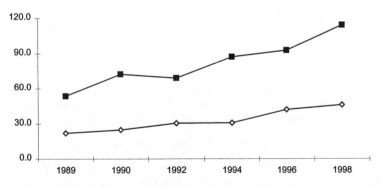

Fig. 6.4   Sales per employee in small and large-sized retail stores, Korea (1989–98)
Source: National Statistical Office, Republic of Korea, "Annual Report on the Survey of Wholesale and Retail Trade," various years.
Note: Numbers are millions of won.

Table 6.9   Price Margin Trends for Supermarkets and Department Stores, Korea: 1995–98 (%)

	1995	1996	1997	1998	Change in 1995–98
Supermarkets	17.8	16.1	15.0	13.6	–4.2
Department stores	24.2	24.8	22.6	21.7	–2.5

Source: Bank of Korea, *Impact of Changes in Distribution Structure on Price Levels* (26 January 2000).

by foreign firms introduced best practices management and challenged domestic retail stores. In addition, changing shopping patterns with the introduction of discount stores may have forced many small stores to specialize their services and existing domestic retail firms to enlarge their size to take advantage of the scale effect.

6.4   Changes in Productivity Growth Rates[5]

This section investigates whether the productivity changes in the service and manufacturing sectors in the 1990s were associated with services liberalization. Even though there are many problems in measuring the output of services, we follow the traditional approach by using the value added as the measure of output. The reason is as follows: first, the production of services covers a wide range of heterogeneous activities. As frequently mentioned by many authors such as Griliches (1992) and Triplett and Bosworth (2000), each service industry contains its own unique problem of measuring output, which makes a treatment applicable to all services almost impossible. Second, recent studies approach the issue with sector-specific data sets by utilizing a wide variety of methodologies. Thus, the productivity changes of different services calculated in various ways are not directly comparable. Third, unlike manufacturing, services are considerably backward in the availability of reliable data, which is particularly the case in Korea. Given the constraints of data, we simply utilize the data from national accounts as a last resort.

We first compare the level of labor productivity of the Korean service industry with that of some of the advanced countries. The growth rates of labor productivity and total factor productivity (TFP) in the Korean service sector since 1970 are then examined. Finally, we will try to see whether productivity growth in the manufacturing sector is associated with services liberalization.

6.4.1   Sectoral Labor Productivity: An International Comparison

In table 6.10, we compare the levels of labor productivity in Korea with those of some developed countries in 1990. Labor productivity is calculated

---

5. The data used for this section are described in the appendix.

as the value added per worker. For the Group of Five (G5) countries, the value added is converted, for comparison, by using the 1985 purchasing power parity exchange rates. Because the purchasing power parity exchange rate for each sector is not available for Korea, we convert the labor productivity of Korea by using the average market exchange rate for the period 1980–90.

Table 6.10 shows that in 1990 the labor productivity of the Korean service sector, except for "utilities," was much lower than that of the United States, the European countries, and Japan. The labor productivity of "construction" and "finance, etc." in Korea was about 40 percent that of the United States. Even worse was the labor productivity of "distribution, etc." and "social services, etc.," which was 18 percent and 15 percent of U.S. levels, respectively.

6.4.2   Productivity Growth in Services

Table 6.11 tabulates the growth rates of labor productivity in the Korean service subsectors since 1970. "Finance, etc.," practically closed to foreign suppliers until the late 1990s, experienced the worst performance, with negative growth rates in labor productivity throughout the period, except for 1985–90. It was during this period that the Korean economy was booming with a large trade surplus. However, "distribution, etc.," which was almost completely liberalized in 1996, and "transport and communication," which was partially liberalized in the 1990s, showed increases in labor productivity in the late 1990s, from 5.09 percent and 0.41 percent in 1990–95 to 7.17 percent and 1.54 percent in 1995–97, respectively.

Because labor productivity is influenced by the magnitude of capital, which is affected by FDI inflows, we next compare changes in TFP in the same period.

Total factor productivity is defined as

$$\text{TFP} = \frac{Y}{K^{\alpha}L^{1-\alpha}}$$

where Y, K, and L are output, capital, and labor inputs, respectively, and α is the elasticity of the production of capital. Thus, TFP growth is calculated as the residual of output growth net of the weighted growth of factor inputs. The underlying assumption is to use the factor shares in total costs as factor weights under constant returns to scale, Hicks neutral technical progress, and the profit maximization of firms in competitive markets. In our study, we consider two inputs, capital and labor.

It is desirable to adjust capital and labor inputs by their quality measures. However, the data on the quality of inputs at the sectoral level are not available. We use gross fixed capital stock for capital inputs and total employment for labor inputs. It is also desirable to have actually utilized

Table 6.10   **Labor Productivity of Selected Countries Relative to the United States in 1990**

Country	Agriculture	Mining	Manufacturing	Construction	Utilities	Transportation, Communication	Distribution, etc.	Finance, etc.	Social etc.	Total Economy
France	74	31	82	86	84	73	100	105	130	93
West Germany	53	19	70	81	63	60	70	166[b]	178	86
United Kingdom	53	n.a.	66	90	85	54	60	257[b]	86	73
United States	100	100	100	100	100	100	100	100	100	100
Japan	25	31	79	89	106	50	70[a]	148	78	70
Korea	16	10	34	42	68	23	18	40	15	26

Source: Authors' calculations.

Notes: "Utilities" denotes electricity, water, and gas; "Distribution, etc." denotes retail, wholesale, restaurants, and hotels; "Finance, etc." denote finance, insurance, real estate, and business services; "Social Services, etc." denotes community, social, and personal services.

[a]The figure is for the comparison of retail and wholesale trade only, excluding restaurants and hotels.

[b]The figure is for the comparison of finance and insurance only, excluding real estate.

Table 6.11   **Annual Average Growth Rates for Labor Productivity in Service Subsectors, Korea: 1970–97 (%)**

Period	Manufacturing	Construction	Utilities	Transportation, Communication	Distribution, etc.	Finance, etc.	Social Services, etc.	Total Economy
1970–75	5.62	-4.60	10.62	11.20	5.13	-2.45	7.64	5.26
1975–80	6.06	2.10	12.54	2.24	-2.87	-6.26	3.16	4.40
1980–85	6.39	6.16	15.95	4.26	3.39	-1.70	7.65	6.96
1985–90	4.76	4.19	3.05	3.49	7.37	1.02	3.65	6.08
1990–95	8.63	-0.14	9.43	5.09	0.41	-1.06	6.46	5.34
1995–97	9.87	1.48	6.32	7.17	1.54	-1.31	0.57	4.83
1970–80	5.84	-1.25	11.58	6.72	1.13	-4.36	5.40	4.83
1980–90	5.58	5.17	9.50	3.88	5.38	-0.54	5.65	6.52
1990–97	8.98	0.32	8.54	5.68	0.74	-1.08	4.78	5.20
1970–97	6.56	1.54	10.02	5.40	2.60	-2.02	5.33	5.55

Source: Authors' calculations.

input levels by using working hours and utilized capital. However, the data on hours worked, both for capital and labor, are limited in their use for our purposes. Regarding working hours, the published data concern the hours paid rather than hours actually worked. Also, the capacity utilization rate at the subsector level is not available, particularly for the service sectors. Therefore, due to the failure to allow for cyclical variations in hours worked and capacity utilization, there is a cyclical bias to our measurements of TFP growth in the short run. However, this problem is lessened in the long run by the booms' being offset by recessions.

Table 6.12 shows that similar patterns can be detected for changes in TFP. As was the case for labor productivity, "finance, etc." recorded negative TFP growth rates throughout the period, except for 1985–90. "Transport and communications" showed a gain in TFP growth in the late 1990s, from 2.2 percent in 1990–95 to 4.12 percent in 1995–97. The trend of TFP growth in "distribution, etc." also improved in the late 1990s, from –0.41 percent in 1990–95 to –0.02 percent in 1995–97.

However, we cannot strictly prove that productivity improvement was caused by liberalization in services from the trend of labor productivity and TFP growth. As already mentioned, the two measures of efficiency considered above are subject to cyclical fluctuations, and there may be a time lag for the liberalization measure to take effect in a sector-wide productivity change. Considering that meaningful liberalization in the Korean service sectors has only been implemented since the mid-1990s, it may be too early to demonstrate any causal relationship between productivity changes and services liberalization.

6.4.3   Contribution of Services Liberalization to Manufacturing

The hypothesis that liberalization in services may increase the productivity of manufacturing subsectors that use liberalized services as inputs can be examined by comparing the productivity growth rates of manufacturing subsectors (table 6.13) and the input coefficients of services to those manufacturing subsectors (table 6.14).

For "nonmetals," which had a negative TFP growth rate of –0.06 percent in 1990–97, we can notice that the input coefficient of distribution services, which were liberalized in the 1990s, was 0.018, relatively lower than the input coefficients of the other service subsectors. Thus, "nonmetals," which uses the liberalized service subsector less intensively, shows poor performance in terms of TFP growth rates when compared with other manufacturing subsectors.

However, it seems to be difficult to extract any consistent pattern from the growth rates of the TFP in the manufacturing subsectors and the input coefficients of the service subsectors. In general, the sum of the input coefficients of services in the manufacturing subsectors is in the range of 0.1 to 0.17, which is not large enough to make a significant impact on their productivity.

Table 6.12 **Annual Average Growth Rates of Total Factor Productivity in Service Subsectors, Korea: 1970–97 (%)**

Period	Manufacturing	Construction	Utilities	Transportation, Communication, etc.	Distribution, etc.	Finance, etc.	Social Services, etc.	Service Total	Total Economy
1970–75	3.58	-2.04	7.52	6.63	4.16	-4.87	5.32	2.17	1.52
1975–80	5.23	-0.64	3.29	-0.58	-3.93	-5.60	2.54	-2.35	-1.13
1980–85	5.81	1.81	3.33	-2.20	1.65	-1.56	7.54	0.77	2.89
1985–90	2.99	3.39	6.56	1.69	3.95	1.27	2.65	3.20	2.65
1990–95	4.90	-2.52	2.39	2.20	-0.41	-2.82	4.44	-0.31	0.99
1995–97	-0.54	-0.01	2.71	4.12	-0.02	-2.72	-1.04	-0.15	0.71
1970–80	4.41	-1.34	5.41	3.02	0.11	-5.24	3.93	-0.09	0.20
1980–90	4.40	2.60	4.94	-0.25	2.80	-0.15	5.10	1.98	2.77
1990–97	3.35	-1.80	2.48	2.75	-0.30	-2.79	2.87	-0.26	0.91
1970–97	4.13	0.00	4.47	1.74	1.00	-2.72	4.09	0.63	1.34

Source: Authors' calculations.

Table 6.13   Annual Average Growth Rates of Total Factor Productivity in Manufacturing, Korea: 1970–97 (%)

Period	Food	Textiles	Wood	Paper	Chemicals	Nonmetals	Metals	Machinery	Other	Manufacturing Total
1970–75	2.57	8.11	6.04	-1.02	-1.42	5.54	23.62	7.96	8.23	3.58
1975–80	9.15	4.67	-3.63	5.43	6.31	-0.09	16.09	2.00	7.52	5.23
1980–85	1.90	4.02	10.32	8.37	3.38	4.64	9.90	8.69	7.51	5.81
1985–90	2.59	1.00	6.39	1.16	3.04	-2.00	6.74	2.97	-5.61	2.99
1990–95	0.69	3.49	4.34	9.14	2.81	0.13	8.18	5.96	-2.24	4.90
1995–97	0.04	-1.02	2.66	-1.47	0.65	-0.55	-0.13	-1.47	1.71	-0.54
1970–80	5.86	6.39	1.20	2.21	2.44	2.73	19.85	4.98	7.88	4.41
1980–90	2.24	2.51	8.36	4.77	3.21	1.32	8.32	5.83	0.95	4.40
1990–97	0.50	2.20	3.86	6.11	2.20	-0.06	5.81	3.84	-1.11	3.35
1970–97	3.13	3.87	4.54	4.17	2.66	1.48	11.94	5.00	2.98	4.13

Source: Authors' calculations.

Table 6.14 Input Coefficients for Selected Manufacturing Subsectors, Korea: 1995

				Output				
Input	Food	Textiles	Wood and Paper	Chemicals	Nonmetals	Metal Products	Machinery	Electronics
Utilities	0.008	0.013	0.023	0.023	0.034	0.015	0.007	0.007
Construction	0.000	0.001	0.000	0.001	0.001	0.001	0.001	0.000
Distribution	0.026	0.029	0.032	0.023	0.018	0.025	0.028	0.026
Restaurants, etc.	0.000	0.000	0.000	0.000	0.000	0.030	0.000	0.000
Transportation	0.013	0.011	0.019	0.014	0.038	0.013	0.013	0.010
Communications	0.001	0.002	0.003	0.002	0.005	0.002	0.002	0.003
Financial services	0.014	0.038	0.033	0.024	0.029	0.025	0.024	0.018
Real estate, etc.	0.028	0.024	0.018	0.032	0.023	0.020	0.022	0.025
Public administration	0	0	0	0	0	0	0	0
Education, health	0.003	0.003	0.002	0.014	0.006	0.004	0.014	0.030
Social services	0.000	0.001	0.001	0.001	0.001	0.001	0.001	0.001
Other services	0.013	0.024	0.014	0.017	0.017	0.022	0.016	0.011
Total	0.108	0.146	0.148	0.152	0.173	0.123	0.129	0.132
	(0.151)	(0.209)	(0.209)	(0.217)	(0.274)	(0.197)	(0.195)	(0.148)

Source: Bank of Korea, *Input/Output Tables* (1995).

Note: The figures in parentheses are the share of service in total intermediate input.

6.5   Concluding Remarks

Due to industrialization that gave priority to manufacturing at the expense of services, the service sector in Korea was grossly underdeveloped prior to the early 1990s. Numerous sector-specific regulations and restrictions on FDI prevented competition and impeded the offering of higher-value services. In 1990, the labor productivity of the Korean service subsectors was much lower than that of the advanced countries. The labor productivity of "distribution services, etc.," in particular, was less than one-fifth that of the United States in 1990.

Since the mid-1990s, the Uruguay Round negotiations and accession to the OECD enabled the Korean government to gradually open the service sector to foreign suppliers. The financial crisis of late 1997 resulted in the Korean service sector's becoming almost completely open, except for a few areas sensitive to national security, culture, and political stability.

The liberalization of services is presumed to bring productivity gains in the service sector and also in manufacturing sectors that use liberalized services as inputs. In searching for some evidence of this in Korea, we examined the changes in productivity of the service and manufacturing subsectors in 1970–97. Because liberalization took place in the 1990s and it takes time to see the full effects of liberalization, it is too early to give a definite answer to whether liberalization in services has caused an increase in productivity in Korea. However, we see a productivity improvement in such sectors as distribution services, which had a large inflow of FDI due to liberalization in the 1990s.

Appendix
Sources of Data

The sectors considered were selected on the basis of their availability for output and factor use. The sector classification used was the International Standard Industrial Classification System. Output was measured as value added in constant prices, the data for which were obtained from the national accounts of Korea collected by the Bank of Korea.

The labor input is measured as total employment. The Annual Report on the Economically Active Population Survey (AREAPS) provides the data on total employment. However, AREAPS does not classify most of the service subsectors for periods earlier than 1991.[6] Thus, we computed the ratio for each service subsector based on the data from the *Statistical Yearbook* of the International Labor Organization (ILO) and the employment tables

---

6. The Annual Report on the Economically Active Population Survey currently classifies service sectors as electricity, gas, and water; retail and wholesale trade, restaurants, and hotels; transport, storage, and communications; financial institutions, insurance, real estate, and business services; and community, social, and personal services.

of the Bank of Korea and estimated the total employment for each subsector by applying the computed ratio to the total employment of the service sector of AREAPS.

Capital input is defined as gross fixed capital stock and was computed by applying the perpetual inventory method (PIM) to the data on the gross fixed capital formation of each industry in constant prices obtained from the national accounts. To use the PIM, we need data on benchmark capital stock and the depreciation rate. However, reliable data on these two variables are not available. Therefore, we extrapolated the gross fixed capital formation back to 1930 based on the time trend from 1953 to 1960 to avoid the problem of benchmark capital stock and accumulated the investment from 1930.[7] For the depreciation rate for each industry, we used the average rates of the corresponding Japanese industries, calculated from the International Sectoral Database (ISDB) published by the OECD, under the assumption that the structure of the Korean economy is most similar to Japan among the developed countries covered by the ISDB.

Finally, to compute the TFP, we need data on the share of labor in value added. The labor share is calculated by dividing the compensation of labor by value added. Because the data on the compensation of employees from the national accounts do not include the compensation of self-employed labor, we adjusted the compensation of employees under the assumption that the compensation of the self-employed is comparable to that of the employed. That is,

The share of labor in value added = (compensation of employees

+ [compensation of employees/total employees]

× [total employment − total employees])/value added.

The data on the compensation of employees and current value added are taken from the National Accounts. The number of total employees is taken from the ILO *Statistical Yearbook* and the employment tables of the Bank of Korea. The share of agriculture and fisheries; community, social, and personal services; retail and wholesale trade; and restaurants and hotels, computed as above, are too high. The employment in these industries shows that a large proportion of unpaid family workers may be underemployed. Thus, when comparing them with some of the advanced countries from the ISDB, we assumed that the unpaid workers were compensated at half the rate of paid workers. After adjustment, the shares of labor in value added for these industries were comparable to the estimate of Kim and Park (1985).

Finally, the data on some of the advanced countries used for international comparison were taken from the ISDB of the OECD, which provides sectoral output and input data of OECD countries from 1970 to 1990.

---

7. We assumed the investment to be zero between 1950 and 1952 during the Korean War.

Table 6A.1   Service Subsectors in which FDI is Restricted (as of January 1990)

Wholly Restricted	Partially Restricted
Production, collection and distribution of electricity	Wholesale of agricultural raw materials, live animals, food, beverages and tobacco
Publishing (newspapers, periodicals and books)	Wholesale of household goods (medical goods and cosmetics)
Collection, purification and distribution of water	Wholesale of nonagricultural intermediate products, waste and scrap (fertilizers)
Drinking establishments	Other wholesale (foreign trade brokers)
Transport via railways	Retail sale of food, beverages and tobacco in specialized stores
Scheduled air transport	Other retail trade of new goods in specialized stores
Nonscheduled air transport	
Post and courier activities	Land transport
Telecommunications	Sea and coastal water transport
News agency activities	Inland water transport
Radio and television broadcasting	Travel agencies
Gambling	General financial intermediation (banking)
	Other financial intermediation (investment, trust, securities)
	Insurance and pension funding
	Real estate rental and development
	Renting of other machinery and equipment (construction equipment)
	Research and experimental development on social sciences and humanities
	Legal, accounting, bookkeeping and auditing activities; tax consultancy; market research and public opinion polling; business and management consultancy
	Advertising
	Other business services (personnel supply services, investigation and security activities)
	Adult and other education (vocational training schools, etc.)
	Human health activities
	Veterinary activities
	Motion pictures and other entertainment activities
	Libraries, archives, museums and other cultural activities
	Sporting and other recreational activities
	Other service activities (barber and beauty shops, wedding chapels, etc.)
	Other recreational activities (parks, beaches, etc.)
	Personal services (tutoring, housekeeping, etc.)

Source: Ministry of Finance and Economy, *Five-Year Foreign Investment Liberalization Plan* (various years)

Note: In KSIC three-digit level.

Table 6A.2   **Service Subsectors in which FDI is Restricted (as of November 1997)**

Wholly Restricted	Partially Restricted
Collection, purification and distribution of water News agency activities Radio and television broadcasting Gambling	Wholesale of agricultural raw materials, live animals, food, beverages and tobacco (meat) Production, collection and distribution of electricity Publishing (newspapers, periodical and books) Other retail trade of new goods in specialized stores (gas stations) Land transport Sea and coastal water transport Scheduled air transport Nonscheduled air transport Telecommunications General financial intermediation (banking) Other financial intermediation (investment, trust, securities) Insurance and pension funding Real estate rental and development Credit information agency

Source: Ministry of Finance and Economy, *Five-Year Foreign Investment Liberalization Plan* (various years).
Note: In KSCI three-digit level.

References

Anwar, Syed Tariq, and Michael A. Taku. 1993. Productivity and efficiency in the Japanese distribution system: A review and developments. *Journal of World Trade* 27:83–110.

Blomstrom, Magnus, and Hakan Persson. 1983. Foreign investment and spillover efficiency in an underdeveloped economy: Evidence from the Mexican manufacturing industry. *World Development* 11:493–501.

Borensztein, E., J. de Gregorio, and J. W. Lee. 1998. How does foreign direct investment affect economic growth? *Journal of International Economics* 45:115–35.

Caves, Richard E. 1974. Multinational firms, competition, and productivity in host-country markets. *Economica* 41:176–93.

Chung, W., W. Mitchell, and Bernard Yeung. 1994. Foreign direct investment and host country productivity: The case of the American automotive components industry. Discussion Paper no. 367. University of Michigan, Institute of Public Policy Studies.

Globerman, Steve. 1979. Foreign direct investment and "spillover" efficiency benefits in Canadian manufacturing industries. *Canadian Journal of Economics* 12: 42–56.

Griliches, Zvi. 1992. Introduction. In *Output measurement in the service sectors*, ed. Zvi Griliches, Chicago: University of Chicago Press.

Ito, Takatoshi, and Masayoshi Maruyama. 1991. Is the Japanese distribution sys-

tem really inefficient? In *Trade with Japan*, ed. Paul Krugman, Chicago: University of Chicago Press.
Kim, J.-D., and S.-I. Hwang. 2000. The role of foreign direct investment in Korea's economic development: Productivity effects and implications for the currency crisis. In *The role of foreign direct investment in East Asian economic development*, ed. Anne O. Krueger and Takatoshi Ito, 267–94. Chicago: University of Chicago Press.
Kim, K.-S., and J.-K. Park. 1985. *Sources of economic growth in Korea: 1963–1982*. Seoul: Korea Development Institute.
Markusen, James R. 1989. Trade in producer services and in other specialized intermediate inputs. *American Economic Review* 79:85–95.
McKinsey Global Institute. 1998. *Productivity-led growth for Korea*. Washington, D.C.: McKinsey Global Institute.
Nickell, Stephen J. 1996. Competition and corporate performance. *Journal of Political Economy* 104:724–46.
Rivera-Batiz, Francisco L., and Luis A. Rivera-Batiz. 1992. Europe 1992 and the liberalization of direct investment flows: Services versus manufacturing. *International Economic Journal* 6:45–57.
Triplett, Jack E., and Barry P. Bosworth. 2000. Productivity in the service sector. Washington, D.C.: Brookings Institution. Manuscript.

Comment   Kazumasa Iwata

The paper investigates the impacts of service liberalization, notably establishment trade through commercial presence, on productivity in Korea. The authors found it premature to show empirical evidence on an increase in productivity. They argue, however, that competition is enhanced, notably in the distribution sector, which may lead to an increase in productivity.

The liberalization of services trade may increase productivity through a number of channels. First, the establishment trade via commercial presence brings about an increase in production and in management know-how and imparts a productivity spillover effect to the economy.

Second, the improved quality and diversified service with lower prices as intermediate inputs (both embodied and disembodied) may improve the productivity of the manufacturing sector. This constitutes the major source of benefit of developing countries that have a comparative disadvantage in service trade.

Third, the liberalization of trade in services may enhance competition. This implies that trade liberalization affects the market structure in an economy. Although it does not exclude the case of foreign monopoly, the Korean economy seems to prove that the domestic economy is large enough to permit the commercial presence of foreign firms. We should recall that some

Kazumasa Iwata is a professor of economics at the University of Tokyo.

developing economies have employed trade restriction as the imperfect substitute for antimonopoly law.

Fourth, the transmission of advanced knowledge through the joint research and development and provision of licenses constitutes the channel-of-productivity increase.

It is interesting to observe that in the case of Korea the establishment trade in services has accelerated after the financial crisis, despite the fact that the massive inflow and subsequent outflow of capital disrupted the financial market. It seems important to pay due attention to the fact that the financial liberalization has been accompanied by over-investment by nonfinancial firms and constituted the precondition to financial crisis. The absence of appropriate prudential policy on financial institutions' financial trade liberalization may have had a destabilizing effect on the economy. The financial disruption seemed to work to deepen the recession.

Table 6.9 provides an international comparison of sectoral labor productivity in 1990, whereas table 6.11 provides an interesting result on changes in total factor productivity (TFP). It indicates an improvement of productivity in the service sector, notably in the transport and communication sector, in 1995–97, despite the recession. However, it is somewhat puzzling that productivity in the financial sector registered a negative rate of increase for a long period of time, from 1970 to 1997. Further, I find it difficult to accept the statement that the distribution sector shows an improvement in productivity when I look at tables 6.10 and 6.11; the negative TFP growth is recorded in 1995–97. In contrast, figure 6.1 points to an increase in sales per employee, but it is not clear whether it implies a productivity gain. It seems necessary to provide additional evidence to confirm the productivity improvement in the distribution sector after 1996.

My first comment is whether the productivity gain is attributable solely to foreign direct investment. I suspect that domestic regulatory reform in the communication sector, in addition to the distribution sector, may have played an important role in increasing the productivity of almost all service sectors. Liberalization of trade in service often takes place simultaneously with domestic regulatory reform. Both promote competition and efficiency in an economy. It is important to discriminate between the effects of reduction in the domestic distortion and in the external distortion. Apparently, services trade liberalization deals with the latter. It seems possible to truncate the impact by employing some econometric methods, although note 2 mentions that it is difficult to differentiate between the impacts of domestic deregulation and external liberalization.

For instance, it may be useful to apply the tariffication to trade in services, as argued by Deardorff (2000), or develop the tax-equivalent measures in assessing external distortion, such as the exercise carried out by the Australian Productivity Council.

My second comment is a question as to whether the observed productivity gain is due to the expansion of the production frontier or the shift of the production point inside the existing production frontier to the more efficient point.

Foreign direct investment (FDI) implies an increase in capital accumulation, coupled with transfer of management and reduction in external distortion, whereas domestic deregulation simply removes barriers and reduces domestic distortion. Thus, the increase in FDI is accompanied by the expansion of production frontiers into the direction that the Rybczynski theorem predicts. The lack of appropriate data may prevent judgment. However, it may be useful to delineate the difference, at least conceptually, because it relates to the issue on the order of liberalization, notably with respect to financial liberalization.

Third, I would like to know why the intermediate service inputs in total intermediate inputs take on such a small share (0.15–0.17) in intermediate inputs of the Korean manufacturing sector. In Japan the share ranges from 0.28 to 0.3 in 1995, whereas in the United States it is 0.32 in 1990. As is shown by the simulation results in a paper presented at this conference by Kun-Ming Chen and others (chap. 4 in this volume), the service inputs are important not only for the development of the service sector but also for strengthening the competitiveness of the manufacturing sector. The smaller share of intermediate service inputs implies a smaller impact (about half) of liberalization of trade in services in the case of Korea, as compared with other countries, such as Japan and the United States.

Fourth, the paper primarily concerns productivity. However, the benefit for consumers appears mainly through price reduction and diversification of services. The price differential between home and abroad and the comparison of cost structures provide additional information on domestic and external distortion. It may be promising to employ cost function, instead of production function, in continuing the analysis on changes in consumer welfare by the liberalization of trade in services.

Fifth, despite the difficulty in the case of abnormal shocks like financial crises, it still seems important to make some cyclical adjustment to the labor and total productivity changes, in order to assess the impact of liberalization.

Finally, with respect to productivity changes in distribution sector, I find it more important to pay attention to the impact of information and telecommunication innovation due to the regulatory reform; it contributes to the expansion of production frontier due to the new innovation and the network externality effect. It may be noted that in the case of Japan the large retail stores like Ito-Yokado intend to make new entry into banking business specializing in payment services by utilizing both the physical and virtual networks. Conversely, many banks attempt to enclose networks of convenience stores. This attempt can enhance competition and bring about an increase in productivity in financial service.

Reference

Deardorff, A. V. 2000. Tariffication of services. Paper presented at the conference organized by Keio University. 19 May.

Comment   Mahani Zainal-Abidin

The paper shows that Korea's service sector liberalization began in the early 1990s with its commitment to the Uruguay Round negotiation and later membership to the Organization for Economic Cooperation and Development (OECD). The pace of liberalization was further accelerated during the East Asian Crisis, and by May 2000 only very few service industries had total or partial restrictions (see table 6.2).

The paper suggests that liberalization of Korea's service sector, which allows for greater participation of foreign direct investment (FDI), will improve productivity of liberalized service subsectors as well as of the manufacturing sector that uses the services of these subsectors.

My comments are as follows.

First, the main channel for productivity improvement put forward in this paper is through higher participation of FDI that can inject technology transfer and overcome economies of scale constraints. Increasing foreign participation is one form of service sector liberalization. An equally important form of liberalization concerns domestic service producers, namely, the lowering of barriers to domestic producers through deregulation and dismantling of regulatory impediments. These may take the form of what is called in Kim's paper barriers to ongoing operation or barriers that are designed to protect incumbent firms.

The paper should describe more the measures of liberalization for domestic service producers—for example, to what extent the Korean services liberalization has removed entry and competition barriers. In other words, to what extent is productivity improvement in the service subsector due to the removal of barriers to ongoing operation of domestic services providers vis-à-vis increased FDI participation?

For example, in Malaysia, productivity gains in service subsectors such as telecommunications are substantially caused by deregulation of domes-

Mahani Zainal-Abidin is professor of applied economics at the University of Malaya.

tic industries through increasing the number of participants and competition in the industry.

Another form of services liberalization is through privatization. Most of the services industries are publicly owned, and privatization of these activities since the late 1980s has allowed these services to be managed in a more efficient manner and has made cost and quality of service key in evaluating their performance. As a result, there was a marked increase in the productivity of these privatized services. For example, parts of the health, education, and water sectors have been privatized, and the ensuing higher productivity has benefited consumers and has attracted investments in these sectors.

Second, I would like to comment on the calculation of service subsector productivity.

The key problem in calculating service productivity is in determining a suitable measure of output over time. This is complicated by two factors: a) market prices may not be observable for publicly provided services, and b) it is often difficult to identify precisely what constitutes a service activity in a particular industry and to account correctly for quality changes in services.

The measurement of output requires identifying whether the output consists of the transaction performed or the outcome achieved through the service. A case in point is the finance subsector. The subsector experienced negative growth rates in labor productivity throughout the 1970–97 period, except for 1985–90. The period 1985–90, in which productivity rose substantially, was when the Korean economy boomed following the 1988 Olympic games. Thus, the increase in output of finance subsector was due to bigger transaction volume and not improvement in services. Further, the finance sector offers a variety of activities that may increase its output (revenue) by expanding the scope of financial transactions, and this higher revenue may not come from improvement in service.

The measurement difficulties are further aggravated by contributions made by technology. In some services, such as distribution, telecommunication, and parts of the financial service industry, technological change has strongly affected the production process and the organization of production. This has contributed to significant improvements in productivity, but this may not always be easy to measure.

To overcome these measurement problems, some studies have used the growth rate of capital as a proxy for services output growth.

Third, I would like to comment on the contribution of services liberalization to the manufacturing sector.

The paper has produced two interesting tables: table 6.8, on annual average growth rates of total factor productivity in the manufacturing sector, and table 6.9, on input coefficients for selected manufacturing subsectors. As contended by the paper, liberalization in service subsectors will benefit the manufacturing industries that use the services of these subsectors through productivity improvements. The paper should explore and formalize these links. These two tables have given the authors a good start, and this work should be expanded to capture the links between service subsectors and manufacturing industries.

Finally, the authors should be commended for initiating a study to estimate productivity improvements in the service sectors, especially for the distribution industry.

The Private Sector's View of Trade Liberalization in Services
A Hong Kong Perspective

Clement Yuk Pang Wong and Anming Zhang

7.1 Introduction

Asia has long been the home of many world-class manufacturers as well as being a major manufacturing centre of the world. Its service industries, however, are still largely shielded from world competition. This is changing as the so-called "Services 2000" round of multilateral trade negotiations under the General Agreement of Trade in Services (GATS) proceeds in the new millennium. Two major recent events have had pronounced effects on service liberalization in Asia: the 1997 Asian currency crisis and China's imminent accession to the World Trade Organization (WTO). The former has led to realization on the part of policy makers that service liberalization will increase the competitiveness of the service sector, which in turn will strengthen the overall economy. As a result, significant additional liberalization has been achieved since the crisis started (*Business Asia,* "Asia's Crisis Reform," 22 February 1999). Commitments to the WTO have in fact been used by some governments as an instrument to lend credibility to domestic liberalization reforms, because failure to honor these commitments

Clement Yuk Pang Wong is an assistant professor at the Department of Economics and Finance, City University of Hong Kong. Anming Zhang is currently an associate professor at the Faculty of Commerce and Business Administration, University of British Columbia commerce department and an associate professor at the Department of Economics and Finance, City University of Hong Kong.

The authors would like to thank W. K. Chan, Steve Ching, Gregory Chow, David Dodwell, Robert Feenstra, Andrew Kao, Anne O. Krueger, W. F. Leung, Chong-Hyun Nam, Pin Ng, Eden Y. Woon, Chung-Shu Wu, participants of the Eleventh NBER Annual East Asian Seminar, and two anonymous referees for their helpful comments. Financial support from the Services Support Fund of the Industry Department of the Hong Kong SAR Government and the Research Grant Council (RGC) of Hong Kong (Competitive Earmarked Grant No. 9040320) is gratefully acknowledged.

would require that compensation be made to countries adversely affected (Mattoo 1999; Low and Mattoo 1998). China's accession to the WTO would create significant opportunities for foreign participation in its service industries and provide the catalyst for domestic liberalization both in China and in other Asian countries. Furthermore, China's accession will pave the way for Taiwan's.

Against this backdrop, multilateral liberalization of services has had an immense impact on the Hong Kong service sector. Hong Kong is not only the most service-oriented economy in the world, but also a "service hub" in the Asia-Pacific region, directing the flow of goods, information, and capital.[1] In 1999, Hong Kong's trade in services reached US$57.53 billion, or 36.5 percent of gross domestic product. Not surprisingly, Hong Kong has evolved into a service hub because of its close proximity to China, excellent telecommunications infrastructure, laissez-faire policy, free flow of information, and the rule of law. As a service hub, Hong Kong plays a key role in providing so-called "trade services" (Deardorff 2001) in the region. Trade services are services, such as financial services, transportation, trading, telecommunications services, and professional services, that serve as inputs in the completion of international trade and investment.

Although the liberalization of services trade would provide substantial benefits for the global trading community, its effects on the Hong Kong economy in general and its service providers in particular are worth understanding. As China and other regional economies undertake service liberalization, Hong Kong's service sector faces both opportunities and challenges. On one hand, market liberalization in other countries, particularly in China, opens up new business opportunities for Hong Kong service providers to expand their operations beyond Hong Kong's boundary. The expanded market can, for example, improve their efficiency through the realization of economies of scale and scope. On the other hand, multilateral trade liberalization diminishes the attractiveness that Hong Kong used to enjoy. Some countries may now bypass Hong Kong, and this would endanger its traditional role as a middleman between east and west, particularly as the gateway to China (Hui et al. 2000; Feenstra and Hanson 2001). Perhaps Hong Kong needs to search for and bolster its identity as a service hub as in the new millennium. What role can the WTO trade forum play? We approach these questions by going directly to Hong Kong service providers themselves because they are the main group (besides consumers) that will be both affected by and affecting the process of service liberalization.

This paper reports the results of a project that studies the private sector's opinions and expectations of service liberalization. It has two objectives. First, it summarizes the project's findings and discusses the implications of

---

1. In 1999, the service sector accounted for 85 percent of Hong Kong's gross domestic product (93 percent if construction and utilities were also included) and 85 percent of employment.

these findings for the upcoming Services 2000 negotiations. Second, it examines the private sector's view of how to maintain and enhance Hong Kong's current status as a service hub in the region. The rest of the paper is organized as follows. Section 7.2 describes the project's background and the methodology used. Section 7.3 provides an overview of the results of a large-scale questionnaire survey and a series of personal interviews with business leaders from Hong Kong service industries. It also discusses the implications of the findings on Services 2000 negotiations. Section 7.4 turns to the issue of how service liberalization can affect Hong Kong's position as a service hub. In this section, we focus on three major service sectors that help shape Hong Kong's position as a service hub—financial services, telecommunications, and transport services. Some concluding remarks are offered in section 7.5.

7.2   Methodology of Study

The main research techniques used in this study are a questionnaire survey and personal interviews. This research approach is quite different from the standard research procedure in economics, which involves theorizing the phenomenon and then subjecting the empirical implications to an actual data test. Economists' traditional skepticism on the usefulness of the survey methodology can be traced back to the famous "billiard player" analogy (Friedman and Savage 1948). However, as argued by Blinder (1991), results from a properly designed survey can provide valuable data that are unavailable to an econometrician.

The service providers covered by this study are members of the Hong Kong Coalition of Service Industries (HKCSI), which is the biggest and most representative association of the service industry in Hong Kong. The primary purpose of the study is to compile a private sector's "wish list" of issues that they would like to be addressed in the WTO negotiations.

7.2.1   Questionnaire Design

The most crucial step in designing the questionnaire was the construction of a proposed wish list for the respondents to choose from. We interviewed fourteen senior executives from leading services companies in Hong Kong. The interviews sought to identify the problems encountered by Hong Kong service providers in market access, the countries in which they would like to see liberalization, and the issues that they would like to be addressed in the Services 2000 Round negotiations.[2] Although the interviewees were from different service sectors and some of their concerns were quite sector-specific, they all alluded to several common concerns. These are restrictions on market access through various forms of establishment requirements, discrimina-

---

2. A summary of the interviews is published in "Barriers Hindering Trade in Services: 'Services 2000,' Preliminary Interview Results" (1999).

tory treatment of foreign service providers, difficulty in accessing trade regulation information, and complex domestic and trade regulations. The feedback collected from the interviews was used, along with our own research, as inputs to construct a proposed list.[3] The list was refined several times after consultations with the members of the HKCSI executive committee, which consisted of representatives from various service industries. The final wish list consists of thirty-eight wishes. Some of the wishes are general concerns (such as applying the same licensing requirements to both local and foreign firms), whereas others are quite specific (such as relaxing restriction on foreign equity ownership). The first thirty-six individual wishes can be classified into one of the following eight categories:

1. Establishment requirements (wishes 1–6)
2. Qualification of professional standard (wishes 7–8)
3. Immigration and visa (wishes 9–11)
4. National treatment (wishes 12–18)
5. Labor market regulations (wishes 19–23)
6. Information flow and transparency (wishes 24–27)
7. Market structure to promote competition (wishes 28–32)
8. Trade facilitation and removal of restrictions in other industries (wishes 33–36)

Wishes 37 and 38 are not classified under any category. They are "relax travel restrictions, durations of stay, and foreign currencies carried abroad" and "relax restrictions on foreign company's profit repatriation." The main parts of the survey are questions 3, 4, and 8 in the questionnaire. Question 3 provides a sector classification scheme for the respondents to indicate their current businesses (Q3) and to define up to two *composite* service sectors called "sector 1" and "sector 2" (Q4) that they want to see liberalized. Our sectoral classification in Q3 follows the GATS classification scheme with only minor adjustments. There are twelve main sectors: (a) business services; (b) telecommunications; (c) construction and related engineering; (d) distribution services; (e) environmental services; (f) educational services; (g) financial services; (h) health-related and social services; (i) tourism and travel and leisure services; (j) recreational, cultural, and sporting services; (k) transport services; and (l) information services. Each sector is further subdivided into subsectors. There are altogether seventy-two subsectors.[4] In defining the two composite sectors in Q4, the respondents

3. We have consulted the literature from United Nations Conference for Trade and Development (UNCTAD), the World Bank, Pacific Economic Co-operation Council (PECC), and Pacific Basic Economic Council (PBEC). See UNCTAD and the World Bank (1994), PECC (1995), and PBEC (1998). We also consulted some surveys that the HKCSI conducted on related topics.
4. If the respondent just selects a main sector without indicating the subsector, we treat this as a "general" subsector. Hence, there are twelve "general" subsectors. This, together with sixty "listed" subsectors, constitute the seventy-two subsectors.

could combine several subsectors from the seventy-two provided. Because the wishes are likely to be (in fact they all are) country-specific, the respondents were also asked in Q8 to identify up to five economies, Hong Kong and four foreign economies, that should undertake market-opening measures to fulfill their selected wishes. Then the respondents could simply build up their own "wish list" by checking the boxes next to the thirty-eight wishes or specifying other wishes in the spaces provided in the questionnaire.

In counting the number of wishes, one must bear in mind that each "wish" can be identified by three fields: a wish code, a subsector (out of the seventy-two subsectors), and an economy the wish is made for. For instance, suppose a respondent selected wish code number 7 in Q8 for "sector 1," which he defines in Q4 as "advertising" plus "public relations," and he put down country codes "A," "B," and "HK" next to the check. Then, this respondent has made a total of six (two subsectors × three economies) wishes.

7.2.2 Background Statistics

In May 1999 we sent the questionnaire to 1,787 members of the HKCSI. A reminder letter followed this two weeks later. We also made random follow-up telephone calls to the respondents approximately one month after the questionnaire was sent out. The calls were made for two purposes. The first purpose was to encourage participation and answer any queries about the questionnaire. The second purpose was to adjust the "effective" size of the sample because the original mailing list might have overstated the effective sample for two reasons. First, it was at least two years old and many of the addresses might no longer have been valid. Second, some firms might not have been the intended targets because their businesses were unrelated to trade in services. Out of the 241 telephone calls made, we found in twenty cases (8.3 percent) the addresses were out of date and could not be updated. Another eighteen cases (7.5 percent) declined to participate in the study because their businesses were not related to trade, either directly or indirectly. Because the calls were made randomly, we project the same percentage of problem cases (15.8 percent) onto the whole sample. This reduced the effective sample size to 1,504 companies.

By August 1999, we had received 114 completed questionnaires; this implied an effective response rate of 7.6 percent. An obvious reason for the low response rate was that the questionnaire itself was lengthy. However, based on our experience from personal interviews, seminars, and follow-up telephone calls, there are also two other factors that we believe are more important explanations. First, there is a general feeling of apathy among business people, especially those from small and medium-sized enterprises, toward the WTO. This can be attributed to a lack of understanding of what the WTO is about or a lack of confidence in what it can do for them. The common perception is that the WTO is too remote to be relevant to everyday business. Second, because market liberalization brings about more intense competition, it is inevitable that some service providers take a cautious attitude and are not keen to see the Hong Kong government open its domestic markets. This tends to discourage companies from making wishes for Hong Kong or even participating in the survey altogether. We would like to note that besides making those 241 calls, we also visited the chairmen of some trade associations, such as the Hong Kong Information Technology Federation and the Institute of Architects, to solicit their participation. We know some of the returned questionnaires came from these associations. This tends to raise the "effective" response rate, because those replies reflect the views of key members of the associations (i.e., from their executive committees).

Table 7.1 provides a summary of the service sectors of our respondents' businesses. The majority of the respondents (38 percent) described the service they provided as "business services." This is followed by financial services (27 percent), transport services (18 percent), distribution services (14 percent), information technology (which consists of telecommunications and information services) (14 percent), and construction and related engineering services (10 percent). Most respondents in our sample appear to be indigenous to Hong Kong. Among the 114 respondents, 75 replied that their parent company's main place of operation was Hong Kong. For the other 39 respondents whose parent company's main place of operation was outside Hong Kong, most are from the United States (10) and United Kingdom (9). Regarding head office locations, 69 of the respondents replied that

Table 7.1   Main Services Provided by Respondents

Sector	No. of Companies	% of Companies
Business services	43	37.7
Financial services	31	27.2
Transport services	21	18.4
Distribution services	16	14.0
Information technology[a]	16	14.0
Construction and related engineering	12	10.5
Tourism, travel, and leisure services[b]	11	9.6
Environment services	8	7.0
Education services	8	7.0
Health-related and social services	2	1.8
Others	15	13.2
Total	114	100.0

[a]Eleven from information services and five from telecommunications.
[b]Nine from tourism and travel-related services and two from recreation, cultural, and sporting services.

their companies had a head office in the Asia-Pacific region, with a vast majority of them (59) located in Hong Kong.

Table 7.2 summarizes the distribution of firm size, based on the 1998 revenue. Most firms (twenty-six firms or 29.5 percent) recorded revenues of between HK$10 and HK$50 million. This, along with the nine cases in which revenues of HK$10 million or below were reported, indicates that most of our respondents (around 40 percent) are relatively small companies. Another 43 percent (thirty-eight firms) are medium-sized companies with reported revenues between HK$50 million and HK$1 billion. Finally, 17 percent of the respondents (fifteen firms) are large firms, with revenue exceeding HK$1 billion in 1998.

A legitimate concern is to what extent the "wishes" are realistic and do not represent wishful thinking. It is possible that a respondent may "wish" that it could access all markets by checking all thirty-eight wish codes in the questionnaire. We trust that the wish list compiled from our survey does represent a realistic picture. First, only two companies checked all thirty-eight wishes. Second, from our personal interactions with business people on various occasions (interviews, follow-up telephone conversations, and meetings), we observe that they are very realistic people. Most of them had strong opinions on certain issues that are related to their business experiences and future expansion plans. The wishes collected in our questionnaire should reflect (a) the business opportunities the targeted economies can offer and (b) the degree of difficulty of fulfilling those wishes. Companies are more likely to have wishes for promising markets and in measures that resolve the practical difficulties that they have encountered in the past. In this sense, the wishes collected from this survey are not "wishful" thinking and should be taken seriously because they are well-grounded in a firm's business plans and past experiences. An indication of this conjecture is that 60 percent of the firms made thirty wishes or less and 75 percent made fifty wishes or less.

Table 7.2   **Business Revenues of Respondents in 1998 by Service**

Business revenues (HK$ millions)	No. of Companies
Less than 10	9
10–50	26
51–100	3
101–200	12
210–500	9
501–1,000	14
1,001–10,000	10
Greater than 10,000	5
Total	88

Note: The total is less than 114 because not all respondents indicated their revenue.

7.3   Overall Results from the Questionnaire Survey and Interviews

The 114 companies that returned the questionnaire recorded a total of 5,555 wishes. Table 7.3 presents the breakdown of the wishes by main service sectors. Six main service sectors account for 88 percent of all wishes. These are, in descending order in terms of the number of wishes, the financial services, business services, distribution services, information technology (which is defined as information services plus telecommunication services), construction and related engineering services, and transport services. Basically, the pattern in this table is consistent with the sectoral distribution of the respondents' businesses depicted in Table 7.1.

Table 7.4 presents the breakdown of the wishes by economies of interest. Most wishes are directed toward Asian economies. The top eight econ-

Table 7.3   **Distribution of All Wishes by Sector**

Sector	No. of Wishes	%
Financial services	1141	20.5
Business services	966	17.4
Distribution services	773	13.9
Information Technology[a]	759	13.7
Construction and related engineering	746	13.4
Transport services	489	8.8
Recreational, cultural, and sporting	174	3.1
Health-related and social services	154	2.8
Environmental services	144	2.6
Education services	137	2.5
Tourism and travel-related services	59	1.1
Others	13	0.2
Total	5555	100.0

[a]This sector consists of information services and telecommunications. There are 465 wishes for information services and 294 wishes for telecommunications services.

Table 7.4   **Distribution of All Wishes by Country**

Country	No. of Wishes	%
China	2104	37.9
Hong Kong	636	11.5
Taiwan	530	9.5
Korea	364	6.6
Singapore	289	5.2
Malaysia	275	5.0
Japan	242	4.4
Thailand	168	3.0
Other countries	947	17.1
Total	5555	100.0

omies for which the respondents have wishes are all located in Asia, and together they account for 83 percent of all wishes. The top three economies are from the Greater China area (China, Hong Kong, and Taiwan) and they account for 59 percent of the wishes. China, in particular, accounts for 38 percent of the wishes. This underscores the role of Hong Kong as the gateway to China for many multinationals trying to gain a foothold there, especially in anticipation of its imminent entry to the WTO. Since Hong Kong is often heralded as one of the most open regimes in the world, it may perhaps be surprising to find that it is the second most mentioned economy in the private sector's wish list. The other one-quarter of the wishes are directed at Korea, Japan, and three Association of Southeast Asian Nations (ASEAN) countries: Singapore, Malaysia, and Thailand. The importance of Asian economies is consistent with our interview experiences and the fact that many multinationals use Hong Kong as a regional base to manage their businesses in the Asia-Pacific region. The notable absence of the U.S. and European Union (EU) economies might be due partly to the fact that many respondents originate from these two areas and therefore have no need to expand in their home markets.

The breakdown of the wishes by the eight wish categories mentioned earlier is provided in table 7.5, where the categories are ranked in descending order of an "adjusted response rate." The "adjusted" response rate controls for the fact that the categories have a different number of wishes.[5] The top ten most mentioned individual wishes are listed in table 7.6. Because the service providers we interviewed had very different wishes for Hong Kong as compared with other economies, we break down the results in tables 7.5 and 7.6 between those for Hong Kong and those for foreign economies. In order to have an idea of which issues should be dealt with in the Services 2000 Round, we also asked the respondents to name the most urgent issues from their wish lists, both for the Hong Kong economy and for foreign economies. The results are presented in table 7.7.

7.3.1   Wishes for Other Economies

The top three types of wishes that Hong Kong service providers demanded from other economies were (a) to relax establishment requirement restrictions, (b) to be accorded national treatment, and (c) to improve information transparency.

5. In comparing the relative importance of the eight wish categories, one must also take into account the number of wish items in each category. Everything else being the same, categories with more wish items would be selected more often than those with fewer items. In order to adjust for this bias, an *adjusted response rate* is computed in addition to the *raw response rate* that is simply the relative frequency. The adjusted response rate of an ith wish category is computed as follows. First, divide the number of votes for a category i by the number of wish items in the ith category. Second, repeat the first step for the other seven categories. Third, add up the answers from steps 1 and 2. Finally, "standardize" the answer in step 1 by dividing it by the answer from step 3.

Table 7.5   Distribution of All Wishes by Wish Category

Category	No. of Wishes	Adjusted Response Rate[a] (%)
A. For All Economies		
Establishment requirement	1092	16.1
Information flow and transparency	689	15.2
Immigration and visas	447	13.2
Market infrastructure	688	12.2
Qualification of professional standard	272	12.0
National treatment	937	11.8
Trade facilitation	456	10.1
Labor market regulations	535	9.5
Total	5116[b]	100.0
B. For Foreign Economies		
Establishment requirement	1032	17.4
Information flow and transparency	610	15.5
National treatment	886	12.8
Qualification of professional standard	241	12.2
Market infrastructure	599	12.1
Immigration and visas	318	10.7
Trade facilitation	391	9.9
Labor market regulations	459	9.3
Total	4536	100.0
C. For Hong Kong Only		
Immigration and visas	129	29.7
Information flow and transparency	79	13.6
Market infrastructure	89	12.3
Trade facilitation	65	11.2
Qualification of professional standard	31	10.7
Labor market regulations	76	10.5
Establishment requirement	60	6.9
National treatment	51	5.0
Total	580	100.0

[a]Adjusted response rate for category $i\,(AR_i) = R_i/(\Sigma R_k)$, where $k = 1, 2, 3, \ldots 8$ and $R_i = X_i/N_i$ where N_i is the number of wishes listed in category i. Note that the sum of all AR_i is equal to 100 percent by construction.

[b]The number of wishes is less than 5,555 because we exclude wishes that do not fall into any of the eight categories.

The use of establishment requirements as a way of restricting market access presents the biggest obstacle for Hong Kong service providers. Four of the wishes under this category are among the top ten wishes in this survey (table 7.7, section B). They are, starting with the most important,

1. Relax restrictions on the scope of business activities.
2. Relax restrictions on foreign equity ownership.
3. Relax restrictions on the number of operating licences.

Table 7.6   **Top Ten Individual Wishes For All Sectors**

	No. of Wishes	%
A. For All Economies		
1. Relax restrictions on the scope of business activities	253	4.55
2. Apply same licensing requirement to local and foreign firms	213	3.83
3. Relax limit of foreign equity ownership	213	3.83
4. Set up inquiry points to disseminate information	210	3.78
5. Relax restrictions on profit repatriation	209	3.76
6. Improve transparency of court ruling	198	3.56
7. Relax restrictions on number of operating licenses	184	3.31
8. Streamline the procedures of obtaining foreign visas	176	3.17
9. Relax restrictions on the composition of local partners	172	3.10
10. Simplify customs clearance procedures	170	3.06
Total	1998	36.0
B. For Foreign Economies		
1. Relax restrictions on the scope of business activities	239	4.86
2. Relax limit of foreign equity ownership	207	4.21
3. Apply same licensing requirement to local and foreign firms	200	4.07
4. Relax restrictions on profit repatriation	199	4.05
5. Set up inquiry points to disseminate information	189	3.84
6. Improve transparency of court ruling	182	3.70
7. Relax restrictions on number of operating licenses	172	3.50
8. Relax restrictions on the composition of local partners	163	3.31
9. Relax discriminatory tax on foreign firms	155	3.15
10. Simplify customs clearance procedures	151	3.07
Total	1857	38.0
C. For Hong Kong Only		
1. Streamline the procedures for obtaining foreign visas	51	8.02
2. Increase the quota of foreign work visas	43	6.76
3. Set up regulations related to e-commerce	36	5.66
4. Increase the duration of stay of foreign workers	35	5.50
5. Set up the electronic data interchange facilities	35	5.50
6. Increase flexibility of hiring and dismissing workers	25	3.93
7. Relax travel restrictions (e.g., durations of stay, FX)	22	3.46
8. Set up inquiry points to disseminate information	21	3.30
9. Enforce intellectual property rights law impartially	19	2.99
10. Simplify customs clearance procedures	19	2.99
Total	306	48.0

Note: Percentage is with respect to the total number of wishes: 5,555.

4. Relax restrictions on the composition of local partners, agents, or board of directors.

In particular, relaxing restrictions on the scope of business activities stands out as the most popular wish in the entire survey. This underscores the structural changes taking place in the service industry. Technological in-

Table 7.7   Most Urgent Wishes

	No. of Wishes
A. For Foreign Economies	
1. Relax restrictions on the scope of business activities	35
2. Apply the same licensing requirements for both local and foreign firms	24
3. Relax restrictions on foreign equity ownership	22
4. Simplify customs clearance procedures	14
5. Relax restrictions on profit repatriation	12
6. Others	150
Total	257
B. For Hong Kong	
1. Increase/eliminate the quota of foreign work visas	14
2. Streamline the procedures of obtaining visas	13
3. Set up e-commerce regulations	11
4. Set up electronic data interchange	8
5. Simplify customs clearance procedures	7
6. Others	74
Total	127

Note: Sixty-four companies responded to this question (Q11).

novations and market liberalization of service sectors have gradually weakened segmentation across service product lines. For example, banks and insurance companies are cross-selling each other products as "bancaasurance"; voice telephone service and facsimiles can now be provided as a package over the Internet; and mobile and fixed-line services are converging. As the product lines blur, the ability of service providers to offer a one-stop, total solution package of services to their customers is absolutely essential. This allows service providers not only to build up customer loyalty but also to benefit from economies of scale and scope. In this respect, restrictions on the scope of businesses would have a negative spillover effect on service sectors that are already open to foreign service providers.

The respondents' second most important concern is to improve information flows and transparency. Two of the wishes in this category are among the top ten individual wishes of the survey. They are (a) to set up inquiry points for dissemination of trade related laws and regulations, and (b) to improve transparency of court rulings on business disputes.

Article III of GATS on transparency stipulates that "Each Member shall publish promptly and, except in emergency situations, at the latest by the time of their entry into force, all relevant measures of general applications which pertain to or affect the operation of this Agreement [i.e., GATS]." Apparently, this overriding principle has not been compiled with by WTO member economies covered in this survey. (China, Taiwan, and Vietnam are yet to join the WTO.) Out of the 210 wishes on the establishment of inquiry points to disseminate trade-related laws and regulations, more than half are directed to WTO member countries. Some interviewees stressed that the problem they faced was not so much whether the regulatory information was available but how government officials implemented it. It was also noted that some countries such as Japan and Korea have not translated all of their regulations in English, which put foreign firms at a disadvantage. The wishes to improve the transparency of court rulings on business disputes underscore another hazard of running a business in foreign economies. China alone accounts for 50 percent of these wishes. The result is consistent with the incidences of arbitrary imprisonment of Hong Kong businessmen in the mainland.

Another major concern is national treatment. This GATS principle is apparently not being followed by the main economies covered in this study. The two main complaints are (a) application of different licensing requirements for local and foreign service providers and (b) discriminatory taxes imposed on foreign firms. Demands to remove them are among the top ten wishes from the survey. In a broader sense, some wishes in other categories may constitute violation of national treatment.[6] For example, "Qualification of professional standard" is the fourth most important wish category in section B of table 7.5. Sectoral distribution shows that the use of unreasonable qualification standards unrelated to the quality of services provided (such as race and residency requirements) are commonly used against foreign professionals, especially in the business service sector. The establishment restrictions on the scope of business activities mentioned above are also a violation of national treatment as they are usually imposed on foreign service providers. For example, many interviewees from the financial service sector complained that foreign brokers are restricted to trading foreign-trenched stocks in many Asian economies. In the case of China, restrictions on renminbi business are a common complaint.

The interviewees also pointed out some measures that put foreign service providers at a disadvantaged position vis-à-vis domestic providers, hence violating the spirit behind national treatment. For example, Singapore requires foreign fund management companies to have a "full-fledged" position before they can bid for business. A full-fledged operation means that the company must establish an office with portfolio managers. This type of restriction discourages foreign fund management companies from operating there because the domestic market is too small to warrant setting up a full-fledged operation. Some interviewees in the courier service business said that many developing countries have laws that limit courier service to commercial documents or the weight of the item delivered. The primary purpose of this type of measure is to protect the local postal service provider.

6. Technically, national treatment is used in the context *after* a foreign service provider has gained market access to foreign markets. However, some measures, such as those on establishment requirements, also constitute unequal treatment although they operate *before* market access is achieved.

The other two wishes in the top ten list are (a) to simplify customs clearance procedure (tenth) and (b) to relax restrictions on profit repatriation (fourth). A cumbersome customs clearance procedure would raise transaction costs and discourage international trade of goods, which support service sectors such as transport and distribution. According to a study by Pacific Basin Economic Council (PBEC 1998), "excessive documentation" has been identified as one of the most reported administrative barriers to trade.

Section A of table 7.7 presents the distribution of the most urgent wishes for foreign economies. The five most urgent wishes are all among the top ten most mentioned wishes in section B of table 7.6 as well. In particular, the issue of capital controls was brought up many times in our interviews. In fact, "relax restrictions on profit repatriation" is the fourth most mentioned wish for foreign economies in this study.

7.3.2   Wishes for Hong Kong

As mentioned earlier, Hong Kong is the second most mentioned economy in the private sector's wish list. Most wishes, including the four most urgent wishes, are targeted at relaxing immigration policy and promoting Hong Kong's high-technology sector. Whether we use the raw or the adjusted response rates, the three main types of wish that Hong Kong service providers demanded from their own government were (a) to relax immigration and visa policy, (b) to improve market infrastructure to promote competition, and (c) to improve information transparency.

Although "immigration and visas" is a relatively small category with only three items, it accounts for one-fifth of all wishes for Hong Kong. Hong Kong also accounts for the lion's share of wishes in this category. Out of the 447 wishes from all our respondents, 29 percent are directed at Hong Kong.[7] All three wishes in this category are among the ten most mentioned for Hong Kong (first, second, and fourth in section C of table 7.6). This result accords with the common complaint we heard from our interviewees. Service providers, especially those from the high-technology sector, are critical of the Hong Kong government's immigration policy. Commercial presence and movement of personnel are the two main modes of supplying services to overseas customers. Both modes of supply, however, assume unrestricted flows of personnel across national borders. The industry practitioners that we interviewed stressed that the most crucial element of a business-friendly immigration policy is the flexibility for people to move in and out of national boundaries frequently with minimum administrative hassle. Employees of many multinational corporations and people working in the trade sector often need to travel to foreign countries frequently for

---

7. From table 7.5, there was a total 447 wishes in the category of "immigration and visas," 318 for other economies and 129 for Hong Kong.

brief stays to attend a meeting or a training program or to meet with business partners. Therefore, being able to obtain a foreign visa quickly and inexpensively is very important. Issuing more foreign work visas with longer durations of stay, such as the Hong Kong special administrative region (SAR) policy of allowing in information technology professionals from China, would not be able to address this kind of "flow" need.

The wishes in the other two wish categories also strongly reflect concerns for the high-technology sector. In the "information flow and transparency" category, one of the most important concerns is "to set up electronic data interchange [EDI] facilities." Some interviewees suggested that Hong Kong needed to bolster its middleman role in the e-commerce age by offering other forms of e-middleman function. They argue that given Hong Kong's excellent telecommunications infrastructure, it can develop into a regional electronic catalog center and authentication center. Electronic catalog is a facility that translates buyers' and sellers' product information into a standardized electronic format understandable by both sides. For example, in the apparel industry, electronic catalogs can help a U.S. purchasing manager pass his product requirements to a supplier in China.

The respondents also wished to see travel restrictions relaxed (seventh in section C of table 7.6). This request is obviously motivated by the restrictions on mainland tourists' duration of stay and the complicated visa application procedures for Taiwanese tourists visiting Hong Kong and mainland China. Due to its excellent location, Hong Kong has the conditions to position itself as a hub destination. Hong Kong can be packaged as part of a China or South Asia itinerary. Another obstacle against Hong Kong's developing as a travel hub is its human resources. According to a recent government estimate ("Editorial: Immigration Policy and Economic Considerations," *Mingpao* 15 August 2000), the tourist and high-technology sectors are identified as the two sectors that are most affected by labor shortage.

Finally, as is the case for other countries, the respondents also recommended that Hong Kong should simplify its customs clearance procedures in order to facilitate trade. This wish is obviously important for Hong Kong to maintain and enhance its status as the major trading port in Asia. We elaborate on this point in section 7.4.

In terms of urgency, the two most immediate issues for Hong Kong (table 7.7, section B) are relaxing the quota of foreign work visas and streamlining the procedures for obtaining visas. The third and fourth most urgent wishes are to set up e-commerce regulations and electronic data interchange.

7.3.3   Implications for GATS Negotiations

Our results above may have some useful implications for GATS negotiations. First, there may be a need to deepen the meaning of transparency in Article III of GATS. This problem is borne out in our survey. "Set up inquiry points to disseminate trade information" and "improve the transparency of court rulings on business disputes" are the fourth and sixth most mentioned wishes in our survey. The feedback from practitioners raised our attention to two dimensions of transparency, namely, the interpretation and implementation of trade regulations. Foreign firms are often put in a disadvantaged position vis-à-vis local firms because the former are less familiar with how and in what context local laws and regulations are implemented or interpreted.

Another dimension of transparency is accountability—an element lacking in Article III of GATS. Because GATS members have no obligation to set up a special information agency to assist foreign companies to obtain trade-related information, it is more difficult to ensure their compliance with GATS Article III.

The importance our respondents assigned to a fair, open, and efficient court ruling procedure would help reduce their political risks. Because many countries are still in the process of adapting their laws and regulations to the requirements of WTO rules, trade disputes are bound to occur. From the service providers' point of view, transparency in court rulings can provide more timely and relevant means to resolve business disputes than the WTO dispute settlement mechanism, which only operates at the country level. Perhaps specifying some minimum level of due process and diligence in trade-related court rulings in GATS Article III would help. For example, court rulings on business disputes involving foreign companies could be published and the rationale of the rulings explained.

Violation of the WTO's national treatment principle is a major complaint in this survey. For example, nondiscriminatory licensing requirements are the second most mentioned wish. At present, GATS Article XVII governing national treatment does not oblige countries to state the public policy objectives behind exemptions from this principle and does not, as in the case of the most-favored nation (MFN) principle, impose a phase-out period on such exemptions.

7.4 Sector-Specific Wishes

In this section we examine the three sectors that make Hong Kong a service hub: financial services, information technology services, and transport services. These three sectors attracted 43 percent of the wishes from the private sector and nearly half of the wishes directed to the Hong Kong government. The hub status Hong Kong currently enjoys is, to a large extent, due to the fact that Hong Kong has adopted a free and open investment and trade policy whereas other Asian economies adopt restrictive policies (see appendix 7.1 for a theoretical analysis). However, as other countries in the region (especially Singapore) gradually open up their service sectors under multilateral liberalization, Hong Kong's status as a hub will be challenged. As our theoretical model in appendix 7.1 shows, competition among service hubs lies in the development of hub infrastructure that facilitates the movement of goods, people, capital, and information through a hub. This includes excellent facilities in container and airport terminals, an efficient and smooth customs clearing process, a flexible and open immigration policy, and a transparent flow of information. The private sector's wishes for these three pillars of hub services seem to strike the same chord as our theoretical model. In our discussion below, we try to minimize repetition by focusing on features unique to each sector. Results that are similar to the overall pattern will only be mentioned briefly.

7.4.1 Financial Services

In the 1990s, the world's financial markets underwent several sweeping changes. The first structural change was large-scale privatization as governments tried to reduce their influence in economic activities. This weakens the segmentation of their financial services industries. Banks, insurance companies, securities brokers, and a new breed of nonfinancial institutions are invading each other's turf. The second phenomenon is an ongoing worldwide consolidation and restructuring of the financial services industries in the industrialized countries such as Japan and the EU as well as in developing economies. The latter also need badly foreign capital inflows to recapitalize their financial sectors. The third major force is Internet technology, which has had a major impact on financial services industries. Trading of financial assets can now be conducted around the clock from anywhere, breaking time and geographical constraints. On-line stockbrokers such as E*Trade and the electronic communication networks are challenging the once-protected markets. These forces have created a stock-market culture among investors worldwide. In Asia, this culture is eroding the once-dominant position of the banking sector. Traditionally, Asian companies relied mainly on bank loans for financing. However, the currency crisis has changed this partnership. Asian companies have discovered that they cannot rely too much on bank loans for financing. To reduce their funding risks, they are gradually turning to the equity and bond markets. Across the region, corporate restructuring and bank recapitalization have accelerated the push to equity markets.

The financial service sector is the most important sector in our survey. There are 1,141 wishes for liberalization in the financial sector, representing 21 percent of the wishes from the private sector. Our questionnaire adopts the GATS classification under which the financial services sector is divided into three main subsectors: insurance, banking, and other financial services. Banking includes all traditional services such as deposits, lending, payment, and money transmission services. Other financial services are mostly fee-based services. These include trading of financial assets (e.g., foreign exchange, equities, and derivatives), underwriting, money broking, financial advising, settlement and clearing, provision of financial information, and fund management.

When the distribution of financial-sector wishes by wish category is decomposed into Hong Kong and all other economies subsamples, we observe a couple of results that are the same as the overall results discussed in section 7.3. First, the top concern for Hong Kong remains "immigrations and visas" and that of foreign economies is "establishment requirements." Four of the top five wishes for foreign economies fall into the establishment requirement category—scope of business activities, number of branches or offices, foreign equity ownership, and licensing requirements. Despite the significant progress made in the Financial Services Agreement, the financial service sector remains rather restrictive in Asia-Pacific economies. The results also shed light on the multitude of obstacles standing in the way of foreign financial institutions' trying to gain a foothold in other Asian countries.

Second, as in the overall result case, both subsample groups continue to place great emphasis on "information flows and transparency" in their wishes. "Set up EDI facilities" is among the top ten wishes for both groups. Without a doubt, financial services are the main engine of growth in e-commerce because they are the most easily adaptable to Internet selling. On-line trading has caught on in Asia, particularly in Korea, Japan, and Australia. It is estimated that almost half of all stock market turnovers in Korea are transacted on-line. Hong Kong, Taiwan, and Singapore are the next biggest markets. One of the main obstacles to e-commerce is security concern. Facilities in public key infrastructure can foster an environment conducive to e-commerce. The introduction of e-certification by the Hong Kong Post in 1999 is a good example of this. The other wish in the category demands that inquiry points be set up for dissemination of trade-related laws and information. There is clearly a need to improve transparency in the financial services sector in the Asia-Pacific region. For example, the restructuring process of Daewoo of South Korea is neither open nor transparent. It neither includes the debts of Daewoo's overseas subsidiaries nor properly addresses transactions between Daewoo's affiliates. The plans were drafted without input from foreign creditors, who complained that they were not given enough access to Daewoo's financial information. Another example of lack of transparency is the Guangdong International Trust and Investment Corporation (GITIC) incident in China, in which lack of communication has led to misunderstandings over the need to register the loans of its overseas subsidiaries with the State Administration of Foreign Exchange. Foreign acquisitions of local financial institutions have slowed down in some countries such as Korea because of the substandard level of disclosure in these economies.

The only result that differs from the overall pattern in section 7.3 is "qualification of professional standards," which is relatively more important in financial services than in overall results. A common complaint from this category is the use of unreasonable qualification requirements as a barrier to entry. These can take the form of requirements on language, race, and length of residency. These measures violate Article VI (4) of GATS (Domestic Regulation) that states that qualification requirements cannot constitute unnecessary barriers to trade in services.

The six main subsectors mentioned in the wish list are "trading of financial instruments," "clearing, settlement, and custodial services," "insurance services," "fund management services," "financial advisory services," and "lending services." With the exception of lending and insurance, the other four subsectors are all from "other financial services." This may suggest that Hong Kong financial services providers have a comparative advantage in these value-added services.

The most important financial service our respondents would like to see liberalized is the trading of financial assets (e.g., securities, derivatives, foreign exchange, bullion). This subsector attracts far more wishes than the rest, accounting for more than one-third of the wishes for financial services. Given the trend of expansion of stock trading in the global financial markets as discussed above, restrictions in this subsector would have a dire consequence for the business opportunities of trade in financial services. In the case of China, this wish was most likely made in response to the restrictions that foreign investors could only trade B-shares. Many countries in Asia also impose similar forms of trading restrictions, such as the maximum foreign ownership of local shares and the prohibition on offering derivative or innovative financial products. Another 43 percent of the wishes requested that national treatment be granted to foreign firms trading financial assets, an improvement in financial market infrastructure to facilitate trading businesses, and a more transparent information disclosure standard. For example, the lack of reliable clearing and settlement services is often cited as one of the main reasons that discourage trading in emerging markets. The lack of information transparency such as lax accounting standards and rampant cross-ownership-motivated trading has also introduced excessive risk to emerging markets.

Given our discussion above, it is not surprising to find that the second most important financial service our respondents sought to liberalize is clearing, settlement, and custodial services (16 percent of the wishes). Other financial sectors that foreign financial institutions find it difficult to enter include insurance, lending, fund management, and financial advisory services. Note that these are all high profit margin businesses. Together, these four subsectors account for 37 percent of financial-sector wishes.

Our respondents also demand to be given national treatment in licensing requirements (fourth and sixth on the top-ten list of Hong Kong and foreign economies respectively). One solution is to require the licensing authority to publish the licensing requirements on the Internet in order to improve the transparency of the selection process. Another major concern in the foreign economies subsample is capital controls. The private sector would like governments to reduce their limitations on the repatriation of profits. Article XI (Payments and Transfers) of GATS stipulates that, except under circumstances envisaged in Article XII (Restrictions to Safeguard the Balance of Payments), a member country shall not apply restrictions on international transfers and payments for current transactions related to its GATS commitments.

Implications for Hong Kong as an International Financial Center

Judging from many criteria, such as stock market capitalization as a percentage of gross domestic product (GDP) and the size of footloose international financial businesses such as foreign exchange and international banking it can attract, Hong Kong is one of the premier financial centers in the Asian time zone. The fact that the financial services sector attracts most of the wishes from the private-sector wish list reflects Hong Kong's competitive strength in this sector. Nevertheless, other Asian economies are embarking on a service liberalization program of their own. Furthermore, as mentioned above, technological innovation is shaking up the financial service industry. Internet technology has made physical distance less important and cross-border supply of services more feasible. As a consequence, financial centers such as Hong Kong, where the costs of doing business are high, will face the threat of being bypassed. As other economies open their financial sectors, some non-personal-intensive, administrative types of services, such as clearing and settlement, custodial services, back-office operations, and, to some extent, even asset trading as technology develops further, could be diverted to less costly countries.

Hong Kong can keep these businesses by, as suggested by our respondents and our theoretical model, improving services trade facilitation and further liberalizing its financial market to bring down the cost of trading. For example, in asset trading, abolishing the minimum brokerage commission and opening exchange floor seats are urgent policies in need of adoption. Hong Kong also badly needs to develop EDI facilities to enable online payment. The lack of a secured and cheap online payment technology is holding back the development of e-commerce. Some financial institutions have started to use the Internet as a tool to provide value-added services. For example, some banks offer electronic bill presentment services to enhance existing cash management services. Another natural extension of their cash management services is to facilitate payment and information flows of business-to-business (B2B) e-commerce.[8]

At the same time, as developing countries start to liberalize their financial markets, services that require close contact with clients such as fund management, insurance and lending may also move closer to the clients'

8. See Citibank's website: [*http://www.citibank.com/singapore/get/english*].

countries. Hence, commercial presence may be a preferable mode of supply to cross-border supply because it can provide more timely information. The relative importance of insurance, lending, and fund management in our wish list for financial services underscores the advantage of gaining better information by getting closer to the source.

Although we cannot deny that there is a substitution effect from trade liberalization, we believe that Hong Kong's status as an international financial center can be maintained. In fact, it is well placed to become the most important financial center in the Asian time zone, given the trends of consolidation we mentioned above. First, a relaxation of foreign equity ownership could bring opportunities to cross-border supply of financial services. In its recent WTO accession agreements with the United States and the European Union (EU), China has made substantial concessions on foreign investment in many sectors. Many mainland companies will seek to use Hong Kong as a base to tap foreign capital. This will bring export businesses by increasing the number of mainland firms listed in the local bourse. The demand for advice on mergers and acquisitions, syndicated bank loans, and related business services will also increase.

Second, as an international financial center, Hong Kong is endowed with a critical mass of talents from a wide range of financial and business services, owing partly to its earlier start as a financial hub. Such a clustering of people could create a pressure for innovation that is important for high valued-added services such as investment banking. This clustering also makes it possible for financial service providers to have access to talents from other related financial services themselves. This allows them to enjoy economies of scope from the input point of view. For example, some financial services such as merges and acquisitions and initial public offerings require talents from many financial services and related business services (e.g., accountants, bankers, and lawyers) to work closely with each other.

7.4.2 Telecommunications and Information Services

In this paper, we use a broader definition of the "telecommunications" sector, which we call the "information technology" (IT) sector. Information technology consists of two related sectors: telecommunications and information services. The telecommunications sector includes basic telecom services (e.g., fax, fixed network, international telecom, mobile, and private leased circuit services) and some simple value-added services such as call waiting and e-mail. The information service sector consists of higher value-added products (e.g., EDI, Internet access, on-line information services), hardware installation and maintenance, and software development. The IT sector is the third most important sector in our survey, after financial and business services. A total of 759 wishes are recorded for this sector. Among them, about 40 percent are for telecommunications, and the remaining 60 percent aim at information services.

Hong Kong's role as a leading business center in the Asia-Pacific region owes much to its advanced telecommunications infrastructure. The role of telecommunications will be especially important to Hong Kong in the age of electronic commerce. Trade liberalization in telecommunications services would stimulate the growth of e-commerce and offer tremendous business opportunities for Hong Kong Internet service providers and software companies. Within the telecommunications sector, China and Taiwan account for 71 percent of the wishes. Such a high concentration of wishes arises from Hong Kong's role as the hub of trade and investment in the Greater China region. At the end of 1999, 50 percent of both Hong Kong's outgoing and incoming telecommunications traffic originated from these two economies. Both China and Taiwan are currently under pressure to open their telecommunications markets in their bids to enter the WTO. As the Greater China hub, Hong Kong stands to benefit, because competition would lower costs and, hence, increase telecommunications traffic in the region.

Hong Kong started to open its fixed telecom-network services (FTNS) market in 1995 by allowing three companies to compete with Cable and Wireless HKT (CWHKT). The number of such "cable-based" FTNS licences is frozen at four until 2003. In the mobile-phone sector, six companies are now competing to serve the city's 6.8 million people. Competition intensified further when number portability was introduced in 1999. In the international services market, CWHKT surrendered its exclusive licence at the beginning of 1999. Since then, seventy companies have entered this market using what is called international simple resale, whereby they lease lines from the CWHKT.

On 26 January 2000, seventeen new telecommunication licences were granted. Five of these are "wireless" FTNS licences that allow licensees to launch broadband Internet access and telephone services, and value-added services such as video conferencing and multimedia services. The other twelve licences are for external satellite-based services. These services allow licensees to bypass the undersea cable gateway of CWHKT. Two weeks later, the Hong Kong Government issued letters of intent to thirteen companies for cable-based external fixed telecom-network services licences. Companies can now build their own submarine cables, hence further eroding CWHKT's position.

Most of the wishes for Hong Kong (65 percent) are concentrated in the international telecom services (35 percent) and fixed telecom-network services (30 percent). For foreign economies, fixed telecom-network service dominates all others subsectors by attracting 40 percent of the wishes. The other 38 percent are roughly split between private-leased circuit services and international telecom services. Mobile services, on the other hand, attract very little attention.

The subsector distribution for information services highlights the importance of software development and implementation and the Internet services markets. Together, they account for 60 percent (Hong Kong subsample) to 67 percent (foreign economies subsample) of the wishes for information services. Deregulation, corporate restructuring, and investment in technology have stimulated a strong demand for Internet services and software development. For example, deregulation of the financial services industry has created a huge market for equity-trading and trading-clearance systems. Another reason is the realization by corporate Asia that a huge profit potential arises from the so-called B2B e-commerce. As this form of business soars, many companies, especially small and medium-sized enterprises that cannot afford major capital investment, are outsourcing their Internet operations. A new breed of Web hosting service providers is springing up, providing business with the essential e-commerce tools from data storage, data management, and business applications software to supply chain management. This creates enormous business opportunities for software companies.

Wishes for Hong Kong: IT Sector

For the Hong Kong market, most of the wishes for telecommunications services are from the "immigration and visas" and "market infrastructure" categories. The concern for a more relaxed immigration regime reflects the extremely tight labor market for the high-tech industry reported by the Hong Kong Government ("Editorial: Immigration Policy and Economic Considerations," *Mingpao* 15 August 2000) and our interviewees. If this is not solved quickly, Hong Kong's lead as a telecommunications hub and its role as a middleman in e-commerce business will be at risk. No wonder our respondents rated "increase/eliminate the quota of foreign work visas" and "streamline the procedures for obtaining visas" as the two most urgent wishes in this survey. Some respondents are also concerned with the enforcement of intellectual property rights law and competition law. Hong Kong's image as an international city has been tarnished by rampant piracy of computer software and brand-name apparel products. Piracy does present a serious threat to Hong Kong's quest to become a high-tech hub. This is an implementation problem that some of our interviewees have raised.

As for competition law, Hong Kong does not have one. Instead, the government has adopted a moral persuasion approach. A policy statement on the objectives of promoting competition and discouraging various forms of anticompetitive practices was adopted in 1997.[9] However, the Hong Kong telecommunications liberalization experience suggests that the current policy leaves much to be desired. First, all six mobile operators raised their rates on the same day (1 January 2000) and by the same extent. The Telecommunications Authority (TA) ruled that there was enough evidence to suggest that the price hike was the result of a tacit agreement. The mo-

9. See Cheng and Wu (1998) for the status of competition policy in Hong Kong.

bile operators backed down and restored their original rates. This event has shown that self-regulation may not be enough to ensure that the six regulatory principles in the reference paper will be implemented effectively. Another issue is cross-ownership, which is common in many Asian economies, as illustrated by the *chaebol* in South Korea and the conglomerates in Hong Kong and Taiwan. Many telecom operators in Hong Kong have close ties with property developers, and some are even wholly owned by them. Therefore, there is a potential for collusion, and consumers might be forced to use the service provided by the fixed-network operator with close ties to developer of their housing units. It is important for the telecommunications authority to ensure that access to facilities must be granted in a nondiscriminatory manner.

For the information service sector, the most-demanded requests for Hong Kong are, as expected, to set up e-commerce regulation and EDI facilities and to streamline application procedure for foreign work visas.

Wishes for Foreign Economies: IT Sector

The biggest obstacles facing Hong Kong telecommunications companies seeking a foothold in foreign telecom markets are the restrictions on establishing a commercial presence. Five of the top six wishes in the telecommunications sector in foreign economies demanded fewer restrictions on establishment requirements. This underscores the fact that many Asian economies still cling to the idea that telecommunications is a strategic sector and various restrictions are imposed on foreign participation. It is interesting to observe that some economies are using the WTO forum as a channel to lend credibility to domestic reform, particularly for China and Taiwan—two economies that are not WTO members yet. For instance, China has agreed to let foreign investors hold 49 percent of telecommunications companies upon China's accession to the WTO; this limit will be raised to 50 percent two years later. At least this is on a par with the level of openness of other developing economies. Taiwan has also recently raised the ceiling of foreign ownership of fixed telecom companies substantially, from 20 percent to 60 percent, and is planning to increase the number of fixed telecom network operating licences.

Surprisingly, "immigration and visas" is also the second most important wish concern for foreign economies. "Streamline the procedure for obtaining foreign visas" is the fourth most mentioned wish. This indicates that restrictions on immigration and visa are fairly common problems in the telecommunications sector.

The main driving force behind government efforts to liberalize telecommunications is the desire to foster its development by attracting the transfer of foreign technological know-how and capital. To this end, relaxation of establishment requirements must be accompanied by a more liberal immigration regime. Although it is certainly desirable to increase foreign work visa quotas and foreign employees' duration of stay, the IT industry practitioners that we interviewed stressed that the most crucial element of a businessfriendly immigration policy is to offer the "flexibility" for people to move in and out of national boundaries with minimum administrative hassle.

As for the information service sector, the respondents are also keen to see, as they do for Hong Kong, foreign economies establish adequate e-commerce regulation. Although the lack of regulation is hailed as the main reason behind the phenomenal growth of e-commerce, the result here does not square with the common perception that the business sector does not want any regulation. Instead, the respondents in our survey demanded that such regulations be set up. Lack of a regulatory framework has been a drag on the development of e-commerce in Asia. Only Singapore and Hong Kong have introduced an electronic transaction bill to give electronic records and digital signatures the same legal status as that of their paper counterparts.

Besides the common concern for e-commerce regulation, our respondents do have many concerns when thinking about expanding into foreign economies. These are familiar concerns mentioned in section 7.3, such as improving the transparency of court rulings (most frequently mentioned) and setting up inquiry points to disseminate information.

Businesses would also like foreign governments to give them national treatment in the licensing applications and to relax local content rules on services. These wishes are borne out by experiences in some countries. For example, China and Singapore have set up government-backed Internet companies to compete with private firms. This poses the risk to foreign firms that they might not receive treatment equal to that of local government-backed companies—a violation of the national treatment principle of GATS. Another form of restriction on marketing channels is the high interconnect fee that Internet firms have to pay telecommunications operators for the "last mile" of connection.

7.4.3   Transport Services

The transport service sector has been traditionally regulated at both national and international levels. There have been concerted efforts, and considerable progress, in deregulating and liberalizing transportation services over the last two decades. For example, countries such as the United States, Canada, the EU states, and Australia deregulated their internal markets (e.g., trucking, rail, and air). At the international level, the Uruguay Round succeeded in applying multilateral trade concepts to three ancillary areas of the air transport sector (aircraft repair and maintenance, the selling and marketing of air transport services, and computer reservation services). However, efforts to bring air carriers' traffic rights under the multilateral purview of GATS have failed. Plans are currently being laid to discuss the inclusion of aviation services within the Services 2000 negotiations.

Although the extended negotiations on maritime services lasted two years beyond the end of the Uruguay Round, they were finally laid to rest in June 1996 when WTO member delegations agreed to suspend negotiations. They deferred them again, this time until Services 2000.

A total of 489 wishes are directed to transport services, representing 9 percent of the entire wishes. The category of "services auxiliary to all modes of transport" receives 170 wishes, air transport 118, and maritime transport 102, whereas the other categories receive far fewer wishes. This is not surprising, given Hong Kong's status as a hub for both aviation and maritime transportation, and given the importance of services liberalization to both these modes of transport and to services auxiliary to them.

The most popular wishes for the transportation sector relate to immigration and visa policy. This is understandable given the nature of transport services, which require movement of personnel across national borders.

The second most popular wishes are concerned with facilitating trade: to simplify customs clearance procedures and to eliminate unnecessary practices for certification and testing of products. These are, in particular, the dominant wishes for the maritime, air cargo, and trucking industries. Currently, transshipments through Hong Kong are treated as imports and re-exports. As a consequence, carriers are required to lodge accurate and complete import and export declarations for transshipped cargo, thereby introducing unnecessary delay and costs in the process. On the other hand, in Singapore transshipped goods avoid customs by being held in a bonded zone during transfer. Companies within the zone can move, break down, store, consolidate, and repack cargo without documentation or incurring customs duties. In terms of our theoretical model in appendix 7.1, N_{AHB} equals to 4 for Hong Kong and 2 for Singapore. We believe that simplifying customs procedure for transshipment services will become increasingly important because of the global trend toward international fragmentation of production processes. More and more raw material and intermediate goods will be in transit from one location to the other to be further processed before reaching their final destinations. Jones and Kierzkowski (1999) argue that as trade liberalization reduces the cost of coordinating activities across national borders (such as transport and communications), it becomes easier for producers to outsource part of their production processes to foreign locales to take advantage of comparative advantages. More production fragmentation, in turn, stimulates more demand for these trade services. They also argue that these coordination costs tend to be relatively fixed in nature. Therefore, the scale of outsourcing tends to be large. Given this kind of dynamic relationship, Hong Kong stands to lose more transport service business to other competing hubs if it delays simplifying its customs procedures.

Other popular wishes relate to establishment requirements and market access. In many economies, current rules restrict both the types of goods that can be shipped and the points from and to which carriers can transport them. The common barriers specific to the courier services industry, mainly found in developing countries, are laws and regulations limiting courier services to commercial documents or cargo below a certain value. The primary purpose of these restrictions is to protect local postal services. There are also cases in which domestic regulatory policies or restrictions are applied to other sectors (e.g., ground transport, telecommunications) that are ancillary to air transport but nonetheless critical to a foreign carrier's ability to provide competitive services in the host economy. In many economies considered in this study, for example, a host of local regulations (such as licensing and ownership restrictions) limits foreign participation in the trucking sector, with the sector sometimes being reserved for nationals. This means express carriers must obtain their road transport through local contracting. The fourth most popular wishes relate to information flow and transparency. In particular, our respondents ask for the establishment of EDI facilities.

7.5 Concluding Remarks

With a "first mover" advantage in adopting an open and free trade policy regime, Hong Kong has evolved to become an important trade services hub in the Asia-Pacific region. Our discussions in this paper show that further liberalization of trade in services by other economies has presented challenges as well as opportunities for Hong Kong in general and for its service providers in particular. There is a need to rethink the role that Hong Kong plays and will continue to play in view of trade liberalization. The fragmentation of supply chains will accelerate as technology improves and barriers to the flow of goods, information, capital, and even human resources are removed. Ironically, fragmentation of the supply chain can come with increased demand for a central location to organize and coordinate activities. These are the "hub" services that Hong Kong should provide. Our study shows that policies or services that can enhance this role, such as a more flexible immigration policy and a faster customs clearance system, are crucial for Hong Kong to continue its role as a service hub in the Asia-Pacific region.

A *flexible* immigration policy is also necessary for Hong Kong to continue to attract multinational corporations to use it as their head office in Asia. This is important for a knowledge-based economy and particularly so for Hong Kong, because its labor market is undergoing changes that might put its future as a service hub in jeopardy. Over the past decade, about 400,000 well-educated Hong Kong people with key skills have emigrated. At the same time, tens of thousands of poorly educated mainlanders with few skills have settled in the territory annually on family reunion grounds. The urgency of this issue is borne out in this survey by the concern about immigration policy from the private sector.

Finally, we reiterate the importance of promoting Hong Kong as a service hub, that is, a concentration of those "hub" services such as project management, product development, risk management, and end-to-end supply chain management. All these activities require a clustering of a *critical mass* of inputs located at the hub. Because many hub services are high value-added outputs, their qualities are extremely dependent on a high level of performance from *all* inputs. The presence of any substandard inputs, no matter how small are their shares, can significantly lower the quality of the final output (Kremer 1993). A deeper question is what the factors are that draw these inputs to the hub in the first place. Further theoretical work is needed here, taking account of multilateral trade liberalization and IT-induced structural changes in the production process of goods and services. We believe that some of the empirical findings arising from our survey will be useful in developing such a theory.

Appendix 7.1
A Multilateral Model of Trade Services Hub

Consider some services known as trade services, such as transportation and telecommunications, that are intermediate inputs to the completion of international trade of some underlying good. The provision of trade services involves three countries: the service provider's country and the exporting and importing countries of the underlying good. If the trade service market is liberalized, the service provider can be from any country. If it is not, the service provider would come from either the exporting or the importing country, with a service provider from each country responsible for the part that falls within its own jurisdiction.

Let us examine the effect of liberalization of this trade service. Because our focus is on a service hub, we assume a *multilateral* setting where a hub country (H) is surrounded by four spoke countries, A, B, C, and D (see figure 7.1). Let us first consider the case in which only country H has liberalized its trade service sector. Following Deardorff (2001), we cast the model in the context of a transport service for the sake of exposition, although the conceptual issues extend to other trade services as well. Let the total cost (S) for a service provider from country J to transport a shipment of Q units from country A to country B be given by

$$(1) \quad S_{AB}^J = c_0^J + c_1^J Q + c_2^J D_{AB} + c_3^J Q D_{AB} + c_4^J Q U_{AB} + c_5^J Q N_{AB}.$$

Because the service market is not liberalized in either country, the service will, as mentioned earlier, be provided by firms from both countries. For simplicity, we assume $J = A$ or B. D_{AB} represents the travel distance between A and B, U_{AB} is the number of loadings and unloadings, and N_{AB} is the num-

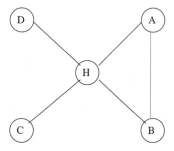

Fig. 7.1   Country H as the trade services hub

ber of customs crossings. The parameters $cJ\backslash i$ are functions of the technology and factor prices of country J, both of which are assumed to be exogenous. The fixed cost per shipment is c_0, c_1 is the cost per unit shipped, c_2 is the cost per unit of distance, c_3 is the cost that depends on both distance and quantity (such as the fuel cost), c_4 is the cost that depends on both quantity and the number of loadings and unloadings, and c_5 captures the cost associated with customs crossings that includes customs duties and the time delay. The last two terms in equation (1) add additional costs by reducing the speed of delivery. Note that U_{AB} and N_{AB} are equal to 2: one loading or customs crossing at A plus one unloading or customs crossing at B. The per-unit service cost is

$$(2) \qquad s_{AB}^J = c_1^J + \frac{c_0^J}{Q} + \frac{c_2^J D_{AB}}{Q} + c_3^J D_{AB} + 2c_4^J + 2c_5^J.$$

On the other hand, the per-unit service cost for this shipment to go via H is

$$(3) \quad s_{AHB}^I = c_1^I + \frac{c_0^I}{\hat{Q}} + \frac{c_2^I(D_{AH} + D_{HB})}{\hat{Q}} + c_3^I(D_{AH} + D_{HB}) + c_4^I U_{AHB} + c_5^I N_{AHB}.$$

Because country H has liberalized its trade service, $I = H, A, B, C, D$, or other countries in equation (3). $\hat{Q} = Q + \overline{Q}$, with $\overline{Q}$ representing all other traffic carried on routes HA or HB. Potentially, $\overline{Q}$ includes, on route HA for example, local traffic between A and H, or connecting traffic between A and C via H, etc. Taking the difference of equations (2) and (3), we obtain

$$(4) \quad s_{AHB}^I - s_{AB}^J = (c_1^I - c_1^J) + \left(\frac{c_0^I}{\hat{Q}} - \frac{c_0^J}{Q}\right)$$

$$+ \left[\frac{c_2^I(D_{AH} + D_{HB})}{\hat{Q}} - \frac{c_2^J D_{AB}}{Q}\right] + [c_3^I(D_{AH} + D_{HB}) - c_3^J D_{AB}]$$

$$+ (c_4^I U_{AHB} - 2c_4^J) + (c_5^I N_{AHB} - 2c_5^J).$$

Equation (4) shows three principal sources of gain from routing the AB shipment through H. The first is comparative advantage: because J is a subset of I, $c_i^I \leq c_i^J$. The second is economies of scale. Since AB goods will now be carried together with goods of other origin-destination pairs ($\hat{Q} > Q$), the fixed cost associated with a shipment will be spread over more quantity. The third is economy of scope. This comes from two areas. First, a service hub has a critical mass of essential inputs for the certain related fragments of the trade service to be provided cost effectively—for example, the sorting, repackaging, and logistics facilities in transport service. The other is the cost saving through eliminating duplication of inputs. Note that as the number of spoke countries increases, the gain from economies of scale and scope will multiply. Equation (4) also illustrates three potential disadvantages of routing AB shipments via H: longer distance ($D_{AH} + D_{HB} > D_{AB}$); additional loadings ($U_{AHB} = 4$); and possibly additional customs crossings ($N_{AHB} = 2$ or 4). All these would reduce the speed of delivery.

Now, consider the impacts on H if *all* countries liberalize their service sector. On the positive side, it allows service providers from H to expand their operations to other countries. The expanded market can, for example, improve efficiency through the realization of economies of scale and scope. Second, an increase in trade activities means more services opportunities for all countries, including H, especially if it could improve its hub infrastructure and remain a service hub. On the negative side, some traffic may be diverted away from H as the comparative advantage enjoyed by country H disappears, that is, $c_i^I = c_i^J$. Consequently, equation (4) becomes

$$(5) \quad s_{AHB}^I - s_{AB}^J = \left(\frac{c_0^I}{\hat{Q}} - \frac{c_0^I}{Q} \right) + \left[\frac{c_2^I(D_{AH} + D_{HB})}{\hat{Q}} - \frac{c_2^I D_{AB}}{Q} \right]$$
$$+ [c_3^I(D_{AH} + D_{HB}) - c_3^I D_{AB}] + (c_4^I U_{AHB} - 2c_4^I)$$
$$+ (c_5^I N_{AHB} - 2c_5^I).$$

Equation (5) shows that now the trade route via H is less attractive than that under unilateral liberalization by H. The shipment may go directly between A and B if the extent of economies of scale and scope is weak, the delivery speed is important, and loading or unloading and customs procedures at H are costly and cumbersome. As long as the hub is well established due to its early liberalization, the benefits from economies of scale and scope at the hub will most likely offset the diversion effect. Furthermore, as discussed below, some activities can only be performed with a critical mass of essential inputs located at the hub.

Perhaps key competitive threats to H under multilateral liberalization come from other competing hubs that spring up after liberalization, such as K (see figure 7.2). The unit cost difference for an AB shipment going through H hub and K hub may be written as

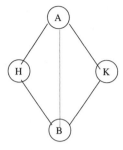

Fig. 7.2   Service liberalization and competing hubs

$$(6)\, s^I_{AHB} - s^I_{AKB} = \left(\frac{c^I_0}{Q_{AHB}} - \frac{c^I_0}{Q_{AKB}} \right) + \left[\frac{c^I_2(D_{AH} + D_{HB})}{Q_{AHB}} - \frac{c^I_2(D_{AK} + D_{KB})}{Q_{AKB}} \right]$$

$$+ \, [c^I_3(D_{AH} + D_{HB}) - c^I_3(D_{AK} + D_{KB})] + c^I_4(U_{AHB} - U_{AKB})$$

$$+ \, c^I_5(N_{AHB} - N_{AKB}).$$

Equation (6) shows that, if both the distance and traffic density of the two routes are about the same, then the first three terms are negligible and the competition between the two hubs lies in the last two terms. They reflect the "frictions" of moving goods, people, capital, and information through a service hub. The more restrictions there are, the less competitive the service hub is. In the context of transport services, these frictions can arise from the fact that goods have to go through port and airport terminal services and customs-clearing procedures. In banking, the frictions can be capital controls. More generally, they can be caused by a restrictive immigration policy, because movement of personnel is often necessary in the provision of many services. Finally, many hub services such as banking and business services are knowledge intensive. A transparent environment for the free flow of information is essential for these services. Restrictions on information flows present another friction that can lower the quality of such services.

References

Barriers hindering trade in services: 'Services 2000,' preliminary interview results. 1999. *Service Economy,* August, 4.

Blinder, A. S. 1991. Why are prices sticky? Preliminary results from an interview study. *American Economic Review* 81:89–100.

Cheng, Leonard K., and Changqi Wu. 1998. *Competition policy and regulation of business.* Hong Kong: City University of Hong Kong Press.

Deardorff, A. 2001. International provision of trade services, trade, and fragmenta-
tion. *Review of International Economics* 9 (2): 233–48.
Feenstra, Robert, and Gordon Hanson. 2001. Intermediaries in entrepôt trade:
Hong Kong re-exports of Chinese goods. NBER Working Paper no. 8808. Cam-
bridge, Mass.: National Bureau of Economic Research.
Friedman, M., and L. J. Savage. 1948. The utility analysis of choices involving risk.
Journal of Political Economy 56:279–304.
Hui, W., L. George, and Lawrence C. Leung. 2000. The trade of an entrepôt: Hong
Kong 1978–1999. National University of Singapore, Department of Economics,
and Chinese University of Hong Kong, Department of Decision Sciences and
Managerial Economics. Unpublished Manuscript.
Jones, Ronald, and Henry Kierzkowski. 1999. Horizontal aspects of vertical frag-
mentation. Paper presented at the International Conference on Global Produc-
tion and Specialization and Trade. 25–27 October, Hong Kong.
Kremer, Michael. 1993. The O-ring theory of economic development. *Quarterly
Journal of Economics* 108 (3): 551–75.
Low, Patrick, and Aaditya Mattoo. 1998. Reform in basic telecommunications and
the WTO negotiations: The Asian experience. Working Paper no. ERAD9801.
Geneva: World Trade Organization.
Mattoo, Aaditya. 1999. Financial services and the WTO: Liberalization in the de-
veloping and transition economies. Working Paper no. 2184. Washington, D.C.:
World Bank, September.
Pacific Basin Economic Council (PBEC). 1998. Administrative Barrier to Trade.
Hong Kong: PBEC.
Pacific Economic Cooperation Council (PECC). 1995. *Survey of impediments to
trade and investment in the APEC region.* Singapore: PECC.
UNCTAD and the World Bank. 1994. *Liberalizing international transactions in ser-
vices: A handbook.* New York: United Nations.

Comment   Chong-Hyun Nam

I enjoyed reading the paper and learned a lot from it. The paper consists of
two major parts, the first being a theoretical model part and the second an
empirical survey part. The relation between these two parts, however, remains very thin.

The first part of the paper asks questions such as how Hong Kong has
been able to evolve and thrive as a "regional service hub," and whether
Hong Kong can continue to be a regional service hub in the future as unilateral or multilateral liberalization in services trade proceeds. These questions were analyzed through a model, and the model suggests that Hong
Kong's geographical location, its ability to reap economies of scale and
scope in services trade, and its relatively earlier liberalization in goods and
services trade could be major factors behind Hong Kong's success as a regional service hub.

Chong-Hyun Nam is professor of economics at Korea University.

The model also suggests that liberalization in services trade, at both unilateral and multilateral levels, can be a mixed blessing for Hong Kong's future as a regional service hub. Certainly, gains are to be made by liberalizing services trade, through expanded markets and a greater possibility of exploiting economies of scale and scope, and through speedier delivery and customs procedures. However, the same gains can accrue to other competing economies in the region, and some traffic can be taken away from Hong Kong.

This kind of model analysis, however, provides little guidance as to what actually might happen to Hong Kong vis-à-vis others in the region as a regional service hub. Some quantitative empirical work is desired to answer this question. The second part of the paper is not designed to answer this question, however.

The second part of the paper investigates what priority wishes Hong Kong's private companies may have in liberalizing services trade at home and abroad. A rather extensive survey was conducted over individual service companies in Hong Kong for that purpose. The survey results seem, indeed, very revealing and useful for policy makers, particularly in facing "Services 2000" negotiations, which were only recently under the auspices of the World Trade Organization (WTO).

I have a couple of questions and a couple of comments on the second part of the paper.

My first question is if there are any other similar surveys done in the past either on Hong Kong or on any other economies. If there is any, a comparative study seems quite interesting.

My second question is how the authors came up with the thirty-eight wishes they adopted in their survey. The thirty-eight wishes can be reclassified into eight more broad categories, including establishment requirements, national treatment, information flows and transparency, and so forth. I wonder how these classifications can square with the format used at the Uruguay Round when governments make their schedule of specific commitments on the General Agreement on Trade in Services. There they used four modes of service supply, namely, cross-border supply, consumption abroad, commercial presence, and presence of natural persons, and they used two types of trade restrictions, namely, restrictions on market access and restrictions on national treatment.

In any case, the thirty-eight wishes seem quite sensibly constructed in the study, and the survey results seem to provide very useful information, as I said before, for policy makers in particular.

Let me now make a couple of comments. My first comment is concerning the survey result that improving information flows and transparency turned out be one of the most prevalent wishes among Hong Kong's service providers. Information gathering and its dissemination regarding individual countries' trade laws and regulations seem to be in high demand. Someone needs to do the job. In fact, the WTO is partly doing such a job at the moment through its Trade Policy Review Mechanism. Obviously, however, that is not sufficient. We may coerce the WTO into playing a greater role in that effort.

The authors argued that there is clearly a need to improve transparency, particularly in the Asia-Pacific region. I agree with that argument. They mentioned the recent case of Daewoo's restructuring process as an example. I may add to that. I believe transparency is badly in need in Korea not only for the business sector but also for the government sector. When the Korean government went for an International Monetary Fund rescue in late 1997, for example, nobody—even in the government circle, except for a very few persons—knew exactly what the actual reserve position at disposal was and what the foreign debt structure was at the time in Korea. The lack of information and transparency in accounting was surely one of the major causes of the recent financial crisis in Korea. Some systematic efforts need to be made on a global basis to improve information flows and transparency, I think.

My second comment is concerned with domestic regulations that may limit market access for many service providers. Such trade-impeding domestic regulations are often found to be country-specific and industry-specific. However, some common elements of trade barriers embodied in those domestic regulations may be drawn from cross-country studies or surveys, as was done in the present study. In the end, again, the WTO may need to be brought in to make something like what I would call a Trade-Related Domestic Regulations Review Mechanism.

8

Korea's Telecom Services Reform through Trade Negotiations

Nae-Chan Lee and Han-Young Lie

8.1 Introduction

It has become a fairly common assertion that deregulation in the telecommunication services market around the world has been a clear trend since a couple of decades ago. Obviously, regulatory reform has been legitimized with the beginning of a multilateral governmental response to the Uruguay Round and World Trade Organization (WTO) negotiations on basic telecommunications. The backgrounds and key factors of deregulation markedly differ across the majority of countries according to different policy objectives, so that no single country can be held up as a model case of successful deregulation. Nevertheless, all agree that the objective of deregulation is to improve social welfare by attaining lower service tariffs, higher service quality, and greater efficiency in the market.

Korea has been actively working on deregulating in its telecommunication services market since the 1990s, after it provided basic telephony to the general public in the 1980s. Recognizing the importance of effective competition for the future growth of its telecommunication services market, the Korean government, acting as policy maker, regulator, and largest stakeholder of the dominant service provider Korea Telecom, has played a central role in restructuring the market. It has established the rules of the game on one hand and has controlled the outcomes of it on the other hand. Although less identified, but of great importance for better understanding the

Nae-Chan Lee and Han-Young Lie are senior research fellows of the Korea Information Society Development Institute (KISDI).

The authors are indebted to Dong-Pyo Hong of the KISDI for valuable comments and discussion on the input-output analysis of telecommunication services and to Ki-Joo Yu and Jung-Woo Ryu for their assistance. They also appreciate Takatoshi Ito and Anne O. Kruger's suggestions. Any shortcomings and errors are the authors'.

motivation and process of Korea's deregulation, is the pressure for market opening that has originated from trade negotiations. In particular, WTO negotiations on basic telecommunications have become a watershed for furthering deregulation and accelerating competition in the market.

The purpose of this paper is to examine the implications and lessons of regulatory reform in Korea's telecommunication services market. The next section (section 8.2) reviews the history of the deregulation process. It explains how Korea has been transforming the market since 1990 by taking advantage of the opportunities given by trade negotiations. This is followed in section 8.3 by an overview of the current marketplace. Section 8.4 explores the impact of the most recent regulatory reforms made in conjunction with WTO negotiations on basic telecommunications on market performance with respect to contestability and competition. We especially focus on the implication of introducing voice resale services into the market and the role of foreign investment in facilitating competition in the mobile-services market. Finally, section 8.5 offers concluding remarks.

8.2   Impact of Trade Negotiations on the Korea's Regulation

Korea has made significant progress in the deregulation of the telecommunication services market since 1990. As a result, it has succeeded in enhancing the overall performance of the market through the promotion of competition. It is natural that adequate credit be given to the government's determination in and efforts toward attaining such progress. In a sense, however, the success of deregulation appears to be a legacy of a series of bilateral and multilateral trade negotiations on telecommunications services. Trade negotiations at least provided good momentum for Korean policy makers to effectively manage domestic pressure against opening the market. In the following sections, we examine the development of the Korean regulatory regime in the 1990s with due attention to the trade negotiations in which Korea has been involved.

8.2.1   The First Stage (mid-1990–mid-1994)

The Korean telecommunication services market in the 1980s was characterized by the construction of the Public Switched Telecommunication Network (PSTN). Extremely low teledensity was a chronic problem during those days, so the top policy priority was simply to satisfy the basic needs of the general public for telephone services. The Korea Telecommunications Authority—predecessor of today's Korea Telecom (KT, henceforth)—was exclusively in charge of providing telephone services under the auspices of the Korean government. By 1988, it had barely managed to attain the policy objective of providing a telephone per household. With the basic demand satisfied, the Korean government began to concentrate on other issues such as quality improvement for telecommunication services and the enhancement of KT's managerial efficiency.

In February 1989, a year after Korea managed to satisfy the basic demand for telephone services, the United States designated Korea as a priority foreign country (PFC) based on Section 1374 of the Omnibus Trade and Competitiveness Act of 1988.[1] The purpose of this designation was to open up Korea's telecommunication services market, especially to value-added and mobile services. In addition, the liberalization of the value-added services market was a major negotiation agenda of the Uruguay Round (UR). The Korean government, after a series of bilateral talks with the United States, decided to carry out structural reform of the existing market in July 1990.

It was clear that the U.S. action provided Korea with a direct motivation for both market liberalization and regulatory policy making at the same time. In the reform process, Korea incorporated the U.S. request to eliminate market access limitations (i.e., to permit foreign ownership in the value-added and mobile services segments). However, Korea kept its own stance in streamlining other regulatory policies such as categorizing service providers and applying market entry conditions to each category of service provider.

A key feature of the reform was to divide service providers into three categories: general, specific, and value-added (see table 8.1).[2] General and specific service providers were differentiated from the value-added by virtue of having their own facilities. The business area of general service providers was wired services, whereas that of specific service providers was confined to wireless services. The requirements for market entry were also different: for the general service providers it was the government's designation; for the special service providers, licensing; and for the value-added service providers, registration. Although foreign ownership was not allowed among general service providers, it was allowed up to 30 percent among special and up to 100 percent among value-added.

The differentiation of entry barriers among the service providers can be interpreted as a strategy to treat foreign entry differently according to the degree of pressure for market opening. Korea could resolve the external issues by fully opening the value-added services market and part of the mobile market. Classification of the service providers was useful from the regulatory perspectives, but it entailed a lot of loopholes. In particular, sticking points were the *positive listing system* for the provision of services and the *request for proposal (RFP) system* for licensing. Through the positive listing system, the government permitted the service providers to supply only the services listed in the Telecommunications Business Act. It took a long

1. Section 1374 entitles the USTR to investigate potential foreign telecommunications trade barriers, identify any trading partner with anticompetitive practices as a priority foreign country, and at any time revoke the identification (taking into account relevant criteria, including progress being made).

2. Before 1990, the only category of the telecommunications service providers in Korea was that of the public telecommunications operators (PTOs).

Table 8.1   **Classification of Service Providers in the First Reform (mid-1990–mid-1994)**

Category	General Service Provider	Specific Service Provider	Value-Added Service Provider
Facilities	Own facilities	Own facilities	Leased facilities
Subservices	Fixed telephony, telegraph, telegram, private leased circuits	Wireless services: cellular, radio paging, TRS, wireless data transmission	Database, data processing, data accumulation and transmission, EDI, e-mail, CRS
Market entry condition	Designation	Licensing	Registration
Foreign ownership	Not allowed	Up to 1/3 of the total shares (not allowed to be the largest shareholder)	Up to 100%
Other regulations on ownership	The largest shareholder: up to 10% Equipment manufacturer: up to 3%	The largest shareholder: up to 1/3 of the total shares Equipment manufacturer: up to 1/3 of the total shares (not allowed to be the largest shareholder) Government invested institution: up to 10% (not allowed to be the largest shareholder)	None

Source: Ministry of Information and Communication.

administrative rationing process for the unlisted services to be introduced into the market, which deterred the introduction of new services. Under the RFP system, a company could make a request for a license only on the condition that the government made public notification prior to licensing. The Korean government retained the RFP system until August 1997 as an important policy tool to set a priori limitation on the number of market entrants into any of the service categories.[3]

After the first reform, duopoly competition for international telephone and regional radio paging service segments was introduced. Competition, however, was managed by government intervention through a prior tariff approval system. Price differentials between incumbent and new entrants were kept constant at a level at which the entrants could secure their market shares without being so drastic as to tip the balance too unfavorably toward incumbents.[4] This managed competition systematically guaranteed excessive profits to all of the service providers once they obtained their entry

3. This is one of the typical limitations to market access, the so-called, economic needs test that is specified in the General Agreement on Trade in Services (GATS), article 16.
4. Under the prior tariff approval system, even the entrant without market power, not to speak of the incumbent, has no autonomy to determine its own tariffs. The system was abolished in late 1995.

ticket, but blurred the original policy objective: enforcement of service providers' competitiveness through competition.

8.2.2   The Second Reform (mid-1994–mid-1997)

The Korean government had maintained the position, ever since the first reform, that it could manage the telecommunications market structure at will. However, in December 1993 when the UR was concluded, the government realized that its policy stance could not last long. Because it had found basic telecommunications to be one of the service sectors left unresolved by the UR, trading partners agreed to extend the period of negotiations on basic telecommunications until 30 April 1996. The government regarded this situation as a strong message that the existing telecommunication services regime would be radically liberalized sooner or later. Therefore, competition became mandatory for the Korean telecommunication services market, which led to the second structural reform of July 1994.

One of the major regulatory changes made in the second reform was the scrapping of the demarcation between general and specific service providers, which were integrated and dubbed *facility-based service providers* (FSPs, henceforth). The other change was the adoption of the negative listing system. As a result, a company licensed to provide facility-based services was legally able to supply any service. In practice, the government still listed the facility-based services in the Telecommunications Business Act, so that FSPs were under regulation in the process of licensing. However, it was a significant improvement because value-added service providers were allowed to supply all types of services except facility-based services. As regards the foreign ownership restrictions, the restriction on telecommunications equipment manufacturers' ownership in facility-based services was abolished so that they could participate in those services on an equal footing.[5] Table 8.2 summarizes the classification of service providers resulting from the second reform.

The second reform, through the introduction of competition into long-distance service,[6] set up a duopoly structure externally in every licensed facility-based services market. However, because the RFP system in licensing served as a major stumbling block to inducing additional entry in the market, it was not possible to reap the fruits of competition. Asymmetric entry barriers also existed between fixed and mobile services, due to the ownership restriction on the largest shareholder in fixed telephony. While the former general service providers could enter the mobile services market without ownership adjustment, the former specific service providers, par-

---

5. Additionally, the ownership limitation on the private leased circuits services—which had been kept tight since the first reform by treating the services as a specific service on the basis of the former classification—was mitigated. Therefore, foreigners formerly prohibited from investment could come to invest up to one third of the total shares.

6. The Korean government designated Dacom as the second long-distance service provider in March 1995.

Table 8.2   **Classification of Service Providers in the Second Reform (mid-1994–mid-1997)**

Category	Facility-Based Service Provider	Value-Added Service Provider
Facilities Subservices	Own facilities Fixed telephony, telegraph, telegram, private leased circuits, wireless services (cellular, radio paging, wireless data transmission, TRS), and other services specified by the minister	Leased facilities Other than those provided by facility-based service providers
Market entry condition	Licensing	Notification
Foreign ownership	Wired line: prohibited Wireless: up to 1/3 of the total shares (not allowed to be the largest shareholder)	Up to 100%
Other regulations on ownership	Wired line: up to 10% for the largest shareholder Government invested institution: up to 10% (not allowed to be the largest shareholder)	None

Source: Ministry of Information and Communication.

ticularly mobile service providers with the largest shareholdings, were obliged to reduce their ownership in order to enter the fixed telephony market. That is to say, there was an asymmetric line of business.

In sum, structural reform in the Korean telecommunication services market was successful to some extent in increasing the number of participants in the market, but not in facilitating the level playing field to enhance the effective competition. However, the practice of managed competition had been continued on the basis of providing appropriate competition in the market, as had been intended at the first stage.

In July 1995, the government announced a blueprint to promote competitiveness in telecommunication services.[7] The main purpose was to establish fair and effective competition in the market, which can be considered a switchover in the policy direction from a managed and progressive competition to a free and full-scale competition.[8] The first step of the action plan was to facilitate the introduction of new service providers in international telephone services, private leased circuits, and various mobile services. The second was to streamline a wide range of existing regulatory measures, including the removal of the RFP system and the reinforcement of an

7. From a political point of view, the announcement of the blueprint might be interpreted as a bandwagon attempt of the newly launched Ministry of Information and Communication (MIC) in December 1994 (the successor of Ministry of Communications [MOC]).

8. Another purpose was to provide KT with greater managerial independence by overhauling existing regulations arising from its status as a government-invested institution, and by permitting participation in the new service markets (such as the mobile market).

independent regulatory body's role.[9] The third was to extend the scope of competition from domestic to international based on the outcomes of the WTO negotiations on basic telecommunications.

At the end of 1995, the government changed its stance dramatically on the regulation of prices by abolishing the prior approval system, under which it had approved all tariffs except local telephony of KT and mobile services of SK telecom, because the two service providers were regarded as assuming market power in each market.[10] As a result, most service providers gained the autonomy to determine their own tariffs, so that a notification alone was enough for any changes in tariffs.

8.2.3   The Third Stage (mid-1997–)

WTO Negotiations

As an initial step toward full-scale competition, the government issued licenses to twenty seven new service providers in 1996 in such services as international telephone (one as the third-service provider), private leased circuits (two), personal communications services (PCS; three), trunked radio services (TRS; six), radio paging (one), wireless data transmission (three), and second generation cordless telephony (CT-2; eleven). It was not until a few months after the conclusion of the WTO negotiations that the government took practical and legal actions for the other regulatory improvements planned in the blueprint. The WTO negotiations on basic telecommunications, after a long series of the consultations, reached its final conclusion on 15 February 1997, with the agreement going into force on 5 February 1998. The main achievements comprised a wide range of binding commitments on market access and a package of procompetitive regulatory principles (the so-called Reference Paper).[11]

Korea made its final commitments on market access for all segments in telecommunication services on 14 February 1997. Foreign ownership was limited to 33 percent in facility-based services by the end of 2000, and was to be raised to 49 percent beginning 1 January 2001. For individual shareholdings, it was limited to 10 percent for wired line services and 33 percent for wireless, respectively.[12] Foreign ownership in KT was limited to 20 percent

9. The Korea Communications Commission (KCC) was originally created as a regulatory body in March 1992. However, being under the auspices of the MIC, KCC played a limited role in regulatory functioning.

10. The Korean government is further considering the introduction into the local telephony of a price-cap system, which is believed to be better than the prior approval system in facilitating cost-oriented pricing and in improving efficiency in the market. In addition, the price-cap system seems advantageous in that it is not subject to non–sector specific consideration and political interference.

11. For details, see Sherman (1998).

12. Limitation on individual shareholdings applied to both domestic and foreign persons in a nondiscriminatory manner.

Table 8.3   Summary of Korea's Existing Regulations and Scheduled Commitments
                   (as of February 1998)

Category	Existing Entry Barriers	Scheduled Commitments
Facility-based services	Aggregate: Wired line: prohibited; wireless: 33%; KT: prohibited Individual: Wired line: 10%; wireless: 33%; KT: 1%	Aggregate: Wired line: 33%, 49% from 2001; wireless: 33%, 49% from 2001; KT: 20%, 33% from 2001 Individual: Wired line: 10%; wireless: 33%; KT: 3%
Voice resale services	Aggregate: Prohibited	Aggregate: 49% from 1999; 100% from 2001
Regulatory principles	Domestic legislation	Reference paper
Numerical restriction	RFP	None

by the end of 2000, and to be raised to 33 percent beginning 1 January 2001, with individual shareholdings limited to 3 percent.[13] In telephone services on a resale basis (so-called voice resale), foreign ownership was allowed up to 49 percent on 1 January 1999, and was to be raised to 100 percent beginning 1 January 2001.[14] Korea also included in its schedule additional commitments to underpin those on market access by adopting the Reference Paper. Table 8.3 summarizes Korea's telecommunication market regulations before and after the WTO negotiations on basic telecommunications, and table 8.4 shows the classifications of service providers resulting from the third reform.

Korea's schedule could be considered insufficient in that it basically phased-in liberalization without allowing foreigners' majority shareholdings in facility-based services, including KT. Nonetheless, it was a significant improvement not only for Korea but also for other trading partners. Aside from its mitigation of foreign ownership restrictions, it was a clear departure from the status quo of the telecommunication services market, particularly in two respects. One was the elimination of the RFP system, which had been a major obstacle to free and full-scale competition in (and a definite limitation on foreign-market access to) the Korean market.[15]

The other enhancement was the introduction and, at the same time, the liberalization of voice resale. Its implication was that the Korean government would systematically induce price competition between the existing FSPs and the newly participating special service providers (SSPs) within

13. It was not permitted for a foreigner to be the largest shareholder in KT.
14. Foreign ownership in the other resale-based services was allowed up to 100 percent as of 1998.
15. It is still legitimate for any government to limit the number of service providers subject to the availability of radio frequencies, in accordance with the consensus made in the WTO negotiations on basic telecommunications.

Table 8.4   **Classification of Service Providers in the Third Reform (as of February 1998)**

Category	Facilities-Based Service Providers	Special Service Providers	Value-Added Service Providers
Facilities subservices	Own facilities Fixed telephony, telegraph, telegram, private leased circuits, mobile services, and other services specified by the minister	Leased facilities Type I (switched reseller): Voice resale, IP-based telephony, international call-back, etc. Type II (switchless reseller): Aggregator, rebiller, etc. Type III: in-building: Communication services	Leased facilities All value-added telecommunication services
Market entry condition	Licensing	Registration	Notification
Foreign ownership	Aggregate: Wired line: 33%, 49% from 2001; wireless: 33%, 49% from 2001; KT: 20%, 33% from 2001 (prohibition of foreign largest shareholding)	Aggregate: 49% from 1999; 100% from 2001	Up to 100%
Other regulations on ownership	Individual: Wired line: 10%; wireless: 33%; KT: 3%	None	None

Source: Ministry of Information and Communication.

the market. It was also a major advancement for Korea to bind itself in fulfilling its regulatory functions through adopting the multilateral obligations of the Reference Paper.

Korea embarked on a third structural reform after the conclusion of the WTO negotiations on basic telecommunications. The motivation was largely twofold: one was to saturate the market with as many domestic providers as possible before the market opening, and the other was to make its laws and regulations conform with the scheduled commitments. In June 1997, the government selected nine new service providers in five service areas such as local, long-distance, private leased circuits, radio paging, and TRS. The meaning of the action was considerably symbolic in Korea, since competition was introduced even in the local service market, which had been exclusively dominated by KT. That was little more than the completion of introducing competition in all segments of the Korean telecommunication services market.

Autonomous Deregulation

The regulatory regime experienced a significant change in August 1997 in accordance with the amendment of the Telecommunications Business Act.

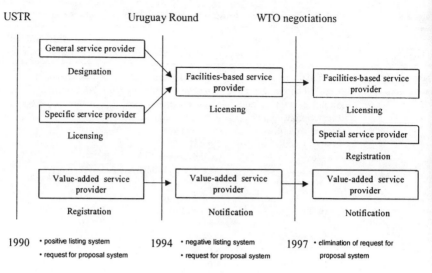

Fig. 8.1   **Deregulation of market entry condition**

Aside from the mitigation of foreign ownership restrictions and the elimination of RFP,[16] one of the most important changes occurred in the classification of service providers. According to the commitments made in the WTO agreement on basic telecommunications, the government newly introduced a category of resale-based services in licensing—so-called *special services*—most of which had never before been legally permitted in Korea. The category was created by a simple decomposition of the former FSPs in the second reform, on the basis of the existence of own facilities and the type of services for provision. The method of classification in Korea became similar to methods in other developed countries. Service providers were classified as facility-based, special, or value-added service providers. They were differentiated by the respective market entry conditions: licensing for facility-based service providers, registration for special service providers, and notification for value-added service providers. (Figure 8.1 shows how the market entry conditions were deregulated.)

Meanwhile, Korea undertook an autonomous liberalization in 1998 and 1999 to facilitate the inflow of foreign capital, the better to cope with its late-1997–98 financial crisis. In accordance with the revision of the Telecommunications Business Law on 17 September 1998, Korea accelerated the removal of other ownership restrictions. The limitation on the foreign ownership of KT was raised from 20 percent to 33 percent as of 17 Sep-

---

16. Nonetheless, the government still has room to improve in that it files license applications only on a periodic basis in March and September. The reason is that the periodic licensing itself imposes a limitation to seasonable market entry.

Table 8.5   **Current Status in Foreign Ownership Regulations (as of April 2000)**

Category	Facility-Based Service Providers	Special Providers	Value-Added Service Providers
Foreign ownership	Aggregate: Wired line: 49% as from July 1999; wireless: 49% as from July 1999; KT: 33% as from September 1998 (prohibition of foreign largest share-holding)	Aggregate: 49% as from September 1998; 100% as from 2001	Up to 100%
Other regulations on ownership	Individual: KT: 15% as from January 1999	None	None

Source: Ministry of Information and Communication.

tember 1998 (the previous date had been 1 January 2001). A 33 percent (10 percent, in the case of wired line services) limitation on individual shareholding for a facility-based service supplier, except KT, was removed as of 17 September 1998. The foreign ownership of a supplier of voice resale services was permitted up to 49 percent as of the same date—which was earlier than the previously scheduled time of 1 January 1999. Furthermore, the limitation on individual shareholdings in KT was expanded from 3 percent to 15 percent as of 1 January 1999. With another revision of the law on 24 May 1999, the limitation on foreign ownership of a facility-based service supplier (except KT) was expanded from 33 percent to 49 percent as of 1 July 1999 (as opposed to the previously scheduled date of 1 January 2001). Table 8.5 summarizes the present state of foreign ownership regulations in Korea.

Competitive Safeguards

It is often said that market entry regulation is justified when the scarce resource of the frequency spectrum should be allocated in an efficient way for wireless services, when economies of scale or scope prevail, and when inefficient duplication of investment may be precluded.

There are several competitive safeguards in Korea for ensuring fair competition between incumbent and new entrants once market entry is accomplished, and for the protection of consumer rights such as interconnection, preselection, telecommunications performance monitoring systems (TPMS), merger and acquisition guidelines, and a universal service fund (see OECD 2000).

Korea Telecom, as an owner of bottleneck facilities such as local loops, is obligated to the mandatory and prompt provision of interconnection, cost-orientation through rate-of-return regulation, and separate accounting. As of 2000, the scheme was extended to mobile FSPs—including SK

Telecom—because of the increasing importance of mobile services and increasing interaction between fixed and mobile networks. Preselection guarantees parity in dialing among long-distance FSPs by allowing users to choose in advance which FSP's service to use. It was first applied in 1997 to Dacom and extended to Onse in 1999. For the purpose of providing users with information about FSPs' quality of services (QoS) and inducing quality competition among FSPs, the government introduced TPMS in 1999, by which several QoS indicators of fixed and mobile services are announced periodically. The scheme was extended to high-bandwidth Internet services including accelerated digital subscriber lines (ADSL) and cable modem in 2000. Mergers and acquisitions (M&A) among FSPs in the telecommunications market are under the auspices of the Ministry of Information and Communication (MIC) and the Korea Federal Trade Commission, which perform separate roles through the M&A guidelines for testing would-be anticompetitive effects of M&A. In addition, a universal service fund to make up for deficits generated from local charges below costs was scheduled to be implemented in 2000.

8.3 Snapshot of Korea's Recent Marketplace

As outlined in section 8.2, harnessed mainly by trade negotiations, the Korean government has enforced step-by-step regulatory reforms to dismantle unnecessary obstacles, especially market entry barriers that deter vigorous market competition. Through a series of deregulation processes, Korea's telecommunication services market has achieved unprecedented growth. This chapter reviews the status quo and dynamics of the marketplace.

8.3.1 Market Participants

The variety of services and the number of market participants are one of the significant barometers for measuring a degree of competition. As of April 2000, 52 FSPs and 229 SSPs were operating businesses in each relevant service market, whereas the value-added service providers totalled 3,729.[17] There are nine major facility-based telecommunication services, including traditional fixed telephony and wireless services. KT, Dacom, and Onse are three major players in long-distance and international service markets, whereas most Type I SSPs have been doing business in the wireless sector ever since voice resale service was permitted in 1998. In April 1999 Hanaro, focusing mainly on the deployment of ADSL for high-bandwidth Internet services, launched its business in the last ten miles of local loops that have long been regarded as an impregnable fortress of KT. Eleven FSPs are operating in the leased-line market. Table 8.6 lists the numbers of FSPs and SSPs as of April 2000.

The mobile service market consists of five competitors: two cellular pro-

17. Different services of the same FSPs and SSPs are double-counted.

Table 8.6   **Number of FSPs and SSPs (as of April 2000)**

		Wired				Wireless			
		Long		Leased			Radio	Wireless Data	
	Local	Distance	International	Line	Cellular	PCS	Paging	Transmission	TRS
Number of service areas	2	3	3	11	2	3	14	3	11

	Type I	Type II	Type III
	Type I	Type II	Type III
Number of classifications of SSPs	30	177	22

Source: Ministry of Information and Communication.

viders (incumbent SK Telecom and Shinsegi Telecom) and three PCS providers (KT Freetel, LG Telecom, and Hansol M.Com). A nationwide radio paging FSP, SK Telecom competes with thirteen regional FSPs in the radio paging service market. Three providers are in the wireless data transmission service market and eleven in the TRS market, where they are in infancy, but about to burgeon.

8.3.2 Subscribers

As of February 2000, the number of fixed telephony subscribers exceeded 21 million and of mobile subscribers amounted to 25 million. KT remains dominant in the fixed service market, whereas the market share of new entrant Hanaro, in its infancy, is less than 1 percent. In the mobile service market, the leading FSP—SK Telecom—accounted for 42.8 percent of the total number of mobile subscribers, whereas the share of Shinsegi Telecom amounted to 13.8 percent. The remaining 43.4 percent belongs to three PCS providers.

The high penetration ratio of fixed telephony service is mainly due to the government's effort to expand the PSTN infrastructures by means of telegraph and telephone bond, and to maintain local-call charges below costs in the context of the general public interest by prior tariff approval systems. Teledensity presently exceeds 50 percent and seems to be reaching a saturation point. In contrast to the fixed telephony market, the growth of the mobile service market has been propelled solely by commercial motives. Ever-diminishing purchase costs of terminal equipment removed entry barriers for customers and accelerated the increase in the number of mobile subscribers exponentially. As of June 1999, the number of mobile subscribers per 100 inhabitants ranked fifth among OECD member countries and surpassed the number of fixed service subscribers in September 1999. The number of mobile subscribers was expected to grow to 29 million by late 2000.[18]

18. The quadratic logistic function is applied to estimate the value. The saturation ratio is assumed to be 65 percent and monthly data from January 1996 to December 1999 are used.

Competition among mobile FSPs for leftover subscribers (consisting mainly of teenagers or individuals in their twenties) is still going on fiercely with the goal of preserving as many customers as possible, aiming at IMT-2000 licenses that were scheduled to be granted in late December 2000.

On the other hand, the number of users of radio paging services reached an apex of 15 million in late 1997, but have been drastically declining since then, and were fewer than 3 million in January 2000. As a result of market shrinkage, origination-only CT-2 service, formerly provided by the nationwide service provider KT and ten regional FSPs, was shut down in January 2000.

Table 8.7 and figure 8.2 summarize the numbers of subscribers in the Korean telecommunication services market.

Table 8.7   **Number and Shares of Subscribers (as of February 2000)**

FSP	Korea Telecom		Hanaro		Total
Fixed subscriber (in millions)	21.43		0.19		21.62
M/S (%)	99.1		0.9		100.0

	SK Telecom	Shinsegi Telecom	KT Freetel	LG Telecom	Hansol M. Com	Total
Mobile subscriber (in millions)	10.89	3.51	4.61	3.02	3.41	25.44
M/S (%)	42.8	13.8	18.1	11.9	13.4	100.0

Source: Ministry of Information and Communication.

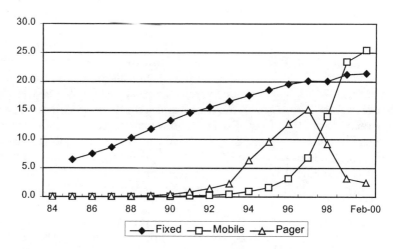

Fig. 8.2   Number of major telecom service subscribers (millions)
Sources: Ministry of Information and Communication, mobile FSPs

8.3.3   Dynamics of the Marketplace

A significant feature of changes in trends in voice services markets is that traditional fixed telephony services markets (including local, long-distance, and international services) have been nearly saturated or have even shrunk, whereas mobile service has grown by leaps and bounds. In 1999, the total revenue for wired services amounted to 6,430 trillion won, and that of wireless services, 9,715 trillion won, adding up to 16,145 trillion won (see table 8.8). The mobile turnover caught up with traditional fixed telephony services in 1999, proving that mobile services are no longer supplemental to fixed telephony services.

The revenue of local services has increased annually by 1.1 percent, on average, since 1996, and amounted to 3,078 billion won in 1999. In order to guarantee universal access for the general public to local telephony services, local-call charges have been maintained at the level below costs incurred. The resulting deficits have been cross-subsidized by such surplus sectors as long-distance and international sectors, which have brought about *de facto* transfers of wealth from heavy users to light users. To address this unfair practice, tariff rebalancing has been executed eight times since 1981 and is expected to continue. Nevertheless, considering that the subscriber market appears to be saturated, with no potential for new demand-inducing packages, market expansion cannot be anticipated.

Worse still is the case of the long-distance service market. Its revenues

Table 8.8   **Revenues of Wired and Wireless Services (billions of won)**

Service Area	1996	1997	1998	1999
Wired				
Local[a]	2,984	3,049	3,072	3,078
Long distance	2,176	2,487	1,573	1,334
International	753	728	590	653
Public pay phone	613	260	502	327
Leased line	622	791	874	1,038
Subtotal (a)	7,148	7,315	6,611	6,430
Wireless				
Mobile[a]	2,254	3,582	5,322	9,118
Paging	1,163	1,511	1,168	550
TRS	3.4	7.2	7.1	12.2
Wireless data transmission	—	0.2	0.7	20.5
CT-2	—	52	43	14
Subtotal (b)	3,420	5,152	6,541	9,715
Total (a + b)	10,568	12,467	13,152	16,145

Source: Computer and Communication Promotion Association of Korea.

[a]Revenues include interconnection revenues.

recorded a peak in 1997, but steeply declined in 1998. It remained on the wane in 1999, accounting for 1,334 billion won. A major reason for the sharp drop was certainly the financial crisis, but what seems more important is that demands for long-distance calls were satisfied by mobile calls. Mobile FSPs, not differentiating services and tariffs by distance, can deliver long-distance calls. Even though long-distance service is generally more economical than mobile service for long conversations, customers appear to prefer using mobile phones partly because they might not recognize this fact, or partly because they are used to the convenience of mobility, resulting in a so-called ratchet effect.

The trend of revenue in international services is similar to that in long-distance service. Although the market size is the smallest of the three traditional fixed telephony services, the market mechanism, including the international settlement regime, is the most complicated, and is an exemplary case in which positive effects of liberalization have been manifested (as will be analyzed in the next section in detail).[19]

The ups and downs of wireless markets, especially mobile and radio paging service markets, reflect trends not only in the number of subscribers but also in amount of revenue. The mobile market has continued to grow from 5,322 billion won to 9,118 billion won in 1999, an increase of more than 40 percent, whereas the radio paging market amounted to 550 billion won in 1999, a 50 percent decrease from the previous year. Radio paging services lost their price competitiveness with mobile services as costs of mobile terminal equipment continued to decline and as the demand for such services as VMS (voice mail service) and SMS (short messaging service) gradually increased, blurring the line of business between them.

Several points in the changing trends are worth noticing. First, although mobile services have grown enough to substitute for some portion of long-distance services, they also have facilitated the expansion of LM (land-to-mobile) calls out-bounding from KT's network to mobile FSPs, to which no one had seriously paid attention only a few years ago. The turnover of LM service in 1999 amounted to 597 billion, twice as much as that of the previous year. Although the mobile-inducing effect could also be observed in public pay-phone service, the ratchet effect of convenience for using mobile service dominated it, resulting in deficits.[20]

Second, traditionally voice-oriented markets are shifting to datacentric ones, which have an impact on leased line services and dial-up Internet interconnection. The revenue of leased line services, quite a portion of which comes from the demand of business customers for constructing intranets and connecting them to the Internet, increased 18.6 percent a year on aver-

19. Numeric values in the table do not contain international settlement deficits.
20. For more details on the interaction between fixed and mobile FSPs from the perspective of the interconnection regime, see Lee (1999).

Table 8.9   Revenues of LM and 014xy Services (billions of won)

Service Area	1996	1997	1998	1999
LM[a]	n.a.	n.a.	1,386	2,510
014xy	n.a.	71	122	242

Source: Korea Information Society Development Institute 1999, 2000.
[a]Revenues include interconnection revenues.

age and recorded 1,038 billion won in 1999.[21] This number was also boosted by the spread of so-called Internet plazas, which are furnished with personal computers providing services such as Internet access and computer games to an unspecified number of users.[22]

On the other hand, residential customers use the dial-up method by which they can connect with the Internet via KT's local switches. A local-call charge is then paid to KT in addition to a flat fee to an ISP (Internet service provider). As a result, some portion of the local service revenue includes dial-up interconnection.[23] There is a peculiar dial-up service in Korea called *014xy service*. This service was introduced by the government in 1994 to facilitate the data services market with a 40 percent discount on local-call charges. For ISPs to qualify as 014xy service providers and obtain identification numbers they must have more than five nodes equipped with routers and relevant facilities nationwide. The increase in demand for data services is reflected in the trend of 014xy revenue, which has nearly doubled annually and accounts for 242 billion won (see table 8.9 for a summary of LM and 014xy revenues, and figure 8.3 for an illustration of how the various telecom services interact).

8.4 Triggering New Games

The liberalization and opening of Korea's telecommunication services market, spurred by a series of trade negotiations, has gradually evolved through the deregulation of market-entry conditions and the mitigation of the foreign ownership ceiling. However, the strongest impact on domestic markets has been the WTO agreement on basic telecommunications. First, it has facilitated contestability of the international service market by allowing voice resellers to enter. Second, it has not only provided a foothold for late-coming mobile FSPs to raise investment funds, thus enhancing competition in the mobile sector, but has also accelerated the privatization of KT by issuing foreign Depositary Receipts.

21. Demands for the leased line also come from FSPs' network construction, e.g., connections of mobile FSPs' switches and base stations.
22. As of January 2000 the number of Internet plazas amounted to about 14,000.
23. Considering that local service revenue maintained the status quo, it might be guessed that the revenue from pure voice use will be diminishing.

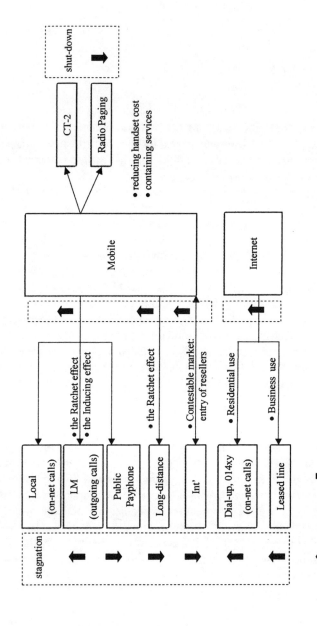

Fig. 8.3   Interaction among telecom services

8.4.1   International Service

Competition in the Outgoing-Traffic Market: Realization of Contestability

Liberalization of the Voice Resale Market. By 1997, Korea's international telecommunications market had been operated by KT, Dacom, and Onse. Since voice resale service was first introduced in January 1998 (pursuant to the revised Telecommunications Business Law, which took in Korea's regulatory commitments made in the WTO agreement), more than thirty Type-I SSPs have entered the international services market.[24]

Type-I SSPs are either affiliates of incumbent FSPs experienced in telecommunications businesses or spin-offs of conglomerates beginning with their large demand pool. SSPs either sell prepaid or postpaid cards with individual access numbers, or provide international telephony services to an unspecified number of the general public with identification numbers in the form of 007xy or 003xy. SSPs' services are provided mostly through mobile FSPs' networks and subscribers' handsets.

Implications of Introducing Resale from the Perspective of Industrial Organizations. Traditionally, the telecommunication services industry has been regarded as having economies of scale or scope and as prone to being monopolized, resulting in monopoly rents for the incumbent monopolist. It is the conventional wisdom of the contestable market theory, however, that if the market is flexible enough that potential entrants who possess the technology and who provide services as homogeneous as the incumbent's can freely enter all or part of market with lower tariffs than the incumbent's whenever they find such cream-skimming behaviors lucrative, and then can exit the market frictionlessly whenever the opportunity for arbitrage vanishes, then the Pareto optimality or at least the second-best (Ramsey pricing) result of the incumbent might be obtained.[25]

What has been going on in Korea's international services market ever since the liberalization ignited by the WTO agreement could be explained in the jargon of this framework: homogeneity of service, frictionless entry and exit, no universal service obligation (enabling cream-skimming by SSPs), and price advantage of resellers over FSPs (see Lee 1998).

First, SSPs' service is much like that of FSPs. To be sure, SSPs' quality of service might be somewhat inferior to that of FSPs because of multinetwork interconnections or technical imperfections, especially in the case of voice on Internet protocol (VoIP) services. However, at least users can achieve

---

24. Resale of international call service is sometimes called *international simple resale.*

25. This paradigm was developed by such Bell Lab economists as Baumol, Panzer, and Willig, implicitly against the Department of Justice's intentions in the break-up of AT&T during the mid-1970s. For more details on the theory, see Baumol, Panzer, and Willig (1982). Ironically, AT&T was divested into long-distance AT&T and seven Regional Bell Operating Companies in 1984, in accordance with the Modified Final Judgment in 1982.

Table 8.10   Price Changes in Fixed Telephony Services of Korea Telecom (won per 3 minutes)

	Feb. 10, 1993	July 1, 1993	Aug. 1, 1994	Jan. 1, 1995	Dec. 1, 1995	Dec. 1, 1996	Sept. 1, 1997	Feb. 16, 1998
Local	30	—	40	—	—	41.6	45	—
Long distance								
Zone 1	100	30	40	—	—	41.6	45	—
Zone 2	360	360	200	—	—	183	172	—
Zone 3	675	675	313	—	—	277	245	—
International[a]	−7%	—	—	−5%	−7%	−15%	−12%	14%

Source: Korea Telecom.
[a]Rate of change in average price.

their general purposes of communicating with foreign residents no matter which services they may use. Even if quality matters, choices between tariff and quality are up to the users. This provides a foothold for SSPs to make inroads into the FSPs' realm. Second, SSPs' frictionless entry and exit are guaranteed institutionally. SSPs can easily enter the market through registration that meets the minimum requirements on financial capability and technical personnel. In addition, SSPs can freely exit simply if the customers' rights are protected, whereas FSPs need authorization for both entry and exit. Third, SSPs sometimes target specific groups of customers through the marketing of pre- or postpaid cards, or by focusing only on those countries with plenty of traffic trading (i.e., those that have no obligation toward ubiquitous services such as FSPs).

Finally, SSPs have a price advantage over FSPs partly due to SSPs' capacity for cost reduction, and partly because of the downward rigidity of FSPs' charges. The main drivers for the cost advantage are twofold. First, SSPs incur relatively negligible sunk costs compared to FSPs because they lease the dedicated lines and use the FSPs' networks.[26] Second, they are able to minimize settlement payments to foreign partners by forwarding traffic via the lease cost routes and thus bypassing the existing international transmission facilities used by FSPs.

Reasons for the stickiness of FSPs' charges could be found in both the prior-tariff-approval system and the FSPs' collusive behavior. KT's collection charges—that is, international tariffs—were lowered several times between 1993 and 1997 under the government's prior-approval system (see table 8.10); this was done with a view toward addressing the unfair practice of cross-subsidizing the deficits of the local telephony sector, in which new entrants' charges had been set below KT's with fixed proportions.

This leader-follower behavior in price differentials, in spite of its abolition in 1995, was maintained even when collection charges for major inter-

26. According to the legal definition, SSPs cannot possess their own facilities.

Table 8.11   **Collection Charges of Korea Telecom and SK Telink as of April 2000 (won)**

| | Korea Telecom | | | | SK TelLink | |
| | Standard | | Discount | | | |
	First Minute	Additional Minutes	First Minute	Additional Minutes	Standard	Discount
United States	14.0	10.5	9.8	7.4	4.8	4.4
China	16.4	12.3	11.5	8.6	6.5	5.9
Japan	24.8	18.6	17.4	13.1	13.0	11.7

Source: Korea Telecom, SK Telink.

national settlement-deficit countries (including, e.g., China) were raised in February 1998 due to ever-worsening traffic imbalances and the devaluation of the exchange rate.

Table 8.11 shows the collection charges of both KT and the leading SSP, SK Telink (an affiliate of the leading mobile FSP, SK Telecom) to the three major countries—the United States, China, and Japan—having the largest volume of traffic with Korea.[27] Note that SK Telink's average charge per minute for these countries is 39 percent lower than that of KT.[28]

Market Performance. At first glance the theory appears to go off the mark, because no tariff reductions of FSPs have occurred in a pure sense since resellers first entered the market. However, evaluation of market performance should occur in the context of enhancing the overall welfare via the widening varieties of choice and demand substitutions. Backed up by price advantages, SSPs have expanded their market shares at extraordinarily high rates and have made inroads on the FSPs' market, as evidenced by the trends in outgoing traffic and revenues.

The outgoing traffic for FSPs has grown annually at the average rate of 18 percent from 1995 to 1998. The volume of outgoing traffic peaked in 1997 with 901 million minutes as a result of the boost in economic growth and increases in trade, but it took a downward turn in 1998. Although there was a decline in the FSPs' traffic between the first half of 1998 and 1999, the SSPs' traffic continued to increase, showing definite evidence that users were switching over from FSPs to SSPs. As of 1999, outgoing traffic during the first half of the year accounted for 430 million minutes, whereas the

27. Note that both providers' charges are billed per second on a usage base and discounted for off-peak hours, but that KT's charges differentiate between the first minute and additional minutes.

28. Average charge per minute for each country is calculated as follows: (a) average standard and discount charges, based on four minutes of use, are calculated separately, and (b) they are weighted by standard and discount hours. Then, each country's average charge is weighted by the corresponding share of outgoing KT traffic in 1999.

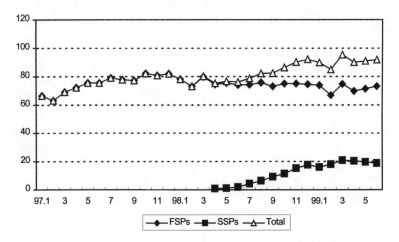

Fig. 8.4   Outgoing traffic of FSPs and SSPs (millions of minutes)
Sources: Evaluation of Competitiveness in Facility-based Telecommunication Services Market (1999, 2000), the Korea Information Society Development Institute, and international FSPs

Table 8.12   Revenues of FSPs and SSPs

Revenue	1996	1997	1998	1999
FSPs (%)	753.0	658.0	593.1	580.5
			(93.3)	(81.7)
SSPs (%)	—	—	42.6	130.0
			(6.7)	(18.3)
Total (billions of won)	753.0	658.0	635.7	710.5

Source: Computer and Communication Promotion Association of Korea.
Note: Numbers in parentheses indicate revenue shares between FSPs and SSPs.

traffic share for SSPs amounted to 21 percent, compared to 0.95 percent in the first half of 1998 (see fig. 8.4).

Regarding revenues (see table 8.12), FSPs recorded 753.0 billion won in 1996, but since then the revenue had been reduced to around 82 percent annually, on average. In contrast, revenues for SSPs have increased from 42.6 billion won in 1998 to 130.0 billion won in 1999. The total decreased in 1998 but by 1999 had fully recuperated above its 1997 level, thanks to the recovery of the economy and market liberalization. In addition, the SSPs' market share had increased from 6.7 percent in 1998 to 18.3 percent in 1999. The net benefits of customers' switching to SSPs from the time of the start-up of resale to June 1999 are estimated to be worth 52.46 billion won, on the assumption that there is a complete substitution of the FSPs' market by SSPs.[29]

29. Average charges per minute of KT and SK TelLink are used as proxies for FSPs and SSPs, respectively, and calculated in the same way as discussed in note 27.

In response to SSPs' elongation, international FSPs have participated in the international simple resale business and have taken on strategies that differentiate classes of customers by introducing new discount-option pricing packages composed of flat-rate and usage-based charges to retain heavy users.

In sum, the introduction of voice resale services has facilitated the contestability of international telecommunication services, and consumers have become the biggest beneficiaries.

Bottleneck in Further Deployment. FSPs' networks are indispensable facilities for SSPs to provide service to an unspecified number of the general public. Most SSPs have provided services mainly through mobile FSPs' networks; KT's PSTN is rarely used. The reason for this can be found in KT's peculiar interconnection arrangement. Whereas interconnection arrangements between FSPs are under the control of the government, those between FSPs and SSPs are left to the parties' own voluntary negotiations. KT had taken a stance of treating SSPs as customers rather than providers and charged user fees higher than interconnection charges available to FSPs with no service of billing. This arrangement inevitably raises costs for SSPs and thus causes a serious bottleneck for FSPs' access to the fixed-line subscribers of KT and for market expansion, although there might be pros and cons in facility-based versus service-based competitions.

Competition in the Incoming-Traffic Market:
Dismantling the International Settlement Regime

Existing International Settlement Regime. Although competition in the domestic outgoing-traffic market is a matter of marketing among service providers, the generation of incoming traffic is irrelevant to domestic competition. Nevertheless, service providers compete with each other to get as much incoming traffic as possible because it can yield settlement revenue as remuneration for forwarding calls to the domestic receiving party and for providing facilities.[30]

Although SSPs have been free to arrange settlement methods, a stereotyped international settlement arrangement has recently been applied to FSPs. An overview of the regime is as follows. International FSPs pay settlement rates to their foreign-counterpart FSPs for utilizing their networks when forwarding calls to the receiving party. FSPs first negotiate the accounting rate and then usually split it evenly to determine the settlement rate, which requires outgoing-traffic-excessive FSPs to settle payments with incoming-traffic-excessive FSPs. However, this system has caused a conflict of interest between developed countries, whose outgoing traffic dominates

---

30. International FSPs regard a deficit in international settlement as a cost driver. Usually, collection charges for deficit countries are higher than those for surplus countries.

Table 8.13   ITU's Recommendatino (by year-end 2001)ᵃ

Teledensity T ≤ 1 (A)	1 < T ≤ 5 (B)	5 < T ≤ 10 (C)	10 < T ≤ 20 (D)	20 < T ≤35 (E)	35 < T ≤50 (F)	T > 50 (G)
0.327 SDR	0.251 SDR	0.210 SDR	0.162 SDR	0.118 SDR	0.088 SDR	0.043 SDR

Source: ITU.
ᵃ1 SDR is equivalent to $1.320999, as of June 2000.

Table 8.14   FCC's Benchmarks and Transitional Period

	Low Income (<$726)	Lower Middle Income ($726 to 2,895)	Upper Middle Income ($2,896 to 8,995)	High Income ($8,956 or more)
Teledensity <1				
$0.23 January 2003	$0.23 January 2002	$0.19 January 2001	$0.19 January 2000	$0.15 January 1999

Source: FCC.

incoming traffic, and developing countries, which are in the opposite situation. The developed countries have urged developing countries to reduce existing accounting rates based on costs, in order to hold down a burden of deficits. Yet developing countries have tenaciously resisted such a proposal because the settlement surplus is a major financial resource for expanding infrastructures and subsidizing universal service.

The possible remedies for this malfunctioning have developed in two directions. One is to commit binding benchmarking to heuristically accelerate the realization of cost-oriented settlement rates, and the other is to create an environment receptive of such alternative means as international simple resale to work as a source of market pressure against the ancient regime. The International Telecommunications Union (ITU), on a multilateral basis, and the United States' Federal Communications Commission (FCC) have independently promoted the first of these directions and issued benchmarks for settlement rates in 1997 and 1999, respectively (see tables 8.13 and 8.14). Korea belongs to class F and the upper-middle-income group, respectively.[31]

WTO negotiations have played a major role in implementing the other direction. The liberalization of resale markets of WTO member countries should be interpreted in this context.

When the presence of SSPs was negligible or resale itself was prohibited, the main focus was on how to protect new domestic entrants from the whip-

---

31. Note that the FCC's benchmarks are those for U.S. partner countries only. Korea extended one year for the fulfillment of the benchmark.

sawing of foreign monopolists in international settlements with foreign FSPs. In most countries, the government or regulatory body has arranged competitive safeguards against whipsawing through the *uniformity of accounting rate*, which sets the rate level equally across domestic FSPs, and for the *proportionate return principle*, which distributes incoming traffic among domestic FSPs in proportion to outgoing-traffic ratios.

Converging to Laissez-faire. As incoming traffic for SSPs has increased, regulatory bodies have become focused on preventing the so-called one-way bypass, an unfair practice in which FSPs reroute calls via foreign SSPs in order to reduce international settlement expenditures to counterparts. Countries such as the United States, Japan, and Korea have reinforced their monitoring functions.

Contests for incoming traffic have been started on the spot market among SSPs in Korea, resulting in declines in interconnection charges.[32] Their total traffic has substituted for part of would-be FSP traffic, distressing the FSPs' settlement revenue despite participating in the SSPs' competition games.

In mid-1999, the Korean government relaxed the existing regulation for FSPs to negotiate a settlement arrangement freely with developed countries in which markets are so competitive that, to a certain degree there is no fear of whipsawing. Currently, in Korea, there are ongoing discussions concerning issues such as the further deregulation of the FSPs' settlement regime and the shift of KT's access arrangement with SSPs to the interconnection regime, striking a balance with measures for FSPs. Figure 8.5 shows the competition between SSPs and FSPs in the international service market.

8.4.2   Mitigation of Foreign Ownership and Growth of the Mobile Services Market

Korea's rapid economic growth over the last two decades has been achieved mostly by the export-centric industrial policy, pushed by the Korean government (see Lee and Lim 1999). The government financed prime-interest-rate loans to target industries through government-controlled financial institutions so as to allocate scarce financial resources effectively, which resulted in the birth of large conglomerates—so-called *chaebol.* Owner-managers and families of *chaebol* exercised exclusive power in managerial decision making as well as controlling subsidiaries. However, as the stock market became an alternative source of external financing for *chaebol*, these owners adhered to the right of management through family ownership of equities and cross-shareholdings among subsidiaries. After all, ownership and management of *chaebol* did not remain separated, and as a

---

32. This is exactly the case in which the market function works, but it also implies losses of international settlement revenue.

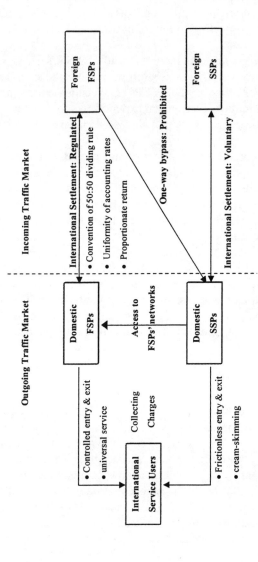

Fig. 8.5 Competition between SSPs and FSPs in international service market

result, managerial efficiency could not be checked through the external stock market or internal institutions. In a sense, the concentration of capital and ownership provided momentum at a time when the Korean economy was about to take off. As economy grew, however, this peculiar corporate governance system in Korea blocked the monitoring function of the capital market and was regarded as one of the major factors that led the economy into the financial crisis between 1997 and 1998. Since then, efforts at regulatory reforms have been to make corporate information and accountability transparent; to strengthen internal monitoring systems such as boards of directors and market disciplines; to mitigate foreign ownership; and so forth. Such measures are expected to accelerate the metamorphosis of the *chaebol's* management system, which in turn will materialize as an effectively competitive structure and promote economic growth.

Until recently, foreign ownership has long been restrained by the government because of the concern that foreign ownership might not only erode the domestic capital market but also deprive domestic corporations of the right of management. Relics of protectionism, however, have been replaced by the more positive notion that foreign ownership could play a significant role in diluting the concentrated ownership of *chaebol* and in securing funding resources. The latter has been highlighted during the period of the financial crisis when the Korean economy was in need of investment funding and foreign currency for recovery. Figure 8.6 summarizes the main points in the mitigation of limitations on foreign ownership.

The telecommunication services market moved ahead of other heavy industries in liberalization and opening even before the onset of financial crisis. As explained in section 8.2, deregulation of foreign ownership kept face with trade negotiations on telecommunication services, including an autonomous lift. Foreign investors, mainly the major telecommunication service providers of developed countries with abundant experience and skills at home focused mostly on the ever-flourishing mobile-service market in Korea, where the three FSPs are subsidiaries of *chaebol*.

At the second stage of the regulatory reform initiated by the UR, there were two cellular FSPs: one was SK Telecom, and the other, Shinsegi Telecom. SK Telecom, starting up as an affiliate of KT (then, Korea Mobile Telecommunication) in 1984, was sold to the SK Group, now one of the biggest four *chaebol*, in 1994, whereas POSCO, the public iron and steel corporation, was a major stakeholder of Shinsegi Telecom.

As of 1997, foreign investors began to take serious interest in Korean telecommunications companies. SBC and AirTouch owned about 18 percent of Shinsegi Telecom's stocks, whereas TEI possessed 6.5 percent of SK Telecom's shares. Although it appears that TEI aimed at realizing stock dividends rather than at participating in management, Shinsegi Telecom at least could raise funds from outside to compete with SK Telecom.

The mitigation of foreign ownership limitations was more influential at

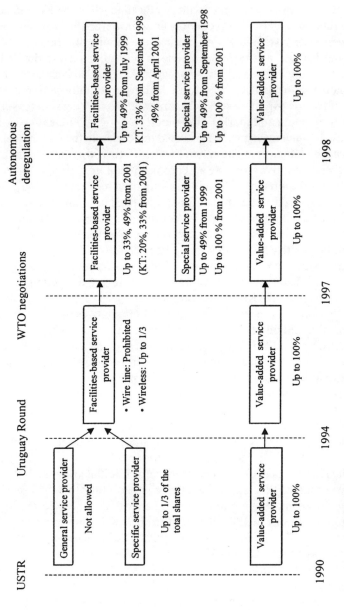

Fig. 8.6 Mitigation of foreign ownership limitations

the third stage of the implementation of the WTO agreement on basic tele-communications. The two late entrants into the FSP market, LG Telecom and Hansol M.Com, affiliates of *chaebol*, have operated their businesses in earnest since early 1998 and have been in need of financial resources to extend coverage and to attract customers to catch up with the front-runners. To make matters worse, this period overlapped with the financial crisis. Under the circumstances, foreign investment provided a source of external funding to PCS providers in return for the concession of some rights to management. This was an opportunity not only to improve their corporate governance system (e.g., through rational decision making on investments) but also to develop the mobile services market, as introduced in section 8.2. As of 1998, BT became a major shareholder of LG Telecom with 24 percent of its stock, whereas BCI and AIG invested in Hansol M.Com and own 9.8 and 6.5 percent of their equities, respectively. Table 8.15 summarizes the foreign ownership structures discussed here.

Foreign investment in the telecommunication markets also had a positive effect on the growth of other industries, in particular, the equipment indus-try. Table 8.16 shows the amount of investment that mobile FSPs spent on

Table 8.15   Foreign Ownership Structure of Mobile FSPs and Korea Telecom

	Major Foreign Investors		
FSP	1997	1998	1999
Korea Telecom	—	—	Foreign DR (14.4%)
SK Telecom	TEI Fund (6.5%)	TEI Fund (6.5%), City Bank ADR (10.31%)	TEI Fund (4.4%), ADR (20.3%)
Shinsegi Telecom	Airtouch (10.7%), SBC (7.8%)	AirTouch (10.6%) SBC (7.8%)	AirTouch (11.4%), SBC (6.5%)
KT Freetel	Motorolar (2.8%)	Motorolar (2.7%)	Microsoft, etc. (7.05%), Motorolar (1.95%)
LG Telecom	—	BT (23.5%)	BT (24.1%)
Hansol M.Com	—	BCI (9.8%), AIG (6.5%)	BCI (20.97%), AIG (13.98%)

Sources: FSPs.

Table 8.16   Mobile FSPs' Investment (billions of won)

	1998	1999
SK Telecom	865	1,460
Shinsegi Telecom	364	483
KT Freetel	511	689
LG Telecom	594	443
Hansol M.Com	592	645
Total	2,926	3,720

Source: Mobile FSPs.

the expansion of networks, including the procurement of switches and base stations.

As of 1998, their investment reached 2,926 billion won in total and increased to 21 percent (3,720 won) in 1999. The far-reaching effect of foreign investment on the value-added production of related industries can be roughly estimated by using table 8.16 with interindustry and fixed capital formation tables based on an input-output analysis issued by the Bank of Korea (BOK).[33] The estimated values of foreign investments are 417 billion won in 1998 and 713 billion won the following year.

The mitigation of foreign ownership limitations also provided a good opportunity for the government to issue foreign DR, and hence, to accelerate the separation of policy and management functions of the government.

8.5 Concluding Remarks

Regulatory reform of Korea's telecommunication services market has been gradually pushed through a series of trade negotiations in two directions: one is to lower entry barriers such as classification systems and licensing processes, and the other is to lift foreign ownership limitations. The trade negotiations that had the most impact on the direction of the telecommunications market were the WTO negotiations on basic telecommunications. Korea introduced voice resale services and mitigated foreign ownership limitations, which were autonomously lifted further in order to attract foreign capital during the economic slump. The inflow of foreign capital had a positive impact on international and mobile service markets. SSPs made inroads primarily on high-profit sectors such as international service, which accounted for about 18 percent of the total minutes of outgoing traffic after less than a year since the beginning of business. As a result, competition in the incoming-traffic market among SSPs and FSPs pushed for decreases in accounting rates and reformed the existing domestic-international settlement regime. However, issues surrounding the deregulation of SSPs' access to fixed networks and how to strike a balance in the international settlement arrangement with FSPs are under discussion, and are expected to materialize within year 2000.

The Korean government in a sense took trade negotiations as an opportunity to effectively manage domestic pressure against market opening. The market regulatory reform provoked conflicts of interest between incumbent

---

33. First, the industry-specific amount of foreign investment is calculated on the assumption that (a) shares of foreign investment in total investment of mobile FSPs are proportionate with shares of equities and (b) proportions of mobile FSPs' investment distributed across related industries are equivalent to those of telecommunications in general and deflated by the constant wholesale price index in 1995. Second, the industry-specific amount of value-added is obtained by multiplying the ratio of the value-added I-O coefficient and the industry-specific amount of foreign investment in real terms, which is summed over related industries. For more details on the analytic scheme, see Hong (1999).

FSPs and late-coming SSPs in the international-services market. However, the ultimate beneficiaries of market liberalization were the consumers, in that they received tariff reductions and a wider scope of choices.

Foreign participation not only provided financial resources to late mobile-FSP entrants, but also contributed to rebuilding their inefficient corporate governance systems. As a result, there was unprecedented growth in the mobile-services market, which influenced the equipment industry in turn.

It might not be easy to provide a clear-cut decomposition of Korea's liberalization regarding the motivations and decisions within Korea, the WTO agreement, and the financial crisis. In the context of market access, WTO negotiations in the UR and Group on Basic Telecom were crucial to the process of Korea's market liberalization. In a sense, however, liberalization could be driven by the government's firm will that competition eventually would be beneficial to consumer welfare. One thorny issue was the political consideration of national sovereignty over telecom networks, as is often the case in developing countries. The phased-in opening of Korea's telecom market could be interpreted as a compromise between economic and political considerations. The financial crisis to some extent played a role, in the sense that Korea implemented its scheduled WTO commitments on foreign ownership eighteen months earlier than originally planned.

In regard to regulatory principles—the so-called, Reference Paper—Korea had no problem making commitments to the WTO because its regulatory framework has undergone progressive reform since 1990. In addition, it was by Korea's own efforts that improvements were made to the regulatory framework.

At an initial stage, foreign entry depends on each country's legal system of licensing processes and equity ownership. Once *de facto* entry is accomplished through either sufficient capital investment or the establishment of on-the-spot corporations, the performance of market liberalization and openness depends on two factors. One belongs to the commercial sphere: the availability of funds necessary for facility investment, market prospects, and the climate of business transactions. The other factor belongs to the solidity of competitive safeguards to guarantee fair competition among service providers, including foreign new entrants: rights of way, access to sea-cable landing stations, interconnection, local loop unbundling with collocation, universal service, and so forth.

While the former factor is a matter of the strategies by service providers, the latter is that of the regulatory regime, which in turn implies that issues of competitive safeguards are no longer confined to the domestic area. Although trade negotiations in the past focused mainly on pulling down foreign entry barriers, those in the future are expected to go forward to details of competitive safeguards as well as further mitigation of entry barriers. In particular, further works on competitive safeguards would have to set forth how these principles should be applied in practice. The reason is that competitive safeguards designed in the previous multilateral trade negotiations are understandably high-level and general enough to accommodate the broad range of different political and legal frameworks among various countries. However, multilateral forums such as the New Round negotiations might not be an appropriate venue for tentatively developing a more complicated set of principles on competitive safeguards. Apparently, an implicit consensus on their role in the previous WTO negotiations has been to ensure a target outcome—i.e., to level the playing field in the telecommunication services market, *not* to provide specific tools whereby those outcomes would be effectively achieved. It would be desirable that bilateral or nonbinding multilateral setting are dedicated over a certain period of time for the sake of building up an understanding and necessity of furthering sufficiently detailed guidelines for competitive safeguards. A better and plausible choice for the New Round negotiations in the telecommunication-services sector is to outline an exemplary list of good and bad practices for each regulatory principle already entered into agreement.

References

Baumol, W. J., J. C. Panzer, and R. D. Willig. 1982. *Contestable markets and the theory of industry structure*. New York: Harcourt Brace Jovanovich.

Federal Communications Commission (FCC). 1997. *Report and order in the matter of international settlement rates*. Washington, D.C.

Hong, D. 1999. An input-output analysis of telecommunication services market investment and its impact on Korean economy (in Korean). *The Korean Journal of Information Society* 11:43–55. The Korea Information Society Development Institute.

KISDI (Korea Information Society Development Institiute). 1999. *Evaluation of competitiveness in facility-based telecommunication services market*. Kwachun.

———. 2000. Evaluation of competitiveness in facility-based telecommunication services market. Kwachun.

———. 1999. *Evaluation of competitiveness in facility-based telecommunication services market*.

Lee, N. 1998. Competition between facility-based service providers and resellers: Are games win-win or zero-sum? (in Korean). *Issues on Information and Communications Policy*. Seoul: The Korea Information Society Development Institute, In-Sung Co Hure.

———. 1999. The advent of new interconnection paradigm: Reciprocal interconnection and interaction between fixed and mobile operators. Paper presented at the Workshop on Interconnection Policies and Framework, OECD/Centro Studi San Salvador. 11–12 November, Venice, Italy.

Lee, Y., and Y. Lim. 1999. Korea's corporate governance: Issues and reforms. Policy Study 99–01. the Korea Development Institute. Seoul: Korea Computer Industry Inc.

Ministry of Information and Communication. 1999. 1999 White Paper. Ministry of Information Communication. Seoul.

Organization for Economic Cooperation and Development (OECD). 2000. *OECD review of regulatory reform: Regulatory reform in Korea.* Paris: Organization for Economic Cooperation and Development.

Sherman, Laura B. 1998. Wildly enthusiastic about the first multilateral agreement on trade in telecommunications services. *Federal Communications Law Journal* 51 (1): 61–110.

Comment   Ramonette B. Serafica

Lee and Lie provide an interesting and insightful discussion of the various stages of market restructuring that Korea's telecom sector has undergone. This is spiced up with a fascinating account of the changing trends in calling patterns or habits as well as of the rise and death of the different services.

The paper not only presents a good discussion of the transformation of the market but also gives concrete evidence of the benefits arising from the more liberal rules on market entry and foreign ownership that resulted from the trade negotiations.

The gains identified in the paper are consistent with what free-market advocates envision. Along with tariff reductions there is now a wider scope of choices available to the consumers along with innovative services, pricing schemes, and technologies.

Another identified gain deals with the role of foreign participation. The fundamental role of the telecommunications sector in building the competitiveness of other sectors of the economy is widely recognized. It is also a prerequisite for international trade in services. However, building the necessary infrastructure and gaining the needed expertise requires a lot of resources, and thus what is at stake in these trade negotiations is really the capacity for most countries to mobilize the investment needed.

For the Korean telecom market, foreign participation addressed the need for additional financial resources particularly at a time when domestic resources were tight. With the aid of foreign investment, latecomers in the mobile market, for example, were able to catch up and attract new customers by expanding their service coverage. Moreover, the paper mentions that foreign participation has also led to the reform of insufficient corporate governance systems, although specifics with respect to telecommunications firms (e.g., by way of anecdotal evidence) were not given.

Finally, reforms in the sector also created positive by-products as the brisk market activity in telecom services spurred growth in the upstream equipment industry.

In total, the authors weave an interesting and compelling story of the liberalization of Korea's telecom services market and of the gains that it has

Ramonette B. Serafica is associate professor of economics at De La Salle University, Manila.

generated. The story, however, focuses on market access concerns, treating an important part of regulatory reform that has to do with procompetitive regulation in less detail. This aspect is critical, especially for the telecom sector, because of its history of monopoly regulation.

Using the traditional Industrial Organization (IO) paradigm, the paper's emphasis is on reforms in market structure and effects on market performance but leaves out the part on market conduct or the rules that govern it, which also affect market outcomes.

The need to have both for successful liberalization is evident in the World Trade Organization (WTO) Agreement on Basic Telecommunications Services, which in fact has two main parts: one is a schedule of commitments with respect to market access and the other is a set of procompetitive regulatory principles.

Individual countries provided detailed schedules on the telecom services to be included, on the scope of commitments, and on the degree of permitted market access. Another significant outcome of the negotiations, however, was the adoption of a procompetitive set of regulatory principles (known as the Reference Paper) covering a range of issues, including anticompetitive practices, interconnection with the major supplier, and regulatory independence and transparency.

A discussion of asymmetric regulation in favor of new players, whether currently in place in Korea or not, and to what extent these have been introduced in conjunction with market access reforms, will be very useful for guiding regulation elsewhere. As an outsider, I would be interested in knowing how the market was opened, to how many players, and to whom; but equally important to understand and appreciate are the rules imposed or relaxed to assist new players. Other interesting questions about aspects of the reform process can be asked as well. What sorts of problems are the new entrants facing vis-à-vis the dominant incumbent? Which among these problems or disadvantages are structural in nature and which are behavioral? Finally, how has the government responded—either proactively or reactively—to ensure that market access reforms are complemented with regulations that assist new entrants?

The need to have both market access and a procompetitive regulatory environment for successful liberalization cannot be stressed enough.

A case in point is the Philippines (see Serafica 2000). Market liberalization in telecoms was introduced in the country before the WTO negotiations on market access for basic telecommunications. Even after the WTO, however, adherence to procompetitive principles remains weak. The previous government effectively demonopolized the industry and competition was introduced even at the local exchange level as early as 1993. Right now we have about ten major facilities-based operators, although industry consolidation is taking place. Thus, by the time of the WTO negotiations on basic telecommunications, the Philippines' response to the negotiations was a schedule of specific commitments based on preexisting policies.

The foreign-ownership limit was set at 40 percent because the Philippines' constitution sets a 40 percent limit on foreign equity in public utilities. Market access was restricted to the provision of services on a facilities basis, with no resale of leased lines, and call-back schemes were not permitted in accordance with our Telecommunications Policy Act of 1995. Although the Philippine commitment acknowledges the objectives of liberalization as embodied in the WTO agreement in broad terms, specific support mechanisms are still lacking.

Sadly, nearly a decade after market liberalization, we still do not have procompetitive rules (e.g., with respect to access to essential facilities of the dominant operator—or major supplier, in WTO parlance).

As we have experienced, lack of support for new entrants in the post-"liberalization" setting threatens the very survival of competition. More-over, it leads to wasteful duplication of facilities or investments made by both the carriers and subscribers, as well as to a lower grade of service as subscribers get caught in the games played by the operators. As such, despite the fact that Filipinos, like the Koreans, have more choices and to some extent lower prices, the full benefits of liberalization have yet to be enjoyed.

Thus, I think policy researchers and other readers can benefit from a balanced discussion or treatment of the reforms in market access and similar adjustments with respect to the procompetitive regulation that Korea's telecom sector has experienced or will need as part of the overall regulatory process. For in the long run, it is the adherence to procompetitive regulation that will determine whether the story of market liberalization in telecoms will end on a sad or happy note.

Reference

Serafica, Ramonette B. 2000. Competition in Philippine telecommunications: A survey of the critical issues. Philippine APEC Study Network Discussion Paper no. DP 2000-15. Makati City: Philippine APEC Study Network.

Comment   Chayun Tantivasadakarn

This comment concentrates on four main points:

1. General comments
2. The lesson learned from Korea's telecommunication reform
3. The possibility of using Korea's experience as a model to facilitate reforms in other developing countries
4. Clarifications and suggestions

Chayun Tantivasadakarn is assistant professor of economics at Thammasat University.

General Comments

In general, this is a well-written paper. It provides readers with not only a systematic chronicle of Korea's telecommunications policy reform but also gives clear explanatory factors behind the change of policy in each stage of reform. The paper also tries to assess the impact of telecommunications on the market structure, industry performance, and foreign ownership. The arguments presented in the paper are well supported with statistics, diagrams, and systematic explanations.

Korea's Lessons

The paper shows that telecommunications reform in Korea has been stimulated by a series of trade negotiations beginning in 1989, with pressure from the United States, then from the Uruguay Round. This, in turn, was followed by World Trade Organization (WTO) negotiations on basic telecommunications. The Asian financial crisis in 1997 generated the final wave of policy changes.

The Korean government took these pressures and turned them into opportunities to maintain the direction of the reform and to keep the momentum going. The reform strategies followed a set of instructions:

1. *At the early stage of reform, open up only the sector that is really pressured by the trade partner.* For instance, the value-added services were opened first in the first stage reform.

2. *Introduce competition to the local service providers gradually.* This process began with an introduction of the second service provider. The government then used the managed competition policy to maintain the excessive profits for the service providers. This policy definitely generated negative impacts on consumer welfare; however, the policy probably facilitated fund raising for further investment of the incumbent firm. The managed competition was maintained throughout the first and second stages.

3. *Introduce full competition later.* For instance, in 1997 the government issued licenses to twenty-seven new service providers. The policy was also intended to saturate the market with as many local service providers as possible before the market was opened to international competition.

4. *Complement the third strategy by utilizing voice-resale services that serve as a tool to stimulate the contestability of international telecommunication services.*

5. *Relax foreign-ownership limitations gradually in the sectors deemed sensitive.* For instance, foreign ownership in the wired line services was not allowed in the first and second stages of reform; however, foreign ownership was increased to 33 percent in the third stage and to 49 percent after 1999.

Can Other Developing Countries Copy the Korean Model?

First, Korea's entire telecommunications reform took more than ten years after a very good start. The liberalization process had begun quite early: in 1989, after the basic need for wired lines had been satisfied. The local monopolist was exposed to competition gradually—initially to competition from private domestic providers, and then to that from foreign firms. This long period of adjustment allowed sufficient time for the incumbent public enterprises and private domestic providers to build up strength to compete with more efficient foreign operators.

Unfortunately, all lower-middle-income countries are still struggling just to satisfy their wired line needs. According to International Telecommunication Union (ITU; 1999) statistics, the average teledensity rate in 1998 for lower income countries was only 8.18 lines per 100 inhabitants. Besides, these countries still have other socioeconomic and political problems that may have higher priority on their government agenda.

Second, the strong will and serious actions of the Korean government were the major driving factors behind the readiness of its domestic service providers for international competition. Such governmental behavior is rare among the developing countries.

As a result, the answer to "Can other countries copy the Korean model?" is probably no. With the pressures of the WTO negotiations on basic telecommunications and the upcoming New Round, the developing countries do not seem to have the necessary conditions and sufficient time to adjust.

Clarification and Suggestions for the Paper

First, it is unclear which government institutions are actually responsible for the determination to carry out the reform policy. Some description of these institutions and their functions should be added to the paper. This will provide very valuable lessons for other countries.

Second, the author has mentioned that in 1995 (in the second stage of reform) the government was considering the introduction of a price-cap system. Economists consider the price cap to be a system that improves price flexibility in an increasingly competitive market while protecting against cross-subsidization, monopoly, and predatory pricing (see Mitchell and Vogelsang 1991). The price-cap system is a very efficient way to ensure that consumers will benefit from the technological progress and the rent from natural monopoly. However, it is unclear in the paper whether it has been implemented, and if it has been, what the result was.

Third, it is possible that the reduction in revenue from the international calls mentioned in the paper was partially caused by new technology, such as the Internet phone, which allows international calls from personal computers (PCs) to other PCs or from PCs to normal telephones at very low

Table 8C.1   Comparative Advantage Index of the Telecommunication Sector:
                    Selected Asia-Pacific Countries

	Relative Main Lines per Employee		Relative Revenue per Employee	
	1991	1996	1991	1996
Malaysia	0.86	1.05	0.72	0.91
The Philippines	0.50	0.75	0.66	0.59
Thailand	0.85	0.98	1.00	0.64
All lower income	0.28	0.61	0.18	0.28
Hong Kong	2.34	0.65	2.93	1.31
South Korea	3.43	2.25	2.01	1.29
Singapore	1.53	1.71	2.58	3.92
Upper income	2.82	2.60	2.09	1.44
Australia	—	—	—	1.65
Japan	2.88	2.11	3.52	4.16
Developed	2.48	1.83	3.06	3.49
Asia-Pacific	1.00	1.00	1.00	1.00

Source: Reprinted from Tantivasadakam (1999).

Notes: Calculated from ITU (1999) data. Relative main lines per employee of country i = main lines per employee of country i divided by the main lines per employee of the Asia-Pacific region. Relative revenue per employee of country i = revenue per employee of country i divided by the revenue per employee of the Asia-Pacific region.

costs or even for free.[1] In fact, this technology will open up an entirely new area of competition to the local service providers.

Fourth, as we know, free trade will generate gains from trade. Some well-known sources of gains are gains from comparative advantage, procompetitive gains, gains from product and input varieties, and gains from average cost reductions due to economies of scale (see Markusen et al. 1995). The paper has shown to some extent the evidence of procompetitive gains due to international competition and gains from the increase of product and input varieties. It might be interesting to explore whether there is any evidence of gains from reduction of average costs because the technology in this sector is generally subject to economies of scale.

Fifth, setting aside the overall benefits of free trade for a country, free trade never guarantees that everyone in the country must also gain. Hence some form of compensation to the loser is needed. In the case of Korea's telecommunication, Korea Telecom (KT) seems to be the loser, at least at the beginning of the reform process. It would be very interesting to know how KT has been affected by the reform and how Korean government has managed the resistance and political pressure.

Finally, International comparison of the Korean comparative advantage

---

1. For example, [http://www.dialpad.com] provides such free international calls for any destination in the United States.

might be something the author should consider to incorporate in the paper. One possible index might be the relative main lines per employee and relative revenue per employee, as in table 8C.1.

The indexes in the table capture the ability of one country's telecommunications employees to handle telecommunications facilities and generate revenue relative to the average ability of the region. The table shows that Korea's indexes for the relative main lines per employee were one of the best in both 1991 and 1996. Similarly, its indexes for the relative revenue per employee were quite high for the same period. More details and similar indexes for each telecommunications facility should be calculated for the periods prior to and after the start of telecommunications reform. The index then can be used to show the impact of the reform on the comparative advantage of Korea's telecommunications sector.

References

International Telecommunication Union (ITU). 1999. *Asia-Pacific telecommunication indicators.* Geneva, Switzerland: ITU.

Markusen, James R., James R. Melvin, William H. Kaempfer, and Keith E. Maskus. 1995. *International trade: Theory and evidence.* New York: McGraw-Hill.

Mitchell, Bridger M., and Ingo Vogelsang. 1991. *Telecommunications pricing: Theory and practice.* A RAND research study supported by WIK. Cambridge: Cambridge University Press.

Tantivasadakarn, Chayun. 1999. Service in the recovery process. Paper presented at the Thailand Update Conference, *Thailand beyond the Crisis.* 21 April, Australian National University, Canberra, Australia.

9

Korea's Liberalization of Financial Services Trade

Sang In Hwang, Inseok Shin, and Jungho Yoo

The financial sector plays a key role in the functioning of a market economy. This was dramatically illustrated by the Asian crisis since 1997. The financial services trade interests us as a subject of international trade and also for its implications on the development of the financial sector.

This paper is about the liberalization of financial services trade in Korea. Following this introduction, section 9.1 briefly reviews the benefits, costs, and risks of the liberalization of financial services trade, as they appear in the literature, and section 9.2 discusses what it takes to realize the benefits of trade in developing countries. Section 9.3 describes Korea's liberalization measures in financial services trade before and after the financial crisis. Section 9.4 discusses the trends in financial services trade in Korea and tries to see what benefits there were from the limited liberalization before the crisis, and section 9.5 concludes the paper.

Among various financial services, the main focus of the paper is on banking services, because banks occupy the central position in Korea's financial sector. Also, two main topics of discussion are two modes of service trade, namely, cross-border and commercial presence, because the other two modes, namely, consumption abroad and the presence of natural persons, are of relatively little importance in the financial services trade.[1]

Sang In Hwang is associate professor at Kangnung National University. Inseok Shin and Jungho Yoo are research fellows at the Korea Development Institute.

1. Mattoo (1999) reports that the financial service trade through commercial presence was two times or more as large as the cross-border trade, whereas the other two modes were insignificant in the case of the United States, the only country that reports trade through commercial presence on a regular basis.

9.1 Benefits and Costs of Liberalization of Financial Services Trade

Banks and other financial service providers offer many different services, such as accepting deposits, lending, underwriting, brokering, and so on, at various rates and fees. In doing so they perform five basic functions, according to Levine (1996): facilitating the exchange of goods and services; facilitating risk management; mobilizing resources; obtaining information, evaluating firms, and allocating capital; and providing corporate control. Obviously, these are critical functions for a market economy, and the externalities involved in the provision of financial services are substantial.

9.1.1 Benefits of Liberalization

Liberalization of financial services trade is supposed to provide certain benefits. Better financial services in terms of cost and variety will be offered to domestic customers across the border and by foreign banks that enter the domestic market. Also, as competition intensifies, the domestic banks will be forced to improve the quality and variety of the services while lowering the costs. Consequently, consumers' welfare will increase, and resources previously employed in the financial sector may shift to more productive employment in other sectors, increasing the economy's output under the international price regime. These benefits closely correspond to the gains from trade in the simple model of international goods trade and will be called in this paper the direct benefits.

Liberalization of financial services trade seems to be advocated as much for indirect benefits as for direct ones. The indirect benefits refer to systemic improvements of the financial sector. They include improvements in the basic functions embodied in the banks' services, especially assessment of borrowers' creditworthiness, risk management, and corporate control. In addition, *foreign banks* are expected to raise the financial sector's standards by which accounts or financial statements are prepared and assets valued, as they follow the home country regulations or international standards (Glaessner and Oks, 1998). The information available about borrowers and, through competition, about financial services providers themselves would also increase in quantity and quality, leading to an improvement of the basic functions.

Also, liberalization of financial services trade can provide an impetus to an improvement in legal and regulatory infrastructure. As financial services trade opens between a developed and a developing country, the banks in both countries would seek to export their services in order to follow and serve their clients. This provides an occasion for the developed country to demand that the developing country's financial regulations and supervision be adequate. The same demand may also come from developing countries' banks, as they seek access to the developed countries' markets. Laws and regulations that have direct bearing on the operations of financial institutions, such as corporate and bankruptcy law, laws regarding negotiable instruments, and the like, may also improve (Levine 1996).

9.1.2   Costs and Risks of Liberalization

Various arguments have been advanced against liberalization of financial services trade. Because few barriers are left in Korea at this moment, as will be explained in section 9.3, some are of little relevance. Examples include the infant industry argument and the argument that too rapid a liberalization could lead to financial distress among domestic financial firms as their profits decline. More relevant would be the argument based on the government's limited ability to properly supervise and monitor a more complex financial system. When combined with the lack of credibility in enforcing prudential regulations and withdrawing (implicit) insurance schemes, the limited ability could encourage banks to take excessive risks at the final expense of the government (Claessens and Glaessner, 1998). This is a risk warranted by the potential for improvements in regulatory infrastructure and strengthening of the government's supervisory capacity.

Other arguments against liberalization seem to be on shaky ground or do not seem strong enough to justify keeping the financial market closed (Musalem, Vittas, and Demirgüç-Kunt 1993). One argument alleges that foreign financial institutions facilitate capital flight out of the host country. However, domestic institutions could play the same role. Another fear is that foreign institutions may drive the local ones out of business and dominate the host country's financial sector. Although this cannot be dismissed as an impossibility, a near or complete domination seems extremely rare. According to Gelb and Sagari (1990), as reported in Levine (1996), foreign banks' median share of total domestic assets in a sample of twenty countries was about 6 percent. A related argument questions the foreign firms' commitment to local markets, worrying about the possibility that they would retreat from the host country in response to difficulties at home or in the local market. Still another argument anticipates difficulty in conducting monetary policy when the presence of foreign institutions is substantial in the financial sector. Claessens and Glaessner (1998) report that in New Zealand, where the financial system is largely in the hands of foreigners, monetary policy is not adversely affected, nor is there evidence of little commitment to the local market.

9.2   Realization of the Benefits

This section discusses what it takes for liberalization to succeed in realizing the benefits. The direct benefits are expected to be realized with little difficulty, as in the goods trade, where the gains from trade arise as the goods are delivered across the border. This would be the case for modes of both cross-border trade and commercial presence.

However, it would not be as easy to obtain the indirect benefits, an improvement of the financial system itself. The reason becomes clear when one recalls the nature of finance. White (1999) says that everywhere finance involves a time-sequencing problem: "Finance always involves an *initial* conveyance of funds—a loan, an investment—and then a *later* reversal of the flow of funds—the loan repayment (plus interest), a stream of dividends, etc." (3; author's emphasis).[2] Even before making a loan, the lender needs to know about the risks and prospects of repayment and, after making the loan, about the borrower's behavior and performance. Information asymmetry exists, however, because the borrower naturally has more information about him- or herself and his or her own work than the lender. If non-repayment is not sanctioned, a moral hazard problem is likely to arise, because the borrower will receive the "upside" benefits from a risky undertaking while not bearing the "downside" costs.

Given these circumstances, financial intermediation evidently requires the basic functions such as information gathering, credit assessment and risk management, and performance monitoring and corporate control. It seems, however, that development of these functions is lacking in many developing countries, including Korea, as the latest Asian financial crises revealed. The lack of development appears to have much to do with the "rules of the game" of a given society. As in North (1990), a society's rules of the game (ROGs) refer to the totality of its formal and informal institutions such as the legal system, government regulations, social customs, and practices. As such, the ROGs refer to the way the written laws and regulations are applied and enforced as well as the laws and regulations themselves.[3] For our discussion of financial services, one important example of the ROGs would be how the decision is actually made regarding who among many applicants gets the bank loans, and another, what kind of penalty the borrower gets when he or she is unable to pay back the loans.

In Korea and some other Asian countries the government extensively and intensively intervened in the workings of the market through formulating and executing development plans, industrial and trade policies, and so-called administrative guidance. In the process, the banks often became the financing instruments of development plans and industrial policies. The government directly or indirectly exercised influence on loan-making decisions. Under the circumstances, the prospects of repayment or performance may not have been a primary concern to the banks. Also, such government involvement usually carried with it an implicit insurance against losses. In this case, banks had little incentive to develop capacities to perform the basic functions. Moreover, banks themselves were prone to moral

---

2. The rest of this paragraph also draws on White 1999.
3. Elsewhere, one of the authors provides a fuller discussion of the rules of the game and the importance in Korean context (Yoo 1998).

hazard, behavior because the owners and managers bore limited responsibilities.

By contrast, development of these capacities will be indispensable if the society's ROGs are fair and transparent.[4] In this case, borrowers will have equal opportunities to obtain a loan and banks to make loans to prospective borrowers, and it will matter how much profit a bank makes. A lending decision will be evaluated and rewarded in an unbiased manner, mainly on the basis of its consequences on profits, both internally within a bank and externally in the financial market. The banks will do their best in gathering and processing relevant information and in examining the prospect of being repaid the principal and interest. This way, high-quality basic financial functions will be embodied in the services they provide.

Hence, how well developed the basic functions of financial intermediation are in a society would critically depend on the fairness and transparency of its ROGs. The existence of factors of production such as skilled manpower and capital would not be sufficient. In addition, the ROGs in the financial sector should be fair and transparent enough so that high international standards of asset valuation and transparency may be put into practice and effective competition reign in the financial market.

Thus, the realization of indirect benefits of financial services trade will depend on the availability of two different kinds of factors. One may be called *internal factors*—factors that render services to production by being employed within the firms producing the services in question. Skilled manpower and capital are typical examples. The other may be called *external factors,* as these render services to production that exist outside firms. In the above discussion such a factor is the society's ROGs. Either one of the two kinds of factors may work as a constraint on the realization of the benefits. Of these two potential constraints, lack or shortage of the internal factors may present less difficulty, because this can be overcome relatively easily at the firm level by training or even by importing from abroad the factors in short supply. However, if unfair and nontransparent ROGs are the binding constraint, the indirect benefits from financial services trade may not be easily realized: What is required is a change in the way the financial market, if not the whole society, operates, although the presence of foreign financial firms may provide an impetus for such a change.

9.3   Liberalization of Financial Services Trade in Korea

The need for financial deregulation and reform has long been recognized, but financial deregulation in Korea has been very cautious and slow, as

---

4. The ROGs are fair when every economic agent enjoys an equal opportunity ex ante and its ability, efforts, and performances are evaluated by the same standard, and when one is held accountable for one's own actions ex post. They are transparent when the rules themselves are known to all economic agents or when a way of knowing them is open to them.

Kang (2000) reports. Only as recently as early 1997, the same year the country was hit by currency and financial crises, the government established the Presidential Commission for Financial Reform.[5] The economic crisis brought about fundamental changes to Korea's economic policies. Regarding the policy on financial services trade, the crisis stands as a defining moment. It revolutionized the policy stance from lukewarm and partial opening to swift and full opening. This section begins with a brief review of the exchange rate policy, and the remainder describes the liberalization of financial services trade, distinguishing the precrisis from the postcrisis period.

9.3.1 Foreign Exchange Rate Policy[6]

Korea's exchange rate policy changed from a nominal anchor approach in the 1970s to a real target approach with the introduction of Multiple Currency Basket Peg System in February 1980. In the wake of the realignment of major currencies in the mid-1980s, Korea's current account showed large surpluses, and the international financial institutions and the U.S. Treasury accused the Korean government of manipulating its exchange rate. This led to an adoption of Market Average Exchange Rate System in March 1990, under which the won-dollar exchange rate was determined in the market within a band that was initially set at 0.4 percent above and below the rate of the previous day. The band was gradually widened over time and was eliminated in December 1997, immediately after the currency crisis. Since then, the policy maintained a floating exchange rate system and limited government intervention in the exchange market to smoothing volatility.

9.3.2 Before the Crisis

Cross-Border Trade

Before the crisis, the policy on cross-border trade was not made on its own merit but was decided as a by-product of capital account liberalization policy. Specifically, for the banking sector, cross-border trade was not allowed (and still is not) under the banking law, since the law followed positive system and had no provisions regarding cross-border trade. In consequence, only limited cross-border trade was allowed under the Foreign Exchange Management Act as a part of permitted capital transactions.

With respect to capital account liberalization, the policy stance was dictated by the concern about the current account balance. Hence, during the 1990s, when the current account recorded large deficits, the Korean government was reluctant to take any policy measures that would ease capital outflows.

---

5. See Kang (2000) for the discussion of financial deregulation before the crisis.
6. This subsection draws on Park (1996).

This policy stance was maintained with regard to the cross-border trade. Liberalization began with those transactions that would lead to capital inflows, although the extent of liberalization was different depending on mode of transactions. Although cross-border trade by individuals was left closed, certain transactions were allowed for corporations as a part of capital account liberalization. Thus, overseas bond issuance by financial institutions and corporations was deregulated in 1991, subject to discretionary quantity control by the government. Also, foreign borrowings by corporations were allowed in 1995, again with attached restrictions on the use of funds and government approval required Liberalization of trade-related short-term financing was made relatively free, with fewer restrictions attached: regulations on deferred import payments and receipt of advance payments for exports were lifted step by step, with few restrictions throughout the 1990s.

Relative to individuals and corporations, banks were given more freedom in transactions with foreign agents, although the allowed transactions were limited to overseas bond issuing and foreign borrowings. For banks, there were no formal restrictions on foreign borrowings, although it is known that the government imposed informal quantity controls. However, according to practitioners, the restrictions were lifted in 1994.

Commercial Presence

In the early 1990s when the current account was still in chronic deficit, the government allowed a number of foreign banks to enter the Korean market in order to help attract foreign capital. The government removed the upper ceilings on foreign bank capital in 1991. The five-year financial liberalization plan announced in 1993 aimed mainly at interest rate deregulation and abolition of the limits on maximum maturity for loans and deposits of banks, among other goals. As a part of the liberalization, the scope of financial activities allowed for foreign banks was broadened to include local branch establishment.

Foreign security companies were authorized to do business in 1992, when the Korean stock market was opened partially. Again, however, they were permitted to open only branches.

Regarding commercial presence, a potentially important development occurred in 1996 on Korea's accession to the Organization for Economic Cooperation and Development (OECD). In order to fulfill its obligations as a member of the organization, the Korean government announced in September a blueprint that gradually would remove barriers to foreign portfolio investment and allow foreign direct investment. The following summarizes the 1996 commitments:

- Foreign banks and securities firms from OECD countries would be permitted to establish subsidiaries in South Korea by 1998.

Table 9.1   Trend in the Expansion of Limits on Equity Investment by Foreigners, 1992–98 (%)

Date	Non–State-Owned Companies	State-Owned Companies
January 1992	10	8
December 1994	12	8
July 1995	15	10
April 1996	18	12
October 1996	20	15
May 1997	23	18
November 1997	26	21
December 1997	55	25
May 1998	100	30

Source: Financial Supervisory Service, Press Release, January 2000.

- Aggregate foreign investment ceilings for investors from OECD countries were to be phased out by 2000.
- Foreign investors from OECD countries would be allowed to establish and hold 100 percent ownership of any type of financial institution by December 1998.
- Foreign investment consulting firms from OECD countries would be able to offer their services without establishing a commercial presence in Korea.

9.3.3   After the Crisis

Since the economic crisis in 1997, the Korean government has started a sweeping reform of the financial sector. We first take a look at the deregulation of the financial market that allows international transactions of stocks, bonds, and other instruments and then discuss the liberalization of the two modes of financial services trade.

Deregulation of the Financial Market

Domestic Stock and Bond Market. In order to promote inflows of foreign capital, Korea opened the domestic stock and bond and money markets.

Ceilings on foreigners' equity investments were completely lifted in May 1998 with the exception of investment in state-owned enterprises, as indicated in table 9.1. Also, the requirement was eliminated that domestic subsidiaries of foreign companies should obtain government approval when introducing more than $1 million from abroad.

The corporate bond and government bond markets were completely opened to foreigners at the end of 1997, as shown in table 9.2. The foreign investment on the bond of non-listed companies was allowed in July 1998.[7]

---

7. A more detailed table on bond market liberalization is provided in appendix table 9A.1.

Table 9.2   Opening of the Bond and Money Markets, 1994–98

Date	Instruments
July 1994	Non-guaranteed convertible bonds issued by small and mid-size companies
June 1997	Non-guaranteed convertible bonds issued by large companies, and non-guaranteed bonds issued by small and mid-size companies
December 1997	Corporate bond and government bond
February 1998	CPs and trade-bill
May 1998	All money market instruments including CDs and Repos

Sources: Ministry of Finance and Economy, Press Release, May 1998; Financial Supervisory Service, Press Release, January 2000.

The opening of the money market, such as the markets for commercial papers (CP) and certificates of deposit (CD), proceeded in steps and was completed in May 1998.

Foreign Exchange Market. As of July 1998, the government liberalized medium-term foreign borrowing in order to help the firms attract foreign capital. In addition, the restrictions on the types of goods and duration of credit were also relaxed for import and export credits.

In April 1999, the government abolished the restrictive Foreign Exchange Management Act and replaced it with Foreign Exchange Transaction Act. The regulation for capital account transactions was changed from a positive system to a negative system. As a result, the following capital transactions were allowed:

- Offshore issuance of securities and foreign borrowing with a maturity of less than one year
- Offshore investment in foreign financial markets, foreign insurance markets, and foreign real estate markets by domestic firms and financial institutions
- Establishment of domestic saving deposits (including trust deposits) by nonresidents with a maturity in excess of one year
- Issuance of won-denominated (maturity exceeding one year) and foreign currency–denominated securities by nonresidents, and transactions of derivatives through domestic financial institutions

From the year 2001, foreign exchange transactions by individuals, such as won-based domestic deposits by nonresidents of maturity less than one year, will be liberalized. The government will also allow individuals to freely deposit their money in banks abroad and buy foreign securities or foreign real estate. At this point, the level of liberalization in Korea will be close to that of the OECD countries.

Table 9.3   The Status of Liberalization in 2000

	Bank	Security	Investment Trust Company	Investment Advisory	Life Insurance	Non-Life Insurance
Branch establishment	Open	Open	Open	Open	Open	Open
Subsidiary establishment	Open	Open	Open	Open	Open	Open
Joint venture	Open	Open	Open	Open	Not open	Not open
Cross-border trade	Partially open	Not open	Partially open	Open	Open	Aviation, hull (open)

Source: Korea Institute of Finance (2000).

Cross-Border Trade

Despite the sweeping reform and deregulation of the financial sector since the economic crisis, the cross-border trade in the banking sector is only partially opened under the Foreign Exchange Transaction Act, because the banking law, which adheres to the positive system, does not deal with cross-border trade. Also, the cross-border trade in securities is not allowed because the security law allows trading only through commercial presence. However, the investment trust companies (ITC) are allowed to trade mutual funds without commercial presence (see table 9.3).

Commercial Presence

Important steps were taken in the spring of 1998 to increase foreigners' access to the Korean financial market. Foreign banks and securities firms were authorized to establish subsidiaries in March. In addition, 100 percent foreign ownership of Korean institutions was allowed in April, and foreign nationals were allowed to become directors of Korean banks in May.

Banks.[8] The branch of a foreign bank is treated as an independent financial institution, and its operations are similar to those of subsidiaries of foreign banks, including retail businesses. There are no restrictions on establishing subsidiaries for foreign banks in Korea. Establishment of a new commercial bank, whether domestic or foreign-owned, requires the permission of only the Financial Supervisory Commission (FSC). The minimum capital required is 100 billion won for establishing a nationwide commercial bank and 25 billion won for a regional bank. In addition, foreign banks in Korea have been able to have local branches in the domestic market since March 1998. The foreign exchange position is regulated for prudential reasons. The maximum oversold position allowed of spot foreign exchange is US$5 million or 3 percent of capital, whichever is greater.

8. A table on Korea's World Trade Organization commitment is available from the authors upon request.

Table 9.4   **Foreign Shares of Domestic Banks as of the end of 1999 (%)**

Banks	Government Share	Main Foreign Investor
Cho Hung	80.05	—
Hanvit	74.65	—
Korea First	49.00	New Bridge (51%)
Seoul	95.68	—
Korea Exchange	35.92	Commerz Bank (23.6%)
Kookmin	6.48	Goldman Sachs (18%)
Korea Housing and Commercial	14.50	ING Group (10%)
Shinhan	—	Korean-Japanese (49.43%)[a]
KorAm	—	BOA (16.8%)
Hana	—	IFC (3.3%)
Peace	—	—

Source: Data supplied by Financial Supervisory Service.
[a]This figure is as of April 12, 2000.

For the ownership of banks, the prior limits of 4 percent in a nationwide bank, 8 percent in a bank converted from a nonbank financial institution, and 15 percent in a regional bank were mitigated by allowing the acquisition of shares in excess of those limits with approval from, or prior notice to, the FSC. Foreign ownership of up to 100 percent was permitted from April 1999, although it is subject to additional review by the FSC in line with the increase in stakes beyond certain predetermined thresholds. Also, laws were enacted to strengthen the powers of banks' boards of directors and to enhance transparency in dealings with shareholders. Foreigners have been permitted to become directors of bank boards since May 1998. Therefore, any foreign bank meeting the conditions, which are applied equally to domestic banks, is allowed to enter the market.[9]

There were sixty one branches of foreign banks and twenty six foreign representative offices as of the end of December 1999. The Korea First Bank was sold in September 1999 to New Bridge Capital in the United States, and there are foreigners participating in management in Housing & Commercial Bank, HanMi Bank, and Foreign Exchange Bank (see table 9.4).

Nonbank Financial Institutions. In the case of security companies, the foreign investment mostly took the form of opening new branches or offices before the crisis. After the crisis, the foreign investments took the form of acquisition of existing firms or share participation. As of the end of 1999, there were twenty branches of foreign securities companies and seven representative offices in Korea (see table 9.5).

With the removal of entry barriers to the security industry, the efforts to

---

9. See appendix table 9A.2 for a greater, detailed explanation of liberalization measures.

Table 9.5   The Trends of Foreign Security Company's Branch, 1991–2000

	1991	1992	1993	1994	1995	1996	1997	1998	1999	2000	Total
Establishment	2	5	2	2	4	5	3	2	2	0	27
Closure					1			3	1	2	7
Total	2	7	9	11	14	19	22	21	22	20	20

Source: Data supplied by Financial Supervisory Service.

Table 9.6   Foreign Participation in Security Firms as of March 2000 (%)

Security Company	Main Foreign Investors	Share
Seoul	QE International (L) Ltd.	21.28
	S.A.C. Capital International Ltd.	3.97
	SR Investment (L) Ltd.	4.29
	SR Global International Fund LP	0.44
	Asian Capital Holdings Fund S/A Berceuse Investment N.V	0.10
Good-Morning	Asia Pacific Growth Fund II	16.7
	KGRF	3.6
	Lombard Korea Ltd.	6.8
	Government of Singapore Investment Corp.	5.9
Daeyu Regent	Regent Pacific Group Ltd.	42.68
KEB Smith Barney	Saloman Brothers Holding Co., Inc.	80.00
Hannuri	Saloman Smith Barney Inc.	25.00
	Saloman Brothers Holding Co., Inc.	24.00
KGI Cho Hung	KGI Korea Ltd.	51.00
Meritz	Trader investment	30.82

Source: Data supplied by Financial Supervisory Service.

bring in foreign investments led to an increase in capital participation and management participation by foreigners, as shown in table 9.6. In addition, in April 1998, due to the liberalization of security firm establishment, foreign firms' merge with and acquisition of domestic security firms became possible.

9.4   Trends and Gains in Financial Services Trade in the 1990s

9.4.1   Trends in Financial Services Trade

Cross-Border Trade: Capital Flow

Policy Stance. As discussed in the previous section, the government pursued financial liberalization throughout the 1990s while varying the speed of liberalization according to concerned economic agents. It seems that the intent was to allow more freedom for banks than for other agents, such as nonfinancial corporations and individuals. Apparently the government preferred gradual liberalization as a way of controlling associated risks and, in particular, appeared to have an intention of utilizing banks as a risk manager. Presumably, the underlying assumption was that banks would make cautious brokers in linking foreign suppliers of financial services with domestic consumers, which was proved to be a gross error by the 1997 crisis.

Characteristics. As a result of the government' strategy, the capital flows in and out of Korea during the period took place mainly through the banks. The Korean economy experienced substantial net capital inflows from 1990 through the 1997 crisis. The magnitude of inflows remained small in the first four years, at 1.2 percent of gross domestic product (GDP) on average, but for the three years from 1994 to 1996 more than doubled to 3.5 percent of GDP (see figure 9.1). Stock investment by foreigners explains only the limited portion of the increase, owing to the quantity restrictions imposed. Thus, debt contracts and debt portfolios were the major carriers of capital inflows, and in consequence the surge in net capital inflows was tantamount to a sharp increase in Korea's external debt.

The Korean financial sector was the major issuer of the debt contracts and portfolios. Of the total increase in external debt during 1994–96, the financial sector explains about 70 percent (see table 9.7). As a matter of fact, the amount of resources provided to the Korean banks by foreign creditors was much larger than represented by external debts, as the Korean banks were allowed to open and expand operations of overseas branches as a part of liberalization measures. The resulting increase in borrowings of the overseas branches from foreign banks was as large as the rise in external debts of the Korean banks (see table 9.8).

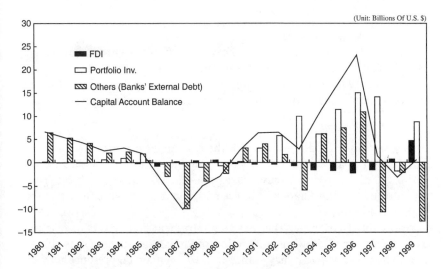

Fig. 9.1   Trend and composition of net capital inflows to Korea
Source: Bank of Korea online service.

Table 9.7   External Debt of Korea, by Sector: 1992–97 (US$100 millions)

	1992	1993	1994	1995	1996	1997
Public sector	56	38	36	30	24	223
Long-term	56	38	36	30	24	223
Short-term	0	0	0	0	0	0
Corporate sector	137	156	200	261	356	462
Long-term	65	78	90	105	136	253
Short-term	72	78	110	156	220	209
Financial sector	235	244	333	493	667	584
Long-term	122	130	139	196	277	310
Short-term	113	114	194	297	390	274
Total	428	439	568	784	1,047	1,268
Long-term	243	247	265	331	437	786
Short-term	185	192	304	453	610	482
Total/GNP (%)	14	13.3	15.1	17.3	21.8	28.6

Source: Bank of Korea, *Annual Foreign Exchange Statistics* (various issues); Ministry of Finance and Economy, Press Release, June 8, 1999.

Table 9.8   Foreign Currency Liabilities of Overseas Branches of Korean Banks, 1992–97 (US$100 millions)

	1992	1993	1994	1995	1996	1997
Domestic branches	157	163	226	363	507	387.9
Foreign branches	201	231	317	413	529	312.5
Sum	358	394	543	776	1,036	700.4
Sum/GNP (%)	13.8	9.9	12.3	16.0	22.2	24.6

Sources: Bank of Korea, *Annual Foreign Exchange Statistics* (various issues); Korea Development Institute, *Major Indicators of the Korean Economy* (various issues).

As is well known, creditors to the Korean banks were foreign banks. Hence, we can sum up that the cross-border financial services trade of Korea for the 1990s had been mainly between Korean banks and foreign banks. Transactions involving final consumers remained limited.

Commercial Presence

Policy Stance. Throughout the 1990s the Korean government continued to be reluctant to allow wider commercial presence of foreign financial institutions, as discussed in the previous section. As a result, the only form of entry into the Korean financial market permitted to foreign banks was the opening of branches.

It was noted earlier that the Korean government preferred banks' acting as risk manager in allowing more inflow of foreign capital and in its allocation. The conservative policy stance toward commercial presence may be explained by the same preference. The rationale must have been that, for the banks to fulfill the job properly intended for them, they needed to be protected from too much protection, which might hamper their soundness. Thus, commercial presence of foreign banks could not be encouraged.

In addition, the policy stance seems to have been strongly influenced by a development in the domestic market that adversely affected the banking sector. Since the 1970s, the Korean government promoted the development of securities markets. Naturally it led to the rapid growth of such players as investment trust companies and merchant banking corporations.[10] That the Korean banks were already facing tough competition from non-banking financial institutions deterred any policy initiatives that might further increase hardship for the Korean banks.

Characteristics. Given the policy stance, the opening of branches of foreign banks in the 1990s was sluggish, to say the least. The number of branches in operation actually declined. In addition, their market share compared to the Korean banks shrank (table 9.9, figure 9.2).

Not only their overall growth but also their scope of business was confined. Among sources of funds for the foreign banks, interoffice borrowing dominated, accounting for over 70 percent of their liabilities. The rest of liabilities were in the form of offshore borrowing (figure 9.3). Reflecting the liability side, foreign currency–denominated lending was the major component of their assets. In particular, the Korean banks were the main borrowers (figure 9.4).

In sum, until the financial crisis, the financial services trade through commercial presence was not much different from cross-border trade. It was only facilitating capital inflows into the Korean economy rather than allowing foreign intermediaries to participate fully in the Korean financial market.

9.4.2   Gains from Trade?

The experience with financial services trade in Korea during most of the 1990s was limited in both the scope of trade liberalization and the length of time. This subsection first tries to see if direct benefits of better financial services were provided and then discusses whether there were any direct benefits of systemic improvement in the financial sector.

Direct Benefits

Liberalization of financial services trade is supposed to provide better financial services in terms of cost and variety and increased output through resource reallocation. We observe the trend in interest rate margin between banks' lending and borrowing rates, because it is relatively straightforward compared to measuring the improvement in variety or estimating the increase in the nation's output.

---

10. Dooley and Shin (2000) argue that the Korean banks were losing their charter values owing to the competition.

Table 9.9   Number of the Korean Branches of Foreign Banks, 1967–98

	1967–75	1976–80	1981–90	1991	1992	1993	1994	1995	1996	1997	1998	Subtotal
Branch												
Newly opened	9	24	44	6	5	2	1	—	—	5	2	98
Closed	—	—	8	5	2	1	3	1	4	4	4	32
Subtotal	9	24	36	1	3	1	–2	–1	–4	1	–2	66
Representative office												
Newly opened	9	17	40	5	2	4	—	2	1	3	2	85
Closed	2	5	35	6	2	3	1	—	2	4	2	17
Subtotal	7	12	5	–1	—	1	–1	2	–1	1	—	23
Total	16	36	41	—	3	2	–3	1	–5	—	–2	89

Source: Financial Supervisory Service, *Bank Management Statistics* (various issues).

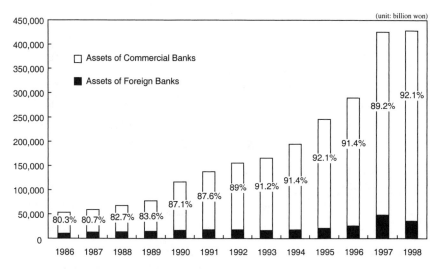

Fig. 9.2   Growth in assets of the Korean branches of foreign banks (in comparison to Korean banks)
Source: Financial Supervisory Service, *Bank Management Statistics* (various issues).

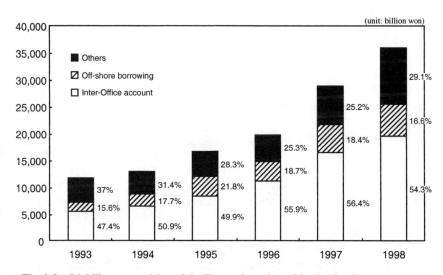

Fig. 9.3   Liability composition of the Korean branches of foreign banks
Source: Financial Supervisory Service, *Bank Management Statistics* (various issues).

The interest margins appear to decline in the 1990s (see table 9.10). When computed by subtracting deposit rate from lending rate, the average over the five years from 1992 to 1996 was 2.60, representing a 27 percent decline from the average margin over the preceding five years of 3.58. Moreover, this may underestimate the actual drop, because the interest rates were un-

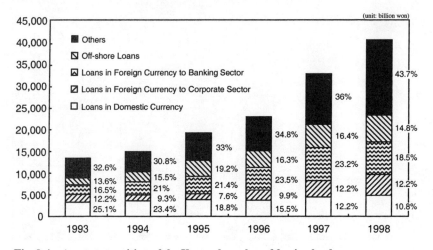

Fig. 9.4   Asset composition of the Korean branches of foreign banks
Source: Financial Supervisory Service, *Bank Management Statistics* (various issues).

der the influence of the government. It should be noted that, despite interest rates' deregulation in the 1990s, banks were not given full freedom to adjust the lending rates. Through moral suasion, the policy makers effectively depressed banks' "official" lending rates as well as deposit rates. As is well known, banks got around this informal control when providing loans by demanding that borrowers deposit some portion of the loans at low interest rates. If the corporate bond rates are used as proxy for the bank lending rates in computing the margins, the magnitude of decline exceeds 40 percent. The cost of capital, when based on corporate bond rates, was clearly declining as well.

However, it does not seem appropriate to attribute these developments to liberalization of the financial services trade alone. Liberalization of domestic financial markets was also under way. Moreover, the scope of financial services trade liberalization before the financial crisis was rather limited, as noted earlier.

Two Hypotheses Regarding Indirect Benefits

Did the Korean economy then obtain some benefits of systemic improvement in the financial sector from the limited liberalization in the 1990s? Hypothetically, we think there could be the following two benefits.

First, there might be gain from better monitoring. We noted that most financial services trade took place between Korean banks and foreign banks. With this, it should be noted that Korean depositors or creditors to the Korean banks did not perform much of a monitoring function on their banks, since they felt safe under the implicit insurance provided by the govern-

Table 9.10   Trend in Interest Rate Margins of Korean Banks, 1987–97 (%)

	1987	1988	1989	1990	1991	1992	1993	1994	1995	1996	1997
Bank lending rate	9.27	9.79	10.48	10.74	10.28	10.82	9.36	9.91	10.82	11.07	11.43
Bank deposit rate	6.54	5.95	5.87	6.21	8.08	8.59	7.45	7.61	7.79	7.55	7.86
Bond yield	12.74	13.58	15.38	18.51	18.98	14.00	12.21	14.22	11.65	12.57	24.31
Real bond yield	9.66	6.44	9.64	10.01	9.66	7.70	7.44	8.00	7.16	7.67	19.83
Margin 1	2.73	3.84	4.61	4.53	2.20	2.23	1.91	2.30	3.03	3.52	3.57
Margin 2	6.20	7.63	9.51	12.30	10.90	5.41	4.76	6.61	3.86	5.02	16.45

Source: Financial Supervisory Service, *Bank Management Statistics* (various issues).

Note: Bank lending rate is the averaged lending rate. Bank deposit rate is the averaged deposit rate. Bond yield is the yield on three-year corporate bond. Real bond yield = bond yield – inflation rate (CPI, 1995 = 100). Margin 1 = Bank lending rate – bank deposit rate. Margin 2 = bond yield – bank deposit rate.

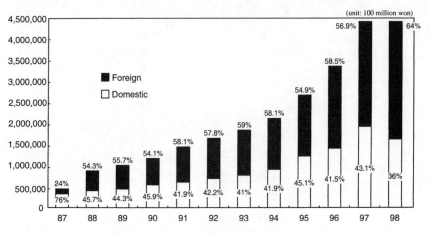

Fig. 9.5   Composition of total assets of Korean banks, by currency
Source: Financial Supervisory Service, *Bank Management Statistics.*

ment. Given the situation, if foreign banks played a market disciplinary role as large creditors, it could be regarded as a systemic improvement of the financial sector.

Second, it is possible that the Korean branches of foreign banks may have played a stabilizing role during the currency crisis. A low local commitment is often pointed to as a risk that might accompany financial liberalization: When crisis hits a country, foreign creditors tend to rush out of the country, leaving domestic companies in a severe foreign currency credit crunch. This would be more likely if creditor-borrower relationships are maintained at arm's length. By contrast, commercial presence of foreign banks might ameliorate the problem because they might then be more committed to relationship banking.

We examine these hypotheses in turn.

Was There Better Monitoring?[11] If foreign banks supplied credit to Korean banks with monitoring, it would have had a rather strong disciplinary effect on Korean banks. Throughout the 1990s more than half of the assets held by Korean banks were in foreign currencies (figure 9.5), and Korean banking regulations prohibited banks from taking net open currency position. Thus, a decrease in the supply of foreign credit would have implied a significant reduction in the opportunity set for asset management.

In order to see if foreign creditors performed any monitoring function, we investigate whether their lending behavior could be explained in terms of creditworthiness of individual Korean banks. If they did monitor Korean banks, we may argue, it should be reflected in their credit policy toward the Korean banks.

11. This part of the paper draws on Dooley and Shin (2000).

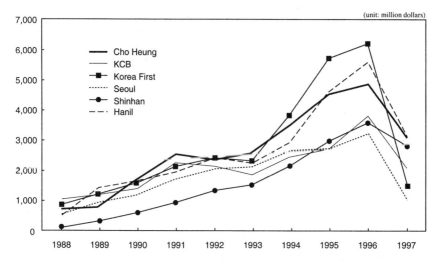

Fig. 9.6   Growth of foreign currency liabilities of the six largest Korean banks
Source: Cho Heung, KCB, Korea First, Seoul, Shinhan, Hanil Bank.

The trend in foreign currency liabilities of the six largest commercial banks in Korea is shown in figure 9.6. As is evident in the figure, all of the banks were increasing their foreign currency liabilities throughout the 1990s. Also evident, however, is an acceleration in the growth rates for the three years from 1994 to 1996 and a sharp drop afterward in 1997. Our examination is focused on development for the 1994–96 period.

Although faster growth than in previous years is common to all six banks, we note two banks, in particular: the Korea First Bank and Seoul Bank. These two banks turned out to be the most troubled banks during and after the crisis development. Capital of both banks was found to be completely eroded, so the government was forced to intervene in December 1997 by injecting public money. However, despite the similarity in terms of capital soundness, in foreign currency liabilities the two banks displayed quite contrasting trends. The Korea First Bank recorded the highest growth rate, whereas Seoul Bank recorded the lowest by a considerable margin. It is questionable, however, if the inability of Seoul Bank to expand foreign currency operation was due to foreign creditors' monitoring. Rather, the bank's expansion was constrained by the supervisory authority. Although there was much to be desired in Korean banking supervision, we are informally told that even under the less-than-satisfactory supervision practice, the supervisory authority considered Seoul Bank to be in serious trouble and felt it necessary to impose discretionary restrictions on its operation. In other words, it appears that in the case of Seoul Bank a different kind of monitoring may have been in operation.

As a way of tentatively gauging whether the individual creditworthiness

Table 9.11   Relationships of Foreign Currency Liabilities of Korean Banks with Other Financial Variables (%)

	Growth Rate	ROA	ROE	Stock Price (won)	Capital Ratio	Net Worth Ratio
Cho Heung	2.78	0.41	5.98	9,701	3.23	6.52
KCB	4.43	0.30	4.80	7,841	3.54	6.23
Korea First	5.77	0.18	2.85	8,588	3.03	6.86
Seoul	1.19	−0.14	−2.09	6,937	4.23	6.84
Shin han	2.25	0.72	7.76	15,702	3.32	9.65
Han Il	5.98	0.33	4.80	9,067	3.36	7.33
Correlation		0.7479	0.2411	−0.224	−0.6298	−0.2592
Coefficient		0.13	0.522	−0.0001	−2.96	−0.41
		(0.28)	(3.48)	(0.0003)	(1.82)	(0.77)
Correlation^2		−0.8255	−0.8668	−0.7051	−0.1542	−0.4697
Coefficient^2		−6.91**	−0.81**	−0.0003	−1.41	−0.58
		(2.73)	(0.27)	(0.0002)	(5.19)	(0.63)

Source: Bank of Korea, Bank Management Statistics (various issues).
Notes: Growth rate = 1993–96 growth rates of foreign currency liabilities ÷ 1990–93 growth rates of foreign currency liabilities. Capital ratio = capital stock ÷ total assets. Net worth ratio = shareholder's equities ÷ total assets. Coefficient is computed by regressing growth rates on each variable. Correlation^2 and coefficient^2 are after excluding Seoul Bank. Numbers in parentheses are standard deviations.
**Significance at the 5 percent level (d.f. = 4).

of Korean banks was one of the determinants for credit policies of foreign banks, we now compute correlation of the growth rates of six banks with their various performance and capital soundness variables. In particular, keeping the above information in mind, we do the same calculations with two sets of banks, one including all of the six banks and the other including all but Seoul Bank.

Inspecting the result, shown in table 9.11, we note the following. First, when Seoul Bank is included, the growth rate of foreign currency liabilities is not significantly related with any variable considered. In addition, it has negative, albeit insignificant, relationships with capital variables. Second, when Seoul Bank is excluded, negative relationships of the growth rate with all the variables are estimated. Moreover, the relationships with performance variables are statistically significant at the 5 percent level, although the degree of freedom was low at 4.

Therefore, albeit tentatively, overall evidences are against the hypothesis that foreign banks were monitoring and concerned about the creditworthiness of individual Korean banks.[12]

12. Surely it begs an answer why there seem to be negative relationships between foreign currency liabilities and their creditworthiness. Dooley and Shin (2000) argue that it reflects incentives for Korean banks to exploit the value of implicit insurance by the government. Also, one may wonder how foreign creditors were pricing their loans to Korean banks. Unfortu-

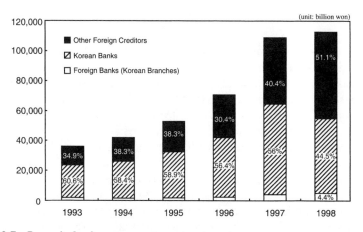

Fig. 9.7   Loans in foreign currency to corporate sector
Source: Financial Supervisory Service, *Bank Management Statistics,* various issues; Bank of Korea, *Annual Foreign Exchange Statistics* (various issues); Ministry of Finance and Economy, press release, 8 June 1999.

Did Commercial Presence of Foreign Banks Mitigate the Credit Crunch? In evaluating the hypothesis on a possible stabilizing role of Korean branches of foreign banks, our answer relies on the following two observations.

First, as shown in figure 9.2, of the total assets held by banks, including both domestic and foreign currency assets, the share of Korean branches of foreign banks changed little after the crisis. If anything, their share decreased a bit in 1998. Hence, based on the changes in total assets, it is hard to argue that Korean branches of foreign banks played a mitigating role in the crisis. Second, narrowing the focus to foreign currency credit to the corporate sector, the share of Korean banks declined and that of the Korean branches of foreign banks increased slightly after the crisis (figure 9.7). This tends to indicate that commercial presence of foreign banks mitigated the crisis. However, their share was too small to be of any material impact. Besides, "other foreign creditors" who provided credit at arm's length also increased their share. Thus, compared to these "other foreign creditors," the commercially present foreign banks do not appear to have played any greater role. Therefore, overall it is hard to conclude either that the commercial presence of foreign banks in Korea rendered soothing effects during the crisis or that it did not.

---

nately, no official data on borrowing rates of individual Korean banks exist. Nonetheless, we are informed by Bank of Korea officials that there was not much discrepancy among Korean banks in terms of international borrowing rates. Finally, as noted by one anonymous referee, the exercise would remain incomplete unless determinants of bank-level foreign currency balances are identified. Therefore, the result of the exercise needs to be taken cautiously.

Discussions

Why Were the Gains from Financial Services Trades Not Evident? In theory, international trade "without distortion" is supposed to increase the welfare of the countries engaged. Thus, given the result that financial services trades did not seem to be welfare improving for Korea before the crisis in 1997, naturally we come to ask what kind of distortions existed. In our view, the answer is obvious: the presence of insured agents and uneven liberalization tilted toward those transactions with insured agents, namely, the banks.

Banks in Korea have been under implicit insurance, as in most other economies. Thus, unless other preventive mechanisms are provided, counterparties of Korean banks in transactions are likely to have incentives to take advantage of insurance. What we have shown is that foreign banks were not exceptions. In addition, by regulation Korean banks were the only effective channels through which financial services trade took place. This, together with the incentives given to foreign banks, implies that most of the financial services trades for Korea were likely to be affected by distortion of insurance.

In retrospect, therefore, one should not be surprised to find that financial services trade brought about little gain to the Korean economy.

Are There Ways to Improve the Situation? The logical remedy to the problem would be to remove the insurance. As has been proven in other countries and by historical episodes, however, it is easier said than done. In fact, some public insurance for banks would be practically unavoidable.

In the presence of certain insured transactions, policy responses to exploit gains from trades should be twofold: regulation and close supervision of insured transactions and liberalization of uninsured transactions. Insured transactions need to be monitored by the insurance provider in order to moderate moral hazard, which is the basic teaching of microeconomics. In the case of financial services trades, it simply points to the necessity of prudential regulation and supervision of banks. While keeping insured transactions under close monitoring, one can make the most gains from trades by allowing relatively free trade of uninsured transactions. It implies that the commercial presence of foreign financial institutions should be more liberalized to obtain gains from trades.

That said, we realize that improved supervision of banks and a more liberalized regime for commercial presence was exactly the policy strategy taken by the Korean government after the 1997 crisis. Thus, as far as the Korean case is concerned, we shall see in the near future if the policy change proves to be sufficient to receive benefits from the liberalization of financial services trade.

9.5   Conclusion

Perhaps it is too early to make a comprehensive assessment of the impact of the liberalization of financial services trade, because many of the liberalization measures currently in place have been taken only after the financial crisis in 1997. Korea's liberalization commitment was rather limited at the conclusion of the Uruguay Round negotiation on financial service in December 1997 (Mattoo 1999). As was mentioned in section 9.3, in 1996 on its accession to the OECD Korea made a substantial commitment to financial reforms and liberalization that were going to be gradually implemented. After the financial crisis, this commitment was bound in the context of the World Trade Organization (WTO), and the reforms and liberalization were quickly carried out. Therefore, the liberalization could not have had much impact, since the variety of financial services to be traded had been rather limited and agents allowed to trade had also been restricted before the crisis, and not enough time has passed after the crisis.

As was discussed in the previous section, the liberalization of financial services trade before the crisis led to a substantial capital inflow via the cross-border mode of trade. Trade via the commercial presence mode also took place, and this also mainly led to an increase in capital inflow, because the main source of the funds of the foreign banks was interoffice borrowing and their main clients were Korean banks. Hence, it was little different from the cross-border mode.

Freer capital flows have the effect of smoothing consumption over time, and this has to be recognized as the benefit of the liberalization. Of greater interest to Korea, however, especially in the aftermath of the financial crisis, would be the quality of five basic functions embodied in the services that foreign banks provide. Section 9.4 considered whether foreign banks played any disciplinary role through better monitoring, given that Korean banks were under government-provided implicit insurance. There was little evidence of such a role. On the contrary, the growth of their lending during the 1990s to domestic banks and a statistically significant, negative correlation with indicators of the banks' performance.

However, it would be too hasty to conclude that liberalization does not provide the benefits discussed in section 9.2. Unlike trade in goods, from which the benefits more or less automatically flow as the goods are delivered across the border, the benefit from services trade is not automatic. The quality of financial services, especially that of the basic functions embodied in the services, depends very much on the ROGs in the domestic financial sector, as discussed in section 9.3. Unless the ROGs become fair and transparent so that information gathering, credit assessment, risk management, and so on may be the essential part of financial intermediation, it seems that liberalization cannot deliver the indirect benefits. The finding in section 9.4 seems to be evidence of this point.

Section 9.4 also examined whether the commercial presence of foreign banks mitigated (foreign currency) credit crunch during the crisis. Related to this was the question of whether foreign banks tend to have low local commitment and flee at the first sign of difficulty in the local economy. No strong evidence was found either for the mitigating role or for low local commitment. However, since the commercial presence of foreign banks in the financial sector was rather limited, the discussion can only be tentative, and no firm conclusion is warranted from Korean experience regarding the effect of the commercial presence.

After the financial crisis, the Korean government drastically widened the scope and quickened the pace of liberalization and carried out reforms, making efforts to rehabilitate the financial sector, strengthening the prudential regulation, and revamping the supervisory infrastructure. This represents a renouncement of its policy on the financial sector before the crisis, the policy that gave rise to "implicit insurance," as discussed in section 9.4. A crucial question for the success of financial reform is whether the renouncement will be adhered to, because the government's strong influence on the financial sector was one of the major factors underlying the crisis. A related development in the process of managing the crisis is that the influence of the government on the financial sector has become stronger after the crisis. It remains to be seen if the government sticks to the renouncement of its past policy, despite its strengthened influence.

Appendix

Table 9A.1   The Trend of Liberalization of Bond Market

Corporate Bond	July 1994	January 1997	June 1997	November 1997	12 December 1997	23 December 1997	30 December 1997	25 May 1998	July 1998
				SMEs					
Nonguaranteed bond									
Straight	Not allowed	Not allowed	Aggregate (50%)	Same as before	No limits				
CB	Aggregate (30%) Individual (5%)	Aggregate (30%) Individual (5%)	Same as before	Same as before	No limits				
BW, EB	Not allowed	Not allowed	Not allowed	Aggregate (50%) Individual (10%)	No limits				
Guaranteed bond	Not allowed	Not allowed	Not allowed	Not allowed	Aggregate (30%) Individual (10%)	Aggregate (30%)	No limits		
				Large Enterprises					
Nonguaranteed bond									
Straight	Not allowed	Not allowed	Not allowed	Not allowed	Aggregate (30%) Individual (10%)	Aggregate (30%)	No limits		
CB	Not allowed	Not allowed	Not allowed	Same as before	Aggregate (50%) Individual (10%)	Aggregate (50%)	No limits		
BW, EB	Not allowed	Not allowed	Not allowed	Not allowed	Aggregate (50%) Individual (10%)	Aggregate (50%)	No limits		
Guaranteed bond	Not allowed	Not allowed	Not allowed	Not allowed	Aggregate (30%) Individual (10%)	Aggregate (30%)	No limits		
Government bond	Not allowed	Not allowed	Not allowed	Not allowed	Not allowed	Aggregate (30%)	No limits		
Listed bond over the counter trading	Not allowed	Not allowed	Not allowed	Not allowed	Not allowed	Not allowed	Not allowed	Allowed	
RP trading	Not allowed	Not allowed	Not allowed	Not allowed	Not allowed	Not allowed	Not allowed	Allowed	
Nonlisted bond	Not allowed	Not allowed	Not allowed	Not allowed	Not allowed	Not allowed	Not allowed	Not allowed	Allowed

Source: Financial Supervisory Service, Press Release, January 2000.

Note: CB indicates convertible bond; BW indicates bond with warranty; EB indicates Eurobond.

Table 9A.2   **Liberalization of Foreign Participation in the Korean Financial Sector**

Sector of Activity	Equity Participation in Existing Korean Institutions	Subsidiary	Branch	Representative Office
Commercial bank	Regional commercial banks, no need to report for up to 15%; commercial banks, no need to report for up to 4%; must report to Financial Supervisory Commission for share between 4% and 10%. F.S.C approval required each time share exceeds 10%, 25%, and 33%.	No restrictions as of April 1998	No restrictions since the General Banking Act was enforced in 1954	No restrictions since the General Banking Act was enforced in 1954
Securities dealing	No restrictions as of April 1998	No restrictions as of April 1998	No restrictions as of November 1990	No restrictions as of January 1981
Investment advisory	No restrictions as of May 1998	No restrictions as of December 1997	No restrictions as of December 1995	No restrictions as of January 1993
Investment & trust company	No restrictions as of December 1997	No restrictions as of May 1998	No restrictions as of December 1996	No restrictions as of January 1993
Trust company	No restrictions	No restrictions	No restrictions	No restrictions
Mutual fund (fund management company)	No restrictions as of September 1998	No restrictions as of September 1998	No restrictions as of September 1998	No restrictions as of September 1998
Insurance[a]	No restrictions	Liberalized before the end of 1996	Liberalized before the end of 1996	Liberalized before the end of 1996
Merchant bank	No restrictions up to 100% as of January 1, 1997	No restrictions as of May 1998	Not allowed	Not allowed
Mutual credit institution	No restrictions up to 100% as of January 1, 1998	No restrictions	Not applicable[b]	Not applicable[b]

Source: OECD (2000).

[a]Auxiliary services were liberalized in April 1998 for insurance brokerage, actuary, and claims adjustment.
[b]Regardless of nationality, branches of mutual credit institutions cannot be established in Korea.

References

Bank of Korea. Various issues. *Annual Foreign Exchange Statistics* (in Korean). Seoul: Bank of Korea.
Claessens, Stijn, and Thomas Charles Glaessner. 1998. Internationalization of financial services in Asia. World Bank Policy Research Working Paper no. WPS 1911. Washington, D.C.: World Bank.
Dooley, Michael P., and Inseok Shin. 2000. Private inflows when crises are anticipated: A case study of Korea. In *The Korean crisis: Before and after,* ed. Inseok Shin, 145–82. Seoul: Korea Development Institute.
Financial Supervisory Service. Various issues. *Bank Management Statistics* (in Korean). Seoul: Financial Supervisory Service.
Glaessner, Thomas Charles, and Daniel Oks. 1998. NAFTA, capital mobility, and Mexico's financial system. World Bank Policy Research Working Paper no. WPS 1984. Washington, D.C.: World Bank.
Gelb, Alan H., and Silvia B.. Sagari. 1990. Banking. In *Uruguay Bond: Services in the world economy,* ed. P. Messerlin and K. Savant, 49–59. Washington, D.C.: World Bank.
Kang, Moon-Soo. 2000. Financial deregulation and competition in Korea. In *Deregulation and interdependence in the Asia-Pacific region,* ed. Takatoshi Ito and Anne O. Krueger, 277–303. Chicago: University of Chicago Press.
Korea Institute of Finance. 2000. Main issues in international financial markets (in Korean). Department of International Finance. Mimeograph.
Levine, Ross. 1996. Foreign banks, financial development, and economic growth. In *International financial markets,* ed. Claude E. Barfield, 224–54. Washington, D.C.: AEI Press.
Mattoo, Aaditya. 1999. Financial services and the World Trade Organization: Liberalization commitments of the developing and transition economies. *World Bank Policy Research Working Paper* no. WPS2184. Washington, D.C.: World Bank, September 30.
Musalem, Alberto, Dimitri Vittas, and Alsi Demirgüç-Kunt. 1993. North American free trade agreement. *World Bank Policy Research Working Paper* no. WPS1153. Washington, D.C.: World Bank.
North, Douglas. 1990. *Institutions, institutional changes, and economic performance.* Cambridge: Cambridge University Press.
Park, Won-Am. 1996. Financial liberalization: The Korean experience. In *Financial deregulation and integration in East Asia,* ed. Takatoshi Ito and Anne O. Krueger, 247–76. Chicago: University of Chicago Press.
Organization for Economic Cooperation and Development (OECD). 2000. Korea: Examination of position under the codes of liberalization and the declaration and decisions on international investment and multinational enterprises. OECD Document no. DAFFE/INV/IME(99)2/REV2.
White, Lawrence J. 1999. The role of financial regulation in a world of deregulation and market forces. Paper presented at IMF Conference, Second-Generation Reforms. 8–9 November, Washington, D.C.
Yoo, Jungho. 1998. The nature of the national economy in a borderless world and the role of the government. In *Korea's choices in emerging global competition and cooperation,* ed. L. Cho and Y. Kim, 65–95. Seoul: Korea Development Institute.

Comment   Kazumasa Iwata

This paper is very useful in assessing the process of liberalization in trade in financial services in Korea. Its analysis centers on cross-border trade and the mode of trade via commercial presence. The financial liberalization was triggered and promoted by Korea's accession to the Organization for Economic Cooperation and Development (OECD) in 1996. To my surprise, it has been accelerated by the financial and currency crisis of 1997.

The authors put emphasis on the importance of high performance of basic functions of financial intermediaries via commercial presence and the improvement of Hayekian legal and regulatory infrastructure (or the "rules of the game" of a given society), aside from the better quality and variety of services provided by foreign institutions. However, they fail to confirm the hypothesis on the expected improvement in regulatory infrastructure; the financial crisis seems to have distorted and prevented the effect of trade in financial services from fully realizing in the economy. However, they testify to the effect of distorted liberalization based on Dooley's hypothesis on insurance models of financial crisis (Chinn, Dooley, and Shresta 1999).

The paper noted that in the 1990s cross-border trade in financial services was largely carried out between Korean banks and foreign banks. Notably, before the crisis the Korean banks were mainly borrowers. Partly due to the regulations on open currency position, the borrowing of overseas branches from foreign banks was as large as the external debt of the Korean Banks. On the other hand, the trade through commercial presence was not much different from cross-border trade, given the limited business scope for foreign banks.

It is indeed interesting to observe that foreign banks as lenders to Korean banks are not concerned about creditworthiness and did not play the disciplinary role of monitoring the individual Korean banks (table 9.11). The increase in borrowing from foreign banks tends to lower the rate of return on assets of the Korean banks. The empirical evidence seems to support the hypothesis on implicit insurance by the government and looting the value of insured assets by foreign and Korean banks (Dooley's insurance model on financial crisis). The insurance model also assumes that the capital flight of uninsured assets will emerge before the crisis and peak at the time of crisis. This paper finds no strong evidence on capital flight through foreign financial intermediaries, as compared with domestic banks.

However, Chinn, Dooley, and Shresta (1999) argue that capital flight started to increase after 1993–94 in Korean as well as other Asian countries. The argument on the International Monetary Fund (IMF) reform revealed that moral hazard on the side of lenders was enhanced by the IMF commitment on the lender of last resort. Thus it seems premature to say that one

Kazumasa Iwata is professor of economics at the University of Tokyo.

should make the most gains from trades by allowing relatively free trade of uninsured transactions by foreign banks with local commitment.

In this context I would like to ask whether capital flight has played an important role in destabilizing the economy. Figure 9.1 demonstrates that portfolio investment continued to increase even in 1997, whereas bank borrowing was reduced after 1997. Does this imply that foreign banks acted as a main conduit of capital flight? We should be cautious, however, about the extremely short-run tradings, notably the derivatives tradings that do not appear in official statistics.

Moreover, my basic question is why foreign banks failed to play a disciplinary role. I believe that financial liberalization in the absence of prudential control, both domestic and international, may result in destabilization of the economy.

The authors argue that the liberalization gain has not been obtained due to the uneven liberalization tilted toward insured agents (domestic banks). However, I would like to ask whether undistorted liberalization can bring about full benefit in the absence of appropriate prudential regulation of financial intermediaries.

From the experience of currency crises, we know that excessive short-term borrowing tends to make the financial structure fragile to shocks. The paper examines the issue from the borrower's (Korean) perspective. However, the risk-weighting scheme embodied in the Bank for International Settlements (BIS) regulation on international banking favored short-term (less than one year) lending to banks incorporated in the non-OECD member countries (20 percent risk weight), while 100 percent risk weight is attached to longer than one year maturity lending to banks. In contrast, the risk weight is 20 percent in the case of lending to banks incorporated in the OECD member countries. Thus it is not excluded that the combination of uneven liberalization on the side of the borrowing country with the deficiency of prudential controls on international banking on the side of the lending country may be the source of the failure to realize the full gain arising from liberalization of trade in financial services.

Finally, the authors present the proposition that the implicit insurance by government distorted liberalization process. Yet the proposition begs further questions: Does it mean that the removal of implicit insurance can stabilize the financial market? How about the role of explicit insurance? Is it different from implicit insurance? More generally speaking, what is the optimal risk-sharing scheme under conditions of uncertainty about contingent liability arising from demand deposit contracts and information asymmetry between lenders and borrowers? Is there a case for an efficient (partial) bankruptcy of banks or the optimal financial crisis, as argued by Allen and Gale (1998)?

Information asymmetry between the regulatory authority and banks points to the importance of self-discipline based on appropriate incentive structure among stakeholders to maintain the soundness of bank management. At the same time, it is interesting to observe that the prudential regulation moves toward the direction of strengthening self- and market-discipline or the self-regulation in risk management; it attaches importance to the establishment of risk management systems within individual banks as well as self-imposed targets. The supervising authority simply checks and approves at each stage of decision-making. The regulation under the information asymmetry becomes more efficient by strengthening self-discipline rather than relying on the discretionary discipline imposed by the authority.

Although we admit that the current BIS regulation is deficient in many respects, it works to make the domestic prudential regulation more efficient to the extent that the best practice embodied in the "core principles" of the BIS promotes the self- and market discipline. Although it is commonly observed among different countries that financial liberalization lacking appropriate international rules and efficient domestic regulation has destabilized the economy, efficient domestic regulation mediated through the implementation of the best practice as common knowledge for borrowing and lending countries may promote liberalization in service trade. The role of the international rules combined with efficient domestic regulation provides useful lessons for the liberalization process in other service sectors.

References

Allen, F., and D. Gale. 1998. Optimal financial crisis. *Journal of Finance* 53 (4): 1245–84.
Chinn, M. D., M. P. Dooley and S. Shresta. 1999. Latin America and East Asia in the context of an insurance model of currency crisis. *Journal of International Money and Finance* 18:659–82.

A Study of Competitiveness of International Tourism in the Southeast Asian Region

Kuo-Liang Wang and Chung-Shu Wu

10.1 Introduction

Along with the continuing growth of national income in most countries, the substantial decline in the real costs of transportation, the liberalization of cross-border movements, and the increasing propensity to travel abroad, the number of total international visitors in the world as a whole kept growing during the period of 1973 to 1994 (figure 10.1). Therefore, most economies in the Southeast Asian region had devoted themselves to international tourism. As a result, total international arrivals to Southeast Asia and the region's share of the total world tourists had been steadily growing (figures 10.2 and 10.3), and all economies in the region had a growing or fairly stable share of the world market (figure 10.4). However, compared with their neighbors in the region, Taiwan, the Philippines, and Singapore had been continuously losing ground in the market for international tourism since the mid-1970s, the end of 1970s, and the beginning of 1980s, respectively; Indonesia had been continuously gaining ground since the mid-1980s; Hong Kong, Malaysia, and Thailand had experienced fluctuations, but still followed a rising trend. Thus, although some economies had been continuously losing ground, other economies had been continuously gaining ground or had experienced fluctuations in international tourism (figure 10.5).

The purpose of this paper is to investigate and identify the factors respon-

The authors are grateful to Anne O. Krueger, Takatoshi Ito, two anonymous referees, and seminar participants, particularly discussants Keiko Ito and Mahani Zainal-Abidin, for constructive comments and useful suggestions. The authors would also like to thank the Tourism Bureau of Taiwan for its kind help with data collection. All remaining errors are the authors'.

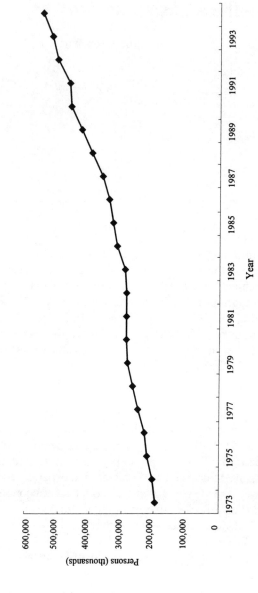

Fig. 10.1 1973–94 total international arrivals of the world as a whole
Source: WTO 1999.

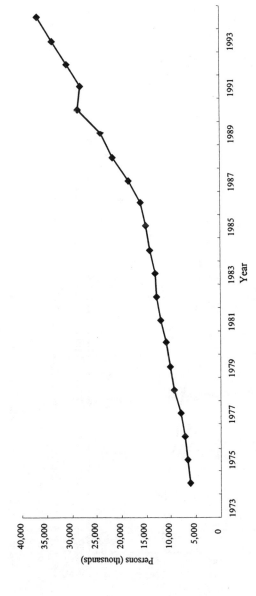

Fig. 10.2   1973–94 total international arrivals to the Southeast Asian region

Source: See note for figure 10.1.

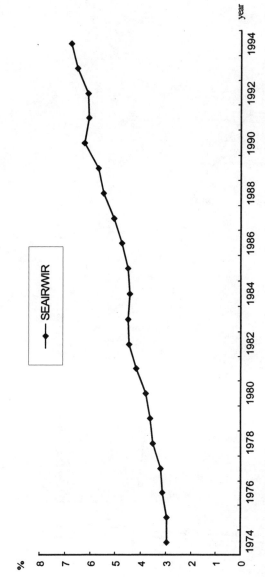

Fig. 10.3 International arrivals of the Southeast Asian region as a percentage of total international tourists in the world for 1973–94

Source: See note for figure 10.1.

Notes: SEAIR and WIR stand for international arrivals to the Southeast Asian region and of the world as a whole, respectively.

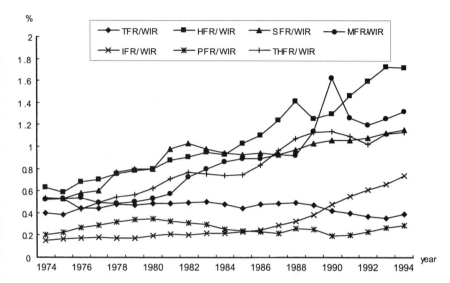

Fig. 10.4   Foreign arrivals of each economy as a percentage of total international tourists in the world for 1973–94

Source: See note for figure 10.1.

Notes: TFR, HRF, SFR, MFR, IFR, PFR, and THFR stand for foreign arrivals to Taiwan, Hong Kong, Singapore, Malaysia, Indonesia, the Philippines, and Thailand, respectively.

sible for each economy's changing competitiveness in international tourism among a group of seven Southeast Asian economies (Taiwan, Hong Kong, and the Association of Southeast Asian Nations (ASEAN) Five, namely, Singapore, Malaysia, Indonesia, the Philippines, and Thailand.[1] The rest of the paper is organized as follows. Section 10.2 sets up a theoretical model, on which our empirical study is based. Section 10.3 discusses the nature and problems of the sample, variables, and data used in the empirical study. Section 10.4 carries out statistical analyses and reports the preliminary findings. The final section concludes the paper.

10.2   Model Building

Consumer demand theory has been the major framework used to study the determinants of demand for tourism services. Empirically, there are two main approaches to modeling the demand for tourism: the single-equation approach and the simultaneous-equations approach. The single-equation

1. These economies are chosen because of their geographical proximity, their similarity in development strategy, and the fact that they have been close rivals in so many product lines on the world markets in recent years.

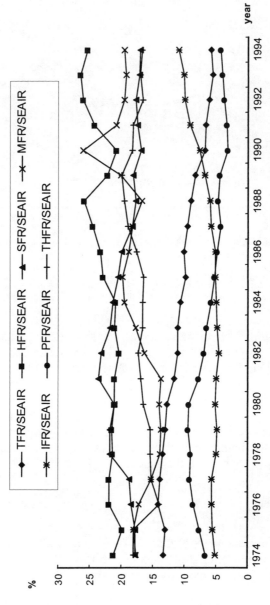

Fig. 10.5   Foreign arrivals of each economy as a percentage of the Southeast Asian region's total international arrivals for 1973–94

Source: See note for figure 10.1.

approach postulates that tourism demand, as measured by tourist expenditures (receipts) or the number of tourist arrivals, depends upon income, relative prices, or exchange rates, and transport costs, as well as other factors like marketing expenditures, political disruptions, or major sporting events. Although studies applying this approach have provided useful insights, the degree of detail that can be included in a single-equation estimation is still limited (Eadington and Redman, 1991). As a consequence, some researchers have shifted to the simultaneous-equations models, most of which are based on Deaton and Muellbauer's (1980) almost ideal demand system (AIDS). These models assume that economic agents have already determined the amount to be spent on foreign tourism and then estimate the sensitivity of a destination's share of the international tourism budget to changes in various determinants, especially income and relative prices among different destinations. The simultaneous-equations approach appears to be perfectly suitable for our purpose of investigating an economy's relative competitiveness in the international tourism market. However, data on each origin's per capita tourism expenditure in Taiwan, Hong Kong, and the ASEAN Five would be needed if the models based on AIDS were to be used in this paper. Unfortunately, the absence of this degree of disaggregated data precludes us from using the existing models. Following Tsai and Wang (1998), we will therefore develop an alternative model in this section to overcome this problem.

An economic agent makes a tourism decision in a number of stages. With given money income and other types of constraints such as time, the agent first decides, via utility maximization, the allocation of expenditure among groups of commodities including tourism. If the economic agent decides to travel, at the second stage, he or she then has to consider whether to travel abroad or domestically. A further stage is the decision to travel to North America, South America, Western Europe, Eastern Europe, Africa, or the Asia-Pacific region. Finally, if the agent does choose the Asia-Pacific region, he or she has to choose which particular country or countries to visit. This decision-making process can be formulated in a two-stage decision.

Let I, L_i, and K_i denote the total number of origins, total economic agents in origin i, and the total potential destinations for visitors from origin i, respectively. Assume that the preferences of an economic agent l in origin i can be described by the following utility function

$$(1) \qquad u_{li} = u_{li}(\mathbf{t}_{li}, \mathbf{x}_{li}), \quad l = 1, \ldots, L_i, \quad i = 1, \ldots, I,$$

where $\mathbf{t}_{li} = (t_{li}^1, t_{li}^2, \ldots, t_{li}^{K_i})$ is a vector of K_i foreign tourism services and $\mathbf{x}_{li} = (x_{li}^1, x_{li}^2, \ldots, x_{li}^{M_i})$ is a vector of M_i commodities other than foreign tourism services consumed by the agent l in origin i.[2] The expression $t_{li}^k, k = 1, \ldots,$

2. Domestic travel is included in the vector $\mathbf{x}_{li}$ because this paper is interested in the relative competitiveness among the competing destinations, not the choice between domestic and foreign tourism.

K_i stands for the consumption of tourism services in destination k by the agent l from origin i. Denoting the given price vectors corresponding to $\mathbf{t}_{li}$ and $\mathbf{x}_{li}$ as $\mathbf{p}_i = (p_i^1, p_i^2, \ldots, p_i^{K_i})$ and $\mathbf{q}_i = (q_i^1, q_i^2, \ldots, q_i^{M_i})$, the money income of the agent l as Y_{li}, and abstracting from other constraints, we can write the utility maximization problem of this agent as

(2) $\max u_{li} = u_{li}(\mathbf{t}_{li}, \mathbf{x}_{li})$,   s.t. $\mathbf{p}_i \cdot \mathbf{t}_{li} + \mathbf{q}_i \cdot \mathbf{x}_{li} = Y_{li}$.

If the group of foreign tourism services is weakly separable from x_{li}, then the utility function (1) can be written as

(3) $u_{li}(\mathbf{t}_{li}, \mathbf{x}_{li}) = u_{li}(v_{li}(\mathbf{t}_{li}), \mathbf{x}_{li})$,

where $v_{li}(\mathbf{t}_{li})$ is a subutility function. If the subutility function $v_{li}(\mathbf{t}_{li})$ is further assumed to be homothetic, then it is well known from consumer theory that a price index r for the "composite commodity" v_{li} can be defined, and the consumption decision can be treated as taking place in two stages.[3] In the first stage, the agent chooses the consumption of the composite commodity v_{li} along with other commodities $\mathbf{x}_{li}$. This in turn determines the expenditure to be spent on the commodity v_{li}, say $Y_{li}^t = rv_{li}$. In the second stage, the agent chooses optimal t_{li}^k, $k = 1, \ldots, K_i$ by solving the following optimization problem:

(4) $\max v_{li}(\mathbf{t}_{li})$ s.t. $\mathbf{p}_i \cdot \mathbf{t}_{li} = Y_{li}^t$.

The homotheticity of $v_{li}(\mathbf{t}_{li})$ implies that the ordinary demand function of t_{li}^k, $k = 1, \ldots, K_i$ is

(5) $t_{li}^k = Y_{li}^t b_{li}^k(\mathbf{p}_i)$.

Based on the popular assumption in the international trade literature that preferences are identical for all consumers in a country, the total demand of the origin i for tourism services in destination k is

(6) $\sum_l t_{li}^k = b_i^k(\mathbf{p}_i)\left(\sum_l Y_{li}^t\right)$.

Multiplying both sides of equation (6) by p_i^k, summing over k, and using the fact that $\sum_k p_i^k \sum_l t_{li}^k = \sum_l Y_{li}^t$ gives us $\sum_k p_i^k b_i^k = 1$. The share of the total foreign tourism expenditure of origin i on the tourism services of destination k can thus be defined as

---

3. Theoretically, the assumption of homotheticity requires an economic agent to consume all of the foreign tourism services available in the model. Since the economic agent normally makes a discrete choice among destinations, the assumption may be inappropriate if he or she does not consume at least one of the foreign tourism services. Nevertheless, the assumption does make sense for a country as a whole. In addition, it warrants mentioning that the homotheticity assumption is a sufficient, but not necessary, condition for the two-stage budgeting argument. Nevertheless, this assumption could greatly simplify the utility-maximizing process and so is widely accepted in both theoretical and applied analyses (Varian, 1992, 150–51).

$$\beta_i^k(\mathbf{p_i}) = \frac{p_i^k b_i^k(\mathbf{p_i})}{\sum\limits_{j=1}^{K_i} p_i^j b_i^j} .$$

The evolution of β_i^k can be used to indicate the change in relative competitiveness of destination k among the K_i destinations competing for visitors from origin i. Noticeably, an increase in the relative competitiveness of a destination must lead to a decrease in the competitiveness of some other destinations because $\sum_k \beta_i^k = 1$.

Although the change in relative competitiveness is important in itself, we are more interested in identifying the factors that determine the time profile of β_i^k. For that purpose, a parameter α_i is introduced to capture the change in preferences, so that $\beta_i^k(\mathbf{p_i})$ is modified as $\beta_i^k(\mathbf{p_i}, \alpha_i)$. Aside from varying with the passage of time, α_i can be "induced" to change by supply-side factors such as improved infrastructure or enhanced promotional campaigns in the destinations. Hence, $\beta_i^k(\mathbf{p_i}, \alpha_i)$ becomes $\beta_i^k(\mathbf{p_i}, \alpha_i(T, \mathbf{z}))$, with T standing for time trend and $\mathbf{z}$ for a vector of supply-side factors. Using the fact that β_i^k is homogeneous of degree zero in $\mathbf{p_i}$, after some manipulation, equation (7) can be expressed in terms of the rate of change from the base year

(8) $$G\beta_i^k = \sum_{j \neq k}^{K_i} \theta_{ij}^k (Gp_i^j - Gp_i^k) + \varepsilon_i^k GT + \sum_n \eta_{in}^k Gz_n$$

where

$$Gx = \frac{x - x^0}{x^0}, \quad x = \beta_i^k, p_i^j, T, z_n;$$

x^0 being the base-year value of x,

$$\theta_{ij}^k = \frac{\partial \beta_i^k}{\partial p_i^j} \frac{p_i^{j0}}{\beta_i^{k0}},$$

$$\varepsilon_i^k = \frac{\partial \beta_i^k}{\partial T} \frac{T^0}{\beta_i^{k0}},$$

$$\eta_{in}^k = \frac{\partial \beta_i^k}{\partial z_n} \frac{z_n^0}{\beta_i^{k0}}.$$

While the prices $p_i^j, j = 1, \ldots, K_i$ are expressed in the currency of origin i, visitors generally have to pay for tourism services whose prices are denominated in the currencies of the destinations. The exchange rate between the currency of destination j and that of origin i is needed to obtain p_i^j. Let the price of tourism services in terms of destination j's currency be p_i^{*j}, and E_i^j is the exchange rate of one unit of currency j in terms of currency i. Then $p_i^{*j} \cdot$

$E_i^j = p_i^j$. Taking log differentiation for any j and $j \neq k$ respectively, and subtracting the latter from the former, yields

$$(9) \qquad (Gp_i^{*j} - Gp_i^{*k}) + (GE_i^j - GE_i^k) = Gp_i^j - Gp_i^k,$$

where the notation Gx denotes the rate of change of the variable x. Equation (9) reveals that, for visitors from origin i, the rate of change in the relative cost of traveling to destination j versus destination k consists of two parts: one from the change in the price of tourism services in each destination, the other from the variation in exchange rates. If visitors have equal access to the information on foreign prices as well as exchange rates and treat them in exactly the same way, then equation (9) can be substituted directly into equation (8).

However, research on the economics of international tourism indicates that visitors generally have better access to information on exchange rates than on prices in the destinations, implying that visitors tend to be more sensitive to changes in exchange rates than those of prices (Artus 1972; Truett and Truett 1987; Crouch 1992; Moshirian 1993).[4] In such circumstances, the variations in exchange rates in the model have to be incorporated as separate variables, and equation (8) becomes

$$(10) \qquad G\beta_i^k = \sum_{j \neq k}^{K_i} \theta_{ij}^k(Gp_i^{*j} - Gp_i^{*k}) + \sum_{j \neq k}^{K_i} \xi_{ij}^k(GE_i^j - GE_i^k) + \varepsilon_i^k GT + \sum_n \eta_{in}^k Gz_n.$$

10.3 Data Description

10.3.1 The Sample

The model developed in section 10.2 describes the way economic agents of a particular origin allocate their foreign tourism budget among various destinations. A decrease (an increase) in the share β_i^k implies that destination k is losing (gaining) the tourism market in origin i. Because the main aim of this paper is to study and compare determinants of competitiveness of international tourism in the Southeast Asian region, the group of destinations in the subutility function is made up of Taiwan, Hong Kong, and the ASEAN Five.

As far as the origins are concerned, one problem is that the destinations, although they compete with each other, may export their tourism services to quite different markets. For example, during 1991–92, Japan was always among the top two origins for all the seven destinations; however, the United States was the second largest origin for Taiwan, and Aus-

---

4. When there is a significant change in exchange rates, the long-run adjustment could mean that both domestic and foreign prices change in the same proportion. Therefore, how prices are quoted may not be important to international tourists.

tralia was the second largest origin for Singapore. Another problem concerns the importance of intragroup tourism; the top origin for Indonesian tourism is Singapore, whereas it is estimated that some three-quarters of foreign visitors to Malaysia are from the ASEAN region. Accordingly, to have a meaningful test on the relative competitiveness, this paper has to (a) limit the origins to those importing a significant proportion of the tourism services from each of the destinations and (b) exclude all the origins that are themselves among the destinations to avoid complications from choosing between domestic and foreign tourism. With these considerations in mind, the United States and Japan are chosen as the two origins.[5] Consequently, for each origin the first sample (hereafter, sample 1) consists of seven destinations. Moreover, on the basis of data quality, three other subsamples will be analyzed for each destination. For the destinations of Taiwan, Hong Kong, and Singapore, sample 2 takes Malaysia, for which the data are the most problematic, out of sample 1; sample 3 further deletes Indonesia and the Philippines from sample 2; sample 4 has only the three newly industrialized economies (Taiwan, Hong Kong, and Singapore) in it. For the destination of Malaysia, sample 2 takes Taiwan out of sample 1; sample 3 further deletes the Philippines and Thailand from sample 2; sample 4 has only Hong Kong, Singapore, and Malaysia in it. For the destination of Indonesia, sample 2 takes the Philippines out of sample 1; sample 3 further deletes Taiwan and Hong Kong from sample 2; sample 4 has only Singapore, Malaysia, and Indonesia in it. For the destination of the Philippines, sample 2 takes Malaysia out of sample 1; sample 3 further deletes Indonesia and Thailand from sample 2; sample 4 has only Taiwan, Hong Kong, and the Philippines in it. For the destination of Thailand, sample 2 takes the Philippines out of sample 1; sample 3 further deletes Singapore and Indonesia from sample 2; sample 4 has only Taiwan, Hong Kong, and Thailand in it (please refer to table 10.1). The period to be covered spans over twenty-two years, from 1973 to 1994, during which continuous time series data are available for all the economies under examination.[6] It is noteworthy that a span of twenty-two years is far above the average time period, thirteen years, in the existing empirical studies of international tourism literature (Crouch and Shaw 1990).

5. In fact, the same empirical procedure has also been tried for other origins (such as the United Kingdom, Germany, and Australia), and the empirical results are somewhere between those of the United States and Japan. For the purpose of simplicity, only two extremes are chosen for this paper. Nevertheless, the empirical results of the other three origins are available upon request from the authors.
6. It might have been desirable to extend the data period up to 1999 in order to include the effect of the Asian financial crisis in the study. Unfortunately, the needed data set has been updated only until 1997 due to some data-collecting problems. Therefore, the investigation of the Asian financial crisis effect must wait for future study.

Table 10.1   **Subsamples' Components for Seven Destinations**

Destination	Sample 4	Sample 3 Add:	Sample 2 Add:	Sample 1 Add:
			Subsample	
Taiwan	Taiwan	Thailand	Indonesia	Malaysia
Hong Kong	Hong Kong		The Philippines	
Singapore	Singapore			
Malaysia	Hong Kong	Indonesia	The Philippines	Taiwan
	Singapore		Thailand	
	Malaysia			
Indonesia	Singapore	Thailand	Taiwan	The Philippines
	Malaysia		Hong Kong	
	Indonesia			
The Philippines	Taiwan	Singapore	Indonesia	Malaysia
	Hong Kong		Thailand	
	The Philippines			
Thailand	Taiwan	Malaysia	Singapore	The Philippines
	Hong Kong		Indonesia	
	Thailand			

10.3.2 The Variables

Dependent Variable

According to the theoretical model in the last section, the dependent variable (β_i^k) represents destination k's share of total foreign tourism expenditure of origin i. Unfortunately, with very few exceptions, this information is simply nonexistent (Moshirian 1993). In practice, most economies estimate tourism receipts by multiplying the number of tourists, average length of stay, and average expenditure per day. As the data on the average length of stay and average daily expenditure are derived from sample surveys, they are not as reliable as the data on visitor arrivals, which are gathered from the arrival and departure cards. In some cases, visitor arrivals are even the only data available. Consequently, each destination's share of the total visitor arrivals at all the destinations in each sample from a given origin will be used as the proxy variable for the dependent variable β_i^k.[7] The sources for the data on visitor arrivals by residence are from various issues of *Annual Statistical Report* by Pacific Asia Travel Association

7. It is admitted that this can hardly be an ideal exercise. Variations in the average length of stay and average daily expenditure are usually too large to be ignored. However, until sufficient and reliable information is available, it might be the best choice. This also explains why the majority of the extant studies use the number of visitor arrivals and departures in the analysis. In addition, because different foreign visitors may come to a destination for different purposes (such as sightseeing, business, visiting relatives, attending conferences, studying abroad, and others), their motivations and economic sensitivities vary considerably. Ideally, the same regression analysis should be run for different groups of foreign visitors. However, data on each different groups of foreign visitors are fragmental for most destinations in this study.

(PATA), which are supplemented by the data from *Yearbook of Tourism Statistics* of the World Trade Organization (WTO) and country-specific sources.

Independent Variables

Two sets of independent variables appear in equation (10). With seven destinations, the first set is composed of six rates of change in relative prices, whereas the other set includes variables affecting tourists' preferences. Since international tourism is an amalgam of goods and services, its price is a much more complex construct than that of any other produce. It is made up of three key elements: the price of transport services to the destination(s), the prices of goods and services purchased in the destination(s), and exchange rates. Although a tourism price defined as a function of the three components for each destination is theoretically desirable, the consumer price index (CPI) is the most popular one used in empirical studies. Some researchers have questioned the legitimacy of using the CPI in this particular context and have attempted to construct a tourist price index (Martin and Witt 1987; Moshirian 1993). However, the results are not as exciting as one would expect. In their conclusions, Martin and Witt admitted, "This study does not provide evidence of clear superiority [of a tourist price index], but rather indicates that the consumer price index, either alone or together with the exchange rate, is a reasonable proxy for the cost of tourism." Therefore, CPI in each destination is used as the proxy for the price of tourism services in that country. Data on the CPI and exchange rate of Taiwan come from the *Taiwan Statistical Data Book* 1995, whereas those of Hong Kong are from the World Bank's World Tables 1992 as well as *Hong Kong Digest of Statistics,* February 1995. The sources for the CPI and exchange rate of all the other destinations are the International Monetary Fund's *International Financial Statistics Yearbook* from 1994 and *International Financial Statistics,* September 1995. All the CPIs are converted so that 1990 is the base year.[8]

The first preferences-related variable is the time trend. It is frequently included in regression analysis to account for exogenous changes in preferences (O'Hagan and Harrison 1984; White 1985; Crouch and Shaw 1990). On the other hand, supply-side factors such as marketing expenditure or a location's innate physical attractiveness or improvement in infrastructure might well induce changes in tastes. Marketing expenditure is of particular

---

8. Transport costs are not taken into account in the analysis, although their importance cannot be denied in any sense. The main reason is the lack of satisfactory information for the whole period of 1973 to 1994 and from some destinations (especially Malaysia, Indonesia, the Philippines, and Thailand). However, this could be justified on the basis of the geographical proximity of the seven destinations. The difference in distance from the United States or Japan to any of the destinations seems not significant enough to affect the relative prices to the extent that a switch in the ranking of tourism costs among the destinations indeed occurs.

importance in a study concerning competitiveness. For instance, it is widely held that the recent boom of international tourism in Southeast Asia has been the result of the very active promotion by member states of ASEAN (Hitchcock, King, and Parnwell 1993). Similarly, a location's innate physical attractiveness for tourists is also an important determinant of an economy's competitiveness in international tourism. However, the fragmented data on marketing expenditure and the dearth of an objective measure of the location's innate physical attractiveness prevent us from including these variables in the statistical analysis. In equation (10), the supply-side factors are formulated to include only what can be reasonably quantified (or, more precisely, differentiable). Nevertheless, there are well-known qualitative (discrete) variables that tend to have extremely strong, though perhaps short-lived, impacts on international tourism. Some typical examples of such disturbances are political or social disorder, travel or foreign exchange restrictions, and special events like the Olympic Games. Following the standard procedure, we will use dummy variables to account for such events. After examining the available information, we will include four dummy variables in our empirical analysis. They are defined as

$$
D79 = \begin{cases} 1, \text{ for year 1979 after the United States terminated its official relationship with Taiwan in late 1978} \\ \\ 0, \text{ other years} \end{cases}
$$

$$
D794 = \begin{cases} 1, \text{ for China's open-door policy, } 1979-94^9 \\ \\ 0, \text{ other years} \end{cases}
$$

$$
D83 = \begin{cases} 1, \text{ for } 1983-94, \text{ referring to the social-political disorder in the Philippines after Aquino's assassination in August 1983} \\ \\ 0, \text{ other years} \end{cases}
$$

$$
D90 = \begin{cases} 1, \text{ for Visit Malaysia Year in 1990} \\ \\ 0, \text{ other years.} \end{cases}
$$

Since there is a high degree of correlation between almost every pair of the rate of change in relative prices, a single relative price for each destination is

---

9. However, an increase in new destinations does not necessarily cause a particular destination to lose its market, since there might be complementarity between the newcomers and the established one.

used instead to avoid multicollinearity. Following White (1985) and O'Hagan and Harrison (1984), we define the relative price of destination k to other destinations faced by visitors from origin i as $rp_i^k = (p_i^k)/[\prod_{j \neq k} (p_i^j)^{w_j}]$, where $w_j = s_j / \sum_{j \neq k} s_j$, and s_j is destinations except for destination k. With this, and with data availability taken into account, equation (10) can be modified to arrive at the following model for empirical studies in this paper:

$$(10)' \ G\beta_i^k = \theta_i^k Grp_i^{*k} + \xi_i^k GrE_i^k + \varepsilon_i^k GT + h_{i79}^k D79 + h_{i83}^k D83 + h_{i90}^k D90$$
$$+ \ h_{iy94}^k D794$$

where $Grp_i^{*k} = Gp_i^{*k} - \sum_{j \neq k} w_j Gp_i^{*j}$ and $GrE_i^k = GE_i^k - \sum_{j \neq k} w_j GE_i^j$.

Although the law of demand leads us to expect a negative value for θ_i^k in equation (10'), the situation is a little bit subtle in the case of international tourism. Consider an increase in the relative price of destination k (relative to alternative destinations) faced by visitors of origin i. Two extreme cases for this to occur are (a) an increase in the price of destination $k(p_i^k)$ with the prices in the alternative destinations $[\prod_{j \neq k}(p_i^j)^{w_j}]$ remaining constant, and (b) a decrease in the prices of alternative destinations with the price in destination k staying intact. In the first case, θ_i^k would as usual be expected to be negative. In the second case, depending on whether destination k and the alternative destination(s) are substitutes or complements, θ_i^k could be negative or positive. It becomes even more difficult to assign θ_i^k when both p_i^k and $[\prod_{j \neq k} (p_i^j)^{w_j}]$ change. However, given that the ASEAN Five tends to be treated as a unit and Hong Kong is closer to Taiwan in many respects, it is hypothesized that visitors regard Taiwan and the alternative destinations as substitutes, so that θ_i^k should be negative. A similar argument is applicable to the sign of ξ_i^k.

As far as the time trend and the dummy variables are concerned, figure 10.5 reveals that Taiwan has been losing its market share compared to other East Asia/Pacific countries since the mid-1970s. Therefore, ε_i^k ought to be negative if there is an autonomous switch in preferences away from Taiwan. The social political disorder in the Philippines since Aquino's assassination in 1983 necessarily discourages visitors from going to that country. As argued above, Taiwan and ASEAN are, if anything, more likely to be substitutes than complements, leading us to expect a positive coefficient for $D83$. The story is just the opposite for the coefficient of $D90$, which marks the extremely successful campaign of the Visit Malaysia Year in 1990. China's open door to foreigners has certainly had a tremendous impact on international tourism industry in East Asia, given its unique natural and cultural heritage. Although all the destinations in this study might well be affected by this policy, Taiwan would doubtless bear the highest cost, for it shares very similar tourism resources with China. As a result, the coefficient for $D794$ ought to be negative. The termination of the official diplomatic relationship with the United States in late 1978 should have adversely affected

U.S. visitors to Taiwan, implying a negative sign for h_{i79}^k when the origin is the United States.

10.4   Empirical Results

10.4.1   Estimation Procedure

The literature of international finance indicates that the linkage between domestic currency and foreign prices may be close. Thus, the pricing of tour packages may change when there is an exchange rate change. Then, there may exist an induced problem from the multicollinearity relationship among independent variables in the regression analysis. To avoid the multicollinearity problem, the variance inflationary factor (VIF) is first used to test the degree of multicollinearity among independent variables. The results show that there is no serious multicollinearity problem in our data because the VIF values of all independent variables for the destinations of Taiwan, Hong Kong, Malaysia, and Thailand are all below 5, those for Indonesia and the Philippines are close to 10, and that for Singapore is little higher than 10 in the U.S. market; those for Taiwan, Hong Kong, and the Philippines are 5.538, 6.214, and 6.116, respectively, and those for other destinations are all below 5.[10] Since the theoretical model of this paper deals with all destinations for a given origin at the same time, the appropriate estimation method is the simultaneous-equations approach. Specifically, the three-stage least squares (3SLS) method is used to estimate the parameters of equation (10') for all destinations simultaneously with respect to each origin.[11] Tables 10.2 and 10.3 report the empirical results for the United States and Japan, respectively. Although the R^2 statistics obtained by the ordinary least squares (OLS) method is presented in each equation, it should be cautioned that specification testing at the system level is more problematic than that for the single equation (O'Hagan and Harrison 1984). Therefore, the corresponding R^2 is added just for the purpose of reference.

10.4.2   Empirical Findings

The U.S. Market

The regression results of seven destinations in the U.S. market for samples 1–4 show that the estimation result of sample 1 is almost the same

---

10. The VIF value is equal to 1 for each independent variable with no correlation with each other, but the VIF value may be even more than 10 for the independent variable highly correlated with others. However, the multicollinearity problem would only increase standard deviations of estimates and reduce their significance levels even if it exists.

11. The full information maximum likelihood (FIML) method has also been used in this study. However, about half of the cases do not converge. Incidentally, the parameter estimates for other destinations are not reported, but they are available upon request from the authors.

as those of samples 2–4; however, the former is better than the latter from the statistical point of view. Therefore, only seven destinations' regression results of sample 1 will be presented in table 10.2.[12] The estimated coefficients of Grp_i^{*k} for Hong Kong, Malaysia, the Philippines, and Thailand are negative as expected and reach at least the 10 percent significance level. Those for Taiwan and Singapore are not significantly different from zero. Available tourism statistics reveal that the majority of U.S. visitors to Taiwan and Singapore are traveling for business purposes or visiting relatives are overseas Chinese. The former group is well known to be less price elastic. In addition, most of the homecoming overseas Chinese stay with their relatives and thus are generally not sensitive to price fluctuations. Consequently, the relative price change has no significant impact on U.S. visitors to Taiwan and Singapore. However, the estimated coefficient of Grp_i^{*k} for Indonesia is positive at the 1 percent significance level. One possible reason for this finding is that, depending on the way relative prices change as well as whether Indonesia and other destinations are substitutes or complements, the coefficient of Grp_i^{*k} could be of any direction. Table 10.2 shows that Indonesia and the Philippines might be complements in the U.S. market.[13] As a consequence, it is reasonable to expect that the relative price changes have a significantly positive impact on the U.S. visitors to Indonesia. By referring to Gujarati (1988, 178–82) as well as Pindyck and Rubinfeld (1981, 128–30), we conclude that the other likely explanation for the perverse result is that some important explanatory variables such as travel costs, which of course are not reflected in Indonesian CPI, are missing from the empirical model in this study. Hence, the significantly positive coefficient of Grp_i^{*k} might just capture the effect of the almost monopolized air fares between Indonesia and the United States. Without doubt, there is no easy way to test this proposition until appropriate data are available.

The estimated coefficients of GrE_i^k for Malaysia and Thailand have the expected (negative) signs. Those for Taiwan, Hong Kong, Singapore, and Indonesia are not significantly different from zero due to the fact that a considerable part of the travel expenses in these destinations is priced in U.S. dollars rather than local currencies. Nevertheless, surprisingly, the coefficient of GrE_i^k for the Philippines is positive and reaches the 10 percent significance level. The perverse exchange rate effect could be due to omission of some relevant explanatory variables, such as the aggregate economic condition or foreign direct investment, that are correlated with GrE_i^k and might have a significantly positive impact on the U.S. visitors to the Philippines. Accordingly, the significantly positive coefficient of GrE_i^k might absorb the effects of omitted variables.

12. The regression results of seven destinations in the U.S. market for samples 2–4 are available upon request.
13. One possible reason for this result is that U.S. tourists visit both countries in the same trip.

Table 10.2   The Regression Results of Seven Destinations in the U.S. Market (1973–94)

	Taiwan	Hong Kong	Singapore	Malaysia	Indonesia	The Philippines	Thailand
Grp^{*k}_i	0.017	-0.405***	-0.028	-0.103*	0.235***	-0.142***	-0.189***
	(0.794)	(-4.979)	(-1.284)	(-2.000)	(5.219)	(-6.980)	(-5.321)
GrE^k_i	0.029	0.242	0.105	-0.808***	0.124	1.152*	-0.817*
	(0.569)	(1.220)	(0.609)	(-3.418)	(0.660)	(2.022)	(-1.922)
GT	0.382	-1.931***	0.041	-3.754***	-5.832***	13.756***	-4.815***
	(1.253)	(-3.463)	(0.061)	(-3.076)	(-3.106)	(5.376)	(-6.795)
$D79$	2.801	-10.786	—	—	—	44.538***	—
	(0.930)	(-1.516)				(3.583)	
$D794$	-25.940***	24.806***	-8.769*	-14.488	-25.662***	45.795***	—
	(-7.725)	(3.767)	(-1.914)	(-1.463)	(-3.890)	(3.578)	
$D83$	-2.233	25.870***	-10.783**	8.455	-19.335***	-34.261*	16.833
	(-0.746)	(3.448)	(-2.680)	(1.030)	(-3.022)	(-2.090)	(1.635)
$D90$	-4.181	-5.733	3.772	50.082***	-6.106	-23.007	21.658*
	(-1.158)	(-0.670)	(0.702)	(4.487)	(-0.723)	(-1.184)	(2.075)
R^2	0.979	0.965	0.589	0.971	0.946	0.969	0.914

Notes: GT indicates exogenous changes in preferences; $D79$ indicates the U.S. termination of its official relationships with Taiwan in late 1978; $D794$ indicates China's open-door policy; $D83$ indicates the social political disorder in the Philippines after Aquino's assassination in August 1983; and $D90$ indicates the Visit Malaysia Year in 1990. The numbers in parentheses are t-statistics. R^2 is from OLS. Dash indicates "not applicable."

***Significant at the 1 percent level.

**Significant at the 5 percent level.

*Significant at the 10 percent level.

The most puzzling findings from table 10.2 are that the estimated coefficients of the rate of change in time trend (GT) for Hong Kong, Malaysia, Indonesia, and Thailand are negative at the 1 percent significance level. This implies that there is an autonomous switch in preferences away from the above four economies. The estimated coefficient of GT for the Philippines is positive at the 1 percent significance level, implying that there is an autonomous switch in preferences for the Philippines. Surprisingly, those for Taiwan and Singapore are not significantly different from zero. One possible reason for the above findings is that figure 10.5 is a synthesized product of multiple factors.

The estimated coefficient of $D79$ is positive and highly significant only for the Philippines, but those for other Southeast Asian economies are not significantly different from zero. This may imply that Taiwan and the Philippines are substitutes for each other in the U.S. market.

The most striking, though not surprising, findings from table 10.2 are that the estimated coefficients of $D794$ for Taiwan, Singapore, and Indonesia are negative and reach at least the 10 percent significance level. This result confirms the belief that the open-door policy adopted by China has had an adverse impact on international tourism in Taiwan, and it may imply that China is also a substitute for Singapore or Indonesia in the U.S. market. As clearly demonstrated in table 10.2, this is definitely the single most important factor responsible for the dramatic decline of Taiwan's market share in the U.S. market during the past fifteen years. Not only does it dominate the diplomatic setback between the United States and Taiwan ($D79$) in determining the relative competitiveness of the latter in the U.S. market, but it also helps clarify the false impression that there are "autonomous" changes in preferences (GT) among the U.S. visitors to switch away from Taiwan. On the contrary, the highly significant positive coefficients of $D794$ for Hong Kong and the Philippines imply that China is a complement for them in the U.S. market, and they are beneficiaries of the Chinese open-door policy.

As to the coefficient of $D83$, it is significantly negative for Singapore, Indonesia, and the Philippines, implying that the Philippines and Singapore (or Indonesia) seem to be complements in the U.S. market; however, it is positive and highly significant for Hong Kong, implying that Hong Kong and the Philippines are substitutes for U.S. tourists.[14] Overall, while the social-political disorder after Aquino's assassination did deter U.S. visitors from going to the Philippines, its single biggest beneficiary is Hong Kong.

The estimated coefficients of $D90$ are positive and significant at the 1 percent and 10 percent levels, respectively, for Malaysia and Thailand, imply-

---

14. This accords with the finding of Chu (1993), which show that the Philippines and Hong Kong are close substitutes for international tourists.

ing that Malaysia did a very successful campaign in 1990, and Thailand is another beneficiary of the event in the Southeast Asian region. In addition, the very successful campaign of the Visit Malaysia Year in 1990 only slightly affected other economies' shares of U.S. visitors, as revealed by the regression results of table 10.2.

The Japan Market

The regression results of seven destinations in the Japan market for samples 1–4 also show that the signs of all corresponding coefficients in sample 1 are the same as those in samples 2–4; nevertheless, the former are statistically better than the latter. Hence, only seven destinations' regression results of sample 1 will be presented in table 10.3.[15] It turns out that the results are very different from those of the United States. The estimated coefficient of Grp_i^{*k} for all the destinations is insignificantly different from zero, except that for Singapore.

Two distinct behavior patterns between U.S. and Japanese visitors might be responsible for this finding: one is the difference in the purposes of visit (including sightseeing, business, visiting relatives, attending conference, studying abroad, and others), and the other is the difference in the patterns of travel (including packaged group tours and personal nongroup tours). The top two purposes for Japanese visitors to the Southeast Asian region are sightseeing and business.[16] By contrast, the first two purposes for U.S. visitors to the region are business and visiting relatives.[17] As far as the patterns of travel are concerned, most Japanese visitors with the purpose of sightseeing came with packaged group tours; most U.S. visitors with the purpose of business or visiting relatives came with personal nongroup tours. The packaged group tours are run by travel agencies, which generally have a better bargaining position for all kinds of tourism expenditures and have a long-term contract with local tourism-related firms in the destinations. Hence, variations in the relative price levels do not have any significant correlation with most of the Southeast Asian destinations' market shares in the Japan market. However, the positive correlation between the market share and the relative price for Singapore is beyond expectation. Yet it is noticeable that their absolute values are all smaller than the corresponding coefficient of GrE_i^k in each sample, implying that the combined price and exchange rate effect might well be consistent with the law of demand. As in the case of U.S. visitors to Indonesia, there are two possible reasons for the significantly positive coefficient of Grp_i^{*k} in table 10.3. The first is

15. The regression results of seven destinations in the Japan market are available upon request.

16. For instance, about 70 percent and 20 percent of Japanese visitors to Taiwan are there for sightseeing and business, respectively, both in 1993 and 1994 (Tsai and Wang 1998).

17. For example, some 40 percent and 30 percent of U.S. visitors to Taiwan are there for business and to visit relatives, respectively, both in 1993 and 1994 (Tsai and Wang 1998).

Table 10.3 The Regression Results of Seven Destinations in the Japan Market (1973–94)

	Taiwan	Hong Kong	Singapore	Malaysia	Indonesia	The Philippines	Thailand
Grp_i^{*k}	0.001	0.044	0.005*	-0.031	-0.048	0.095	0.060
	(0.032)	(1.189)	(2.033)	(-0.289)	(-0.633)	(1.173)	(1.307)
$Gr_rE_i^k$	-0.335*	0.165	-0.010	0.461	0.324	-2.264*	1.238**
	(-2.051)	(1.044)	(-0.603)	(1.112)	(0.715)	(-1.770)	(2.207)
GT	0.167	-0.803**	0.210***	-1.325	2.975	-4.644	1.620**
	(0.390)	(-2.370)	(2.970)	(-0.692)	(0.853)	(-0.828)	(2.343)
$D79$	13.390***	-6.799**	—	—	—	—	—
	(3.292)	(-2.191)					
$D794$	-9.836*	5.380	0.095	6.990	—	-18.155	—
	(-1.956)	(1.165)	(0.162)	(0.503)		(-0.521)	
$D83$	10.150**	9.602**	-1.823***	1.258	-2.007	-43.902	-9.518
	(2.147)	(2.514)	(-3.529)	(0.107)	(-0.194)	(-1.189)	(-1.454)
$D90$	-7.549	-0.510	-0.339	124.069***	11.317	-26.595	-2.426
	(-1.421)	(-0.133)	(-0.482)	(7.632)	(0.810)	(-0.661)	(-0.310)
R^2	0.979	0.965	0.589	0.971	0.946	0.969	0.914

Notes: See table 10.2.

that, depending on the way relative prices change as well as whether Singapore and other destinations are substitute or complements, the coefficient of Grp_i^{*k} could be of any direction. Although table 10.3 shows that Singapore and the Philippines might be complements in the Japan market, no further convincing evidence could be given to support that Japanese visitors to Singapore have treated Singapore and the other destinations under study as complements.[18] The second likely explanation for the perverse price effect is that some important explanatory variables such as air fares, which account for more than 50 percent of the tourism expenditures of the Japanese visitors to Singapore and are not reflected in Singapore's CPI, are missing from the empirical model. As a result, the positive coefficient of Grp_i^{*k} might capture the effect of the almost monopolized air fares between Singapore and Japan. Again, this proposition cannot be easily tested without appropriate data.

With reference to the estimated coefficient of GrE_i^k in the Japan market, it is significantly negative for Taiwan and the Philippines, but insignificantly different from zero for Hong Kong, Singapore, Malaysia, and Indonesia. The reason for the former is that the tourism expenditures in Taiwan and the Philippines are priced in terms of U.S. dollars only. Therefore, in contrast to the case of U.S. visitors, the variation in exchange rates affects Japanese visitors to Taiwan and the Philippines adversely. This result corroborates those widely found in the international tourism literature (Truett and Truett 1987; Crouch 1992). On the other hand, one possible reason for the latter finding is that the tourism expenditures in Hong Kong, Singapore, Malaysia, and Indonesia may also be priced in terms of Japanese yen in addition to U.S. dollars. As a result, just as in the case of U.S. visitors, the variation in exchange rates does not significantly affect Japanese visitors to Hong Kong, Singapore, Malaysia, and Indonesia. Beyond expectation, the estimated coefficient of GrE_i^k for Thailand is positive at the 5 percent significance level. Again, the likely explanation for the perverse exchange rate effect is the same as that in table 10.2.

Table 10.3 shows that Hong Kong lost some of its market share and Singapore and Thailand gained some market shares during the past two decades because of the changes in preferences among Japanese visitors. By examining the tourism statistics, we find that there is a clear trend for Japanese tourists to switch from Hong Kong to Singapore after 1980. There is some evidence showing that the change in the age structure among Japanese visitors might have played a crucial role in this matter. Japanese visitors have been getting younger and younger all the time. However, to what extent the preferences of the younger generations differ from those of the

18. One possible piece of evidence to support the complementarity between Singapore and the Philippines is that most Japanese visitors come to both countries in the same trip. However, available data on the above evidence have been incomplete.

older ones and thus contribute to the gain of Singapore's and the loss of Hong Kong's competitiveness in the Japanese market is certainly a topic worth further investigation.

As far as $D79$ and $D794$ are concerned, the estimated coefficient of $D79$ for Taiwan in the Japan market is positive and highly significant, but that for Hong Kong is significantly negative. Two possible reasons might be responsible for the above finding. One is that Taiwan and Hong Kong are substitutes in the Japanese market because of cultural similarity. The other is that the promotion effort of Taiwan was focused on the Japanese market in 1979 after the United States terminated its official relationship with Taiwan in late 1978. These two factors might explain why Taiwan gained its share and Hong Kong lost its share of the Japanese market in 1979. The estimated coefficient of $D794$ for Taiwan is negative at the 10 percent significance level in samples 1–2, implying that, just as in the case of U.S. tourists, the Chinese open-door policy has had an adverse impact on Taiwan's market share in Japan. Nevertheless, the effect of China's open-door policy is insignificant in samples 3–4 and much less significant than that in the case of U.S. visitors to Taiwan. It is well known that China's attractiveness is mainly based on its abundant cultural and natural heritage. For most visitors from the Western world, such as the United States, ancient Chinese civilization is certainly mysterious and fascinating. Conversely, due to the geographical proximity and long-lasting historical connection with China, the Japanese have a better understanding of and thus less curiosity about mainland China. This, along with the fact that the United States is one of the targeted countries with tourism subsidized by the Chinese government, whereas there is still widespread xenophobia toward Japanese visitors in China, might explain the different reactions to China's open-door policy.

Table 10.3 indicates that the social-political disruption in the Philippines after 1983 has affected Japanese visitors to Taiwan and Hong Kong positively. That is, Taiwan and Hong Kong are beneficiaries from the chaos in the Philippines. Table 10.3 shows that the estimated coefficient of $D83$ is negative and highly significant, implying that Singapore and the Philippines might be complements for Japanese tourists.[19] Finally, the estimated coefficient of $D90$ for Malaysia is positive and extremely significant. This result indicates that the successful tourism promotion by Malaysia in 1990 did appear to have attracted the Japanese visitors away from other economies in the Southeast Asian region. It is noteworthy that the positive impact of $D90$ is much more significant for Japanese visitors than U.S. visitors. There are two possible reasons for this. First, most Japanese visitors came to the region for sightseeing and with packaged group tours; most U.S. visitors came to the region for business and visiting relatives, and their patterns of travel

---

19. One possible reason for this result is that Japanese tourists visit both countries in the same trip.

are individual. Second, the packaged group tours are operated by travel agencies, which generally have more information about promotional campaigns than individual tourists do. Because the Visit Malaysia Year campaign in 1990 aimed chiefly at pleasure-oriented visitors, it should not be surprising to find that it had more effect on the Japanese visitors than on the U.S. visitors.

10.5 Summary and Conclusions

In response to the continuing growth of national income in most countries, the substantial decline in the real costs of transportation, the liberalization of cross-border movements, and the increasing propensity to travel abroad, most economies in the Southeast Asian region devoted themselves to international tourism and experienced tremendous growth in terms of foreign visitors during the period of 1973 to 1994. Although the Southeast Asian region's share of the total world tourists had been steadily growing, and all economies in the region had a growing or fairly stable share of the world market, some economies had been continuously losing ground, and some other economies had been continuously gaining ground or had experienced ups and downs in the market for international tourism vis-à-vis the neighboring Southeast Asian economies in the past two decades. The main purpose of this paper is, therefore, to examine and identify the factors determining the relative competitiveness of each economy's international tourism industry among a group of rivals consisting of Taiwan, Hong Kong, Singapore, Malaysia, Indonesia, the Philippines, and Thailand.

Based on a two-stage budgeting decision, a theoretical model has been developed. With the assumption of homotheticity in preferences, it is derived that the change in a destination's share of the international tourism budget of a given origin depends on changes in relative prices among alternative destinations and factors affecting preferences. Using each economy's share of the total visitors from an origin to all the destinations as the proxy for the dependent variable, and with the CPI in each destination as the proxy for the price of tourism services, an empirical analysis by 3SLS has been performed. Two origins (the United States and Japan) and seven destinations (Taiwan, Hong Kong, and the ASEAN Five) are considered for four different samples. The major findings are, first, that the relative price effect depends critically on the purposes of visit and the patterns of travel. U.S. visitors are not sensitive to price fluctuations in Taiwan and Singapore due to the fact that the majority of them come to the two economies for business or to visit relatives. Japanese visitors are not sensitive to the relative price variations in the Southeast Asian destinations, except Singapore, because the majority of them come to these destinations with packaged group tours, which are run by the better-bargaining-power travel agencies. Second, the exchange rate effect depends on whether the tourism expenditures in the destinations are priced in local currencies or the currency of the origin country. International tourists are not sensitive to the variations in exchange rates if a considerable part of the travel expenses in the destinations are priced in the currencies in their origin countries. Third, supply-side factors could indeed have decisive influences on the market share. However, their effects appear to be origin- and destination-dependent. China's open-door policy stands out as the single most important factor responsible for the loss of Taiwan's market share and a major factor in the gain of Hong Kong's market share in the United States. It seems, however, to have a less significant impact on Japanese visitors to Taiwan and Hong Kong. The promotional campaign of Malaysia in 1990 did have a significant impact on the Malaysian competitiveness in the United States and Japan. Social-political disorder in the Philippines had a negative impact on U.S. visitors, and it also led to the gain of Hong Kong's and the loss of Singapore's and Indonesia's relative competitiveness in the United States. In contrast, the social-political disorder in the Philippines has not had any significant impact on Japanese visitors to the Philippines, but it did lead to the gain of Taiwan's and Hong Kong's, and the loss of Singapore's, relative competitiveness in the Japan market.

The empirical results in this paper yield two important lessons. First, different promotional strategies may be produced for different countries of origin. The relative price effect depends on the purposes of the visit and the patterns of travel; the exchange rate effect depends on whether the tourism expenditures in the destinations are priced in local currencies or the currency of the origin country; and the supply-side effects are origin- and destination-dependent. As a consequence, it could be a serious mistake to treat all foreign visitors as homogeneous in designing any policy to attract foreign visitors. In other words, a successful promotional policy for any destination demands an in-depth study of the characteristics of visitors from various origins and the destination itself. Secondly, it is suspected that the perverse price effect and the perverse exchange rate effect obtained in tables 10.2 and 10.3 could be due to the omission of important explanatory variables. Therefore, more data on potential supply-side variables are needed if the determinants of international tourism are to be understood more thoroughly. Only in this way can more meaningful policy implications be obtained.

References

Artus, J. R. 1972. An econometric analysis of international travel. *International Monetary Fund Staff Papers* 19 (3): 579–614.
Chu, F. L. 1993. Elasticities of substitution in Pacific Rim tourism. National Taiwan University, Graduate Institute of National Development. Mimeograph.

Crouch, G. I. 1992. Effects of income and price on international tourism. *Annals of Tourism Research* 19:643–64.

Crouch, G. I., and R. N. Shaw. 1990. Determinants of international tourist flows: Findings from thirty years of empirical research. In *The tourism connection: Linking research and marketing.* Proceedings of the Twenty-First Annual Conference of the Travel and Tourism Research Association, 45–60. Salt Lake City, Utah: TTRA.

Deaton, A., and J. Muellbauer. 1980. An almost ideal demand system. *American Economic Review* 70:312–26.

Eadington, W. R., and M. Redman. 1991. Economics and tourism. *Annals of Tourism Research* 18:41–56.

Gujarati, D. N. 1988. *Basic econometrics,* 2nd ed. New York: McGraw-Hill.

Hitchcock, M., V. T. King, and M. J. G. Parnwell. 1993. *Tourism in South-East Asia.* London: Routledge.

Martin, C. A., and S. F. Witt. 1987. Tourism forecasting models: Choice of appropriate variable to represent tourists' cost of living. *Tourism Management* 8:223–45.

Moshirian, F. 1993. Determinants of international trade flows in travel and passenger services. *Economic Record* 69 (206): 239–52.

O'Hagan, J. W., and M. J. Harrison. 1984. Market shares of U.S. tourist expenditures in Europe: An econometric analysis. *Applied Economics* 16 (6): 9119–31.

Pindyck, R. S., and D. L. Rubinfeld. 1981. *Econometric models and economic forecasts,* 2nd ed. New York: McGraw-Hill.

Truett, D. B., and L. J. Truett. 1987. The response of tourism to international economic conditions: Greece, Mexico, and Spain. *Journal of Developing Areas* 21 (2): 177–90.

Tsai, P. L., and K. L. Wang. 1998. Competitiveness of international tourism in Taiwan: U.S. versus Japanese visitors. *Applied Economics* 30:631–41.

Varian, H. R. 1992. *Microeconomic analysis,* 3rd ed. New York: W. W. Norton.

White, K. J. 1985. An international travel demand model: U.S. travel to Western Europe. *Annuals of Tourism Research* 12 (4): 529–45.

World Tourism Organization (WTO). 1999. *Yearbook of Tourism Statistics.* Madrid, Spain: WTO.

Comment   Keiko Ito

First of all, we should underline the fact that this paper deals with a topic that is not yet very developed in economics literature. In development studies, it is widely recognized that tourism makes a very significant contribution to the acquisition of foreign currencies for developing economies. According to the World Bank's *World Development Indicators,* the share of travel services in total exports is relatively high in these countries. For example, it was about 10 percent in Indonesia, Thailand, and the Philippines in 1995. Although there are several conceptual difficulties when considering tourism and a lack of appropriate data for empirical analyses, much more attention should be paid to international tourism as a major export industry.

Keiko Ito is a research assistant professor at the International Centre for the Study of East Asian Development in Kitakyushu, Japan.

Confronting these difficulties, the authors try to estimate the sensitivity of a destination's share of the international tourism budget to changes in relative prices among different destinations. This paper is certainly an original work with respect to the following points:

1. It is probably the first work that simultaneously estimates the tourism demand model for seven Southeast Asian countries and compares the travel patterns of American and Japanese tourists.
2. The estimated results clearly show that the factors affecting preferences are very different between American and Japanese tourists. In particular, they confirm that social or political events and marketing promotion have significant impacts on tourists' preferences. The results, therefore, provide some relevant policy implications.
3. This study finds that American tourists are relatively price sensitive, whereas Japanese ones are not so sensitive to price variation. In addition, the effect of relative foreign exchange rates variation is different among their destinations.

The last point provides the central theme and focus of analysis in the paper. The authors aim to investigate relative price and exchange rate variations' effects on tourist demand for each destination. However, the estimated sign and significance of the coefficients are very different among the countries of origin and of destination, and it is quite difficult to induce some persuasive interpretations. I summarize the sign and the significance of the relative price and exchange rate variations for the countries in question in table 10C.1. The authors obtain fairly reasonable results for relative price changes for American tourists, who are price sensitive in most of their seven destinations. However, for Indonesia, the coefficient of relative price changes is positive and significant. The authors give two interpretations for this result. The first one is that Indonesia and the Philippines seem to be complements in the U.S. market, and, therefore, it is reasonable to expect a significantly positive relative price change effect. However, I am not quite sure from where this interpretation derives. I think that without more thorough investigation, we cannot determine which country is a substitute for, or a complement of, any particular country. Rather, I agree with the authors' second interpretation, that some important explanatory variables such as transportation costs are missing from the empirical study. As the authors mention, the local government monopolizes the Indonesian airline industry, and the consumer price index (CPI) does not reflect adequately the price for tourists. Although the authors explain that some previous studies did not provide evidence of clear superiority of a tourist price index to the CPI, I still have some questions on using CPI as a proxy for the price index. When the CPI is used, the estimated parameters do not necessarily reflect the true elasticity of demand for travel services with respect to the relative price of travel services. Particularly in developing countries, the price of travel services, which is mostly for foreigners and richer people, might be

Table 10C.1   Summary of Sign and Significance of Coefficients

	Origin			
	United States		Japan	
Destination	Price Change	Exchange Rate	Price Change	Exchange Rate
Taiwan	+	+	+	_*
Hong Kong	_***	+	+	+
Singapore	–	+	+*	–
Malaysia	_*	_***	–	+
Indonesia	+***	+	–	+
The Philippines	_***	+*	+	_*
Thailand	_***	_*	+	+**

Source: Summarized by the discussant using the estimated results of tables 10.2 and 10.3.
***Significant at the 1 percent level.
**Significant at the 5 percent level.
*Significant at the 10 percent level.

different from the CPI. Therefore, I think that it is still worth constructing the tourist price index and investigating its validity.

As for the price sensitivity of Japanese tourists, the authors explain that the Japanese are not price sensitive because the majority of them travel in packaged group tours. This interpretation seems quite reasonable, but I still wish to suggest using the tourist price index. Moreover, careful investigation is further required for the effect of exchange rates. As shown in table 10C.1, the signs of the coefficients are different among the origins and the destinations, and there are no clear rules for the variation of signs. I partially agree with the authors' view that international tourists are not sensitive to the variations in exchange rates if a considerable part of the travel expenses in their destinations are priced in the currency of their own countries. However, for example, airfares, which are a major proportion of travel expenses, usually reflect exchange rate fluctuations.

Therefore, I still expect that the exchange rate variation has a more significant impact on both American and Japanese tourists. Again, the most likely reason for these ambiguous results for the exchange rate variations might be that some other important explanatory variables are missing. To investigate the relative price change effects, one possible alternative is that one could use the exchange rate–adjusted tourist price index instead of using the price index and the exchange rate separately. Moreover, one could use a better proxy for the supply-side factors rather than using the time trend.

So far, I have commented on the price and exchange rate sensitivity. Although there are some setbacks in constructing dependent and independent variables, I should emphasize that the major findings from this study provide some interesting evidence and implications. That is, the travel patterns of American and Japanese tourists are very different, and so are the price sensitivities of both. This implies that policy makers of each destination are required to consider the most effective strategy for each origin.

Last but not least, I would like to confirm again this paper's important contributions to the field of empirical analysis of tourism in Southeast Asian countries. Even if there are many limitations, by using considerably longer time series data, this paper represents a pioneering piece of work and raises an important issue that is definitely lacking detailed data of the tourism industry in this region. Further research, incorporating more supply-side and country-specific factors, should be encouraged.

Comment   Mahani Zainal-Abidin

The paper investigates the relative competitiveness of international tourism among a group of seven East Asian economies. The study shows that some destinations did indeed lose their competitiveness, and it highlights some interesting findings, namely:

1. Tourists are not sensitive to relative price variations if their purposes for visiting are business or to visit relatives or if they come in a packaged tour.

2. Tourists are not sensitive to relative variations in exchange rates if most of the tourism expenditures at the place of destination are priced in their home currencies.

3. Qualitative supply-side factors have a decisive influence on a country market share, but they are origin- and destination-dependent.

Tourism trade is an important gross domestic product (GDP) contributor to many developing economies. Many of these countries have offered tourism in their General Agreement on Trade in Services (GATS) liberalization commitment because they feel that they have a comparative advantage in this area. Therefore, it is important that East Asian economies maintain or increase their competitiveness to raise the GDP contribution of this industry in their respective economies. In this regard, this paper has dealt with an important issue of relative competitiveness between countries and the factors that influence the demand for tourism services.

I enjoyed reading the paper because it gives an extensive exposition on the topic of demand for tourism services. My comments follow.

Why Europe is Not Included in the Set of Origins

The paper (which only considers the United States and Japan) assumes that tourism markets are interchangeable, whereas tourism products are

Mahani Zainal-Abidin is professor of applied economics at the University of Malaya.

homogeneous and substitutable. In reality, tourism markets are segmented and sometimes product specific. For example, tourists from the United States are likely to visit Taiwan, Hong Kong, the Philippines, and Thailand because of historical ties and political links. The Philippines has very close political links with the United States, and Thailand was the military base for the United States during the Indo-Chinese war. On the other hand, European tourists feature significantly in Malaysia, Singapore, and Indonesia, again due to historical links.

Therefore, in evaluating the relative competitiveness in the tourism industry among East Asian countries, it is essential that the universe of the origin be as comprehensive as possible. It is unlikely that Malaysia, for example, develops competitiveness in the U.S. market when most of the time the majority of its tourists are from European countries.

In part, this is due to the nature of tourism, which is such that information and marketability of a destination take a long time to be developed. Furthermore, the tourism market is relatively segmented and product specific: for example, some Association of Southeast Asian Nations (ASEAN) destinations, such as Malaysia, Indonesia, and Thailand, have developed a market niche in eco-tourism and consequently are almost noninterchangeable with other destinations. Similarly, Hong Kong is a getaway to China and attracts tourists when China opens its doors to foreign tourists.

Perhaps in analyzing the issue of relative competitiveness, it is useful to consider the rate of growth in tourist arrivals and tourism contribution in each destination in addition to the present analysis of relative share of tourism expenditure. In this way, we can incorporate arrivals from new markets—for example, Australian tourists are emerging as a significant source of tourism for ASEAN destinations.

Tour Packages and Intragroup Tourism

The paper acknowledges the importance of packaged or intragroup tourism but decided to limit or exclude their role. In ASEAN, at least, a substantial portion of tourists, particularly from Japan, come in this form. They generally do not make a single destination trip but conduct a tour of ASEAN destinations—starting perhaps with Singapore, followed by Malaysia, and ending up in Thailand. It is important to include this form of tourism if we want to evaluate the relative competitiveness of the various destinations because it constitutes a large part of tourism in these destinations.

Relative Price Variations

In most of the estimated equations, the price coefficients met the expected signs: namely, a price reduction will increase the demand for tourism. However, there are instances in which the estimates are contrary to expectation; that is, the price coefficient is positive.

Once again, this is the case of a segmented market where a price increase indicates quality improvement, which has resulted in higher demand for tourism services. Not all increases in tourism services are generated by lower prices. For example, prices in Singapore have been increasing, but tourist arrival has not dropped. Singapore has increased the quality of its services to compensate for higher costs, and now it targets the higher end of the tourist market.

Thus, the paper may wish to reevaluate the competitiveness of such markets to include improvement in quality of services.

Variation of Samples

The paper gives the variations of samples used in the estimation. It would help the readers if these variations were put in a table for easier reading. The paper should also explain these variations. For example, for the destination of Indonesia, sample 3 deletes the Philippines, Taiwan, and Hong Kong.

Supply-Side Factors

Relative competitiveness is significantly influenced by supply-side factors. I do agree with the authors that marketing efforts are critical, but they are difficult to incorporate in econometric equations.

In the last two years, marketing has featured very prominently in the drive to attract tourists—Thailand with its hugely successful *Amazing Thailand* campaign and Singapore with its shopping and art festivals. Perhaps this factor should be included in the estimation of competitiveness in view of its importance.

Another important supply-side factor is the price of airfares and airlines' availability and connections. Each destination has used its national airline as a key channel to attract tourists by offering steep discounts in airfares. As with marketing, this factor is difficult to qualify but worth considering.

Effects of the Crisis

The East Asian Crisis has caused sharp currency devaluation in most of the destinations. What are the effects of this depreciation on the demand for tourism services and the relative competitiveness of these destinations? For example, it would be very interesting to compare the effects of currency devaluation between Thailand and Malaysia during the Crisis, because the former enjoyed a tourism boom during the crisis, whereas the latter did not when they both experienced a similar level of depreciation.

This brings to attention the question of whether prices or supply-side factors are more important in influencing the demand for tourism services.

Globalization and Harmonization
The Case of Accountancy Services in Japan

Fukunari Kimura

11.1 Introduction

The inclusion of the General Agreement on Trade in Services (GATS) in the Marrakesh Agreement in 1994 was an epoch-making step to expand the scope of international discipline on commercial policies. Among professional services, accountancy services are a particularly promising field for harmonization, and the negotiation in the World Trade Organization (WTO) has set the framework of how to treat domestic regulations. We can draw a number of important lessons from the ongoing story of accountancy service liberalization, particularly in case of Japan.

An important contribution of GATS was to define the scope of services trade and set the framework of liberalization procedure. As policy principles, the GATS first imposes the most-favored-nation (MFN) principle with explicit exemptions and transparency requirements. Then the market access requirements and the national treatment (NT) principle are promoted in the form of each country's specific commitments. Because most of the service sectors are subject to complicated domestic regulations and institutional arrangements, liberalization will inevitably step into the traditional territory of domestic policies.

The borderline between "pure" domestic policies and international commercial policies under the international policy discipline is not clear-cut in many cases. The framework of domestic regulations is deeply rooted in his-

Fukunari Kimura is professor of economics at Keio University.

The author would like to thank the East Asian Seminar participants, particularly Ann Krueger, Takatoshi Ito, Aaditya Mattoo, and Edwin Lai, for constructive comments and encouragement.

tory-dependent country-specific institutions and has been regarded as being under the policy discretion of each country. The policies on service industries were mostly treated as a part of domestic indigenous institutions. On the other hand, the wave of globalization of economic activities calls for international policy discipline, and the momentum toward institutional harmonization among countries has intensified. The scope of international policy discipline is not given a priori but depends on the balance between the benefits from having country-specific institutions and the advantage obtained from globalization. The process of liberalization, therefore, is inevitably accompanied with complicated politico-economic conflicts.

Accountancy services are an area with particularly strong domestic regulatory arrangements. The business accounting system is based on the country-specific legal framework and is not often open to foreign firms or natural persons. The accountants qualification system tends to work as an unintentional barrier to foreign penetration. At the same time, the globalization of economic activities has recently called for the international convergence of accounting systems. In particular, the construction of international accounting standards has proceeded in the effort of the International Accounting Standards Committee. The global restructuring of the accountancy service sector has also advanced through megamergers.

In Japan, drastic institutional changes called the "accounting Big Bang" are going on. The disillusion of the Japanese economic system and the wave of globalization accelerate the convergence of the accounting system with international standards. A series of institutional reforms relating to business accounting is likely to change a wide range of economic institutions, including corporate governance, in the near future.

The purpose of the paper is to review the transition of the Japanese business accounting system and to examine the function of market forces affecting the relationship between domestic institutions and international policy discipline. The next section provides the definition of accountancy services and lists their special characteristics. Section 11.3 summarizes the historical legal and regulatory background of the Japanese business accounting system. Section 11.4 examines the wave of globalization with which Japan has been washed out, and section 11.5 reviews the recent reforms in Japan. Section 11.6 summarizes the liberalization effort in the WTO, and the last section draws lessons from the case.

11.2   Characteristics of Accountancy Services

Because the legal and regulatory framework for accountancy services widely differs across countries, it is not even easy to define accountancy services in the international context. The Provisional Central Product Classification (CPC) of the United Nations has "accounting, auditing, and bookkeeping services [CPC 862]" under the category of "business services." The category CPC 862 is further subdivided as follows:[1]

1. Accounting and auditing services (CPC 8621)
   Financial auditing services (CPC 86211)
   Accounting review services (CPC 86212)
   Compilation of financial statements services (CPC 86213)
   Other accounting services (CPC 86219)
2. Bookkeeping services, except tax returns (CPC 8622)
   Bookkeeping services, except tax returns (CPC 86220)

However, the range of services provided by accountants or accounting firms has substantially expanded beyond the traditional definition. In addition to accounting, auditing, and bookkeeping services, they often provide merger audits, insolvency services, tax advice, investment services, and management consulting. Differences in the regulatory environment result in different definitions from country to country.

In case of Japan, at least in a colloquial usage, accountancy services mean professional services provided by accountancy firms that conduct legally required audits as a core service. The government authorizes Japanese Certified Public Accountants (CPAs) under the CPA law. Only accountants with the CPA qualification can conduct legally required audits.

Accountancy services have several special characteristics, which have generated peculiar responses to globalization. First, accountancy services considerably differ across countries. The accounting system is an important component of legal and economic institutions that support market functioning and thus has a strong path-dependent nature. It has historically developed in the country-specific environment of corporate governance. Accountancy services are also prone to be interlocked with other economic institutions. Moreover, there are considerable differences across countries in the form and degree of government involvement. For example, in Japan, legal auditing practices are specified in detail by the government and are conducted by accountants with publicly authorized licenses. On the other hand, in the United States, accountancy services are primarily based on private qualification and follow formats that differ across states while government agencies, particularly the Securities and Exchange Commission (SEC), are strictly monitoring accounting practices.

Second, although accounting practices by themselves are provided on an individual or small-group basis, the operation of accounting firms has strong economies of scale. Particularly in the globalization era, accountancy firms must be ready to provide a wide range of services requested by globalizing client firms, such as tax advice, management consulting, merger

---

1. The following information is obtained from WTO (1998a).

and acquisition arrangements, investment consulting in both domestic and international operations, and others. The current business environment often makes "one-stop-shopping" services convenient for clients. In addition, typically in the United States, accounting firms must strengthen their financial bases to defend themselves against increasing massive legal claims.

Third, the globalization of economic activities makes international convergence of accounting practices increasingly attractive. It has become troublesome for private companies with international operations to prepare multiple forms of financial statements for country-by-country requirements. In addition, differences in regulatory frameworks across countries have sometimes been an obstacle to raising funds in the international financial market. Of course, the reform of economic institutions requires a lot of energy and momentum. However, the benefit of international convergence has gradually come to outweigh its cost for a large number of globalizing firms. Such a market environment also makes services provided by large accountancy companies with international alliances increasingly attractive.

Fourth, the Anglo-American accounting method has a long history and is now regarded as the most advanced and sophisticated in the world. As a background, we observe that Anglo-American-type corporate culture as well as the system of corporate governance has gradually penetrated into countries all over the world as the globalization of economic activities has proceeded. The technological dominance together with network externalities allows U.S. and British accountancy firms to successfully formulate the global network of alliances.

Fifth, since the accounting system is a part of basic economic institutions, the international convergence inevitably triggers fundamental changes in market functioning in lagged-behind countries. The modern accounting system is an essential part of modern (Anglo-American-type) capitalism. Introducing it results in the emergence of modern corporate culture and corporate governance, which possibly causes serious conflicts with local traditional values. In the process of policy reform, various types of politico-economic pressure may come up.

These characteristics make the case of accountancy services extremely insightful and their relevance to the liberalization of other service sectors worth seeking. The wave of globalization from the market side comes into wild collision with traditional, indigenous values and redefines the borderline of "pure" domestic policies. The institutional convergence or harmonization in the field possibly triggers drastic institutional reform including regulatory framework and corporate governance.

11.3   Historical Background in Japan

The modern business accounting system in Japan has gradually been formed since the end of WWII and has acquired a strong path-dependent nature.[2] The Japanese business accounting system has a "triangle" structure, in which three lines of the accounting system, based on securities and exchange law, commercial law, and corporation tax law, coexist. The original purpose of each law is different; commercial law focuses on the protection of stockholders and creditors, whereas securities and exchange law works to facilitate investment. Commercial law was based on the old German-law tradition, whereas securities and exchange law was written under the heavy influence of the U.S. laws in the occupation period. Commercial law and corporation tax law apply to all firms, whereas securities and exchange law only covers companies participating in the stock market. The required accounting formats as well as accounting procedures are different. The history of the accounting system has been a complicated evolutionary process, coordinating the three systems.

Another important feature is that in Japan the governmental sector has directly conducted the formulation of the accounting system. The starting point of business accounting standards was the Financial Accounting Standards for Business Enterprises, written by the Economic Stabilization Board in 1949. The Business Accounting Deliberations Council has conducted a number of revisions since then. Although the council contains members from the private sector, including professional accountants and others, it is a part of the governmental sector under the supervision of the finance minister.

Business accounting consists of financial accounting (accounting for external reporting), managerial accounting (accounting for internal reporting), and tax accounting as well as external audits. The legally obliged audits in Japan have a dual structure: one is for securities and exchange law, and the other is for commercial law.[3] The Special Law of the Commercial Law legislated in 1974 tried to harmonize the contents of legally required audits for both laws, although there are still a number of differences in required documents, the way of publicizing, and other detailed procedures. Commercial law audits by CPAs are required only for large companies with capital of more than 500 million yen or with liabilities of more than 20 billion yen. Securities and exchange law audits with CPAs, on the other hand, are compulsory only for large companies raising funds in the security market.

The CPA law authorizes the qualification of Japanese CPAs. Only accountants with the CPA qualification can conduct legally required audits. The registration for the Japanese Institute of Certified Public Accountants (JICPA) is virtually compulsory for CPA activities. An auditing firm can be

2. Arai (1999) provides a detailed review of the historical formulation of the Japanese accounting system.
3. In addition, there are special sorts of legally required audits, such as ones for educational institutions, labor unions, and so on. Some companies also have voluntary audits even if no legal audit is required.

Table 11.1   **Membership of the Japanese Institute of Certified Public Accountants (as of March 31, 2000)**

Regions	CPAs	Members Foreign CPAs	Members Auditing Firms	Members Total	Submembers (Junior CPAs Registered)	Total	Junior CPAs Not Registered
Hokkaido	177	0	2	179	39	218	2
Tohoku	188	0	2	190	25	215	3
Tokyo	7,694	9	89	7,792	2,585	10,377	158
Tokai	905	0	13	918	225	1,143	14
Hokuriku	147	0	2	149	16	165	2
Kyoto/Shiga	203	0	1	204	61	265	6
Kinki	1,691	0	25	1,716	506	2,222	22
Hyogo	336	0	3	339	80	419	9
Chugoku	234	0	2	236	52	288	7
Shikoku	135	0	1	136	18	154	2
North Kyushu	319	0	0	319	81	400	4
South Kyushu	113	0	2	115	12	127	2
Okinawa	26	1	0	27	7	34	0
Total	12,168	10	142	12,320	3,707	16,027	231

Source: [http://www.jicpa.or.jp].

established only with five or more qualified CPAs. Table 11.1 presents the number of members of the JICPA, which indicates high geographical concentration to the Tokyo region. The number of CPAs has increased over time, but a shortage of supply has long been claimed (figure 11.1).[4] The size of auditing firms as well as the number is also small; only five firms (as of February 1999) have more than 100 partners (figure 11.2).[5] The qualifying exam for CPAs is provided by the Ministry of Finance and is known to be one of the most demanding exams, being a match for a qualifying exam for lawyers. Table 11.2 shows the recent pass ratios in the Japanese CPA exams.

The system of CPA qualification in Japan has been regarded as an obstacle for foreigners. The qualification of CPAs was granted by special examination for foreign accountants in the past, but the exam has not been held since 1975. When the Accountants Law was established in 1948, accountants with foreign CPAs did not need to take any exam (Article 23). In 1950, the article was abolished, and instead the Foreign CPA System was introduced (Articles 16–32), which granted qualification to a foreign CPA with the authorization of the CPA Management Committee and without qualifying exams. The number of foreign CPAs increased throughout the 1960s. In 1977,

4. Japan has only 12,000 accountants, which is a very small number compared with the size of the Japanese economy. The International Accounting Standards Committee (IASC) claims to cover more than 2 million accountants in the world (see [http://www.iasc.org.uk]).
5. The merger between Ota Showa and Century in April 2000 reduced the number of large auditing firms into four. See below for details.

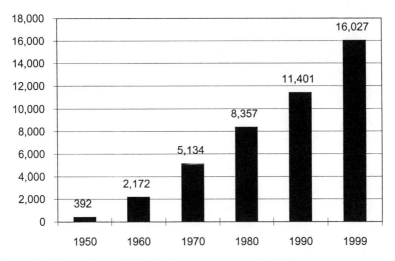

Fig. 11.1   The number of CPAs in Japan including junior CPAs
Source: http://www.jicpa.or.jp.

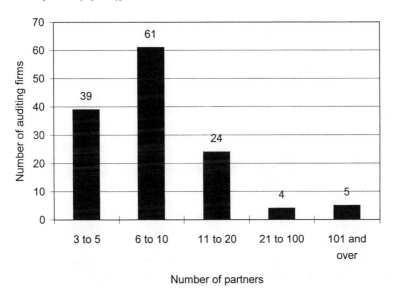

Number of partners

Fig. 11.2   Size distribution of auditing firms in Japan as of 28 February 1999
Source: Data provided by JICPA.

however, JICPA issued a statement. It claimed that foreign accountancy firms should not be legally qualified as auditing firms under CPA law and that it is not appropriate for foreign accountancy firms to conduct tax-related activities that CPA law prohibits a CPA to conduct. In response to it, foreign accountancy firms then in Japan made auditing activities indepen-

Table 11.2   The Results of Japanese CPA Exams

Year	Number of Applicants	Number of Passes	Pass Ratio (%)
1987	4,635	394	8.5
1988	5,205	378	7.3
1989	5,735	596	10.4
1990	6,449	634	9.8
1991	7,157	638	8.9
1992	8,102	798	9.8
1993	9,538	717	7.5
1994	10,391	772	7.4
1995	10,414	722	6.9
1996	10,183	672	6.9
1997	10,033	673	6.7
1998	10,006	672	6.7
1999	10,265	786	7.7

Source: [http://www.jicpa.or.jp].
Note: The results of the second-round exam are shown. Successful applicants become junior CPAs and proceed to the internship period (at least three years) to prepare for the third-round exam.

dent of other services. The Aoyama Audit Corporation with Price Waterhouse was the first approved foreign auditing firm (1983). Foreigners, particularly Americans, have criticized the Japanese CPA system as a barrier.[6]

11.4   The Wave of Globalization

Accountancy services in the world have a peculiar industrial organization. There are several giant firms with a wide range of services operating all over the world. On the other hand, just as in other professional services, there are numerous small firms and individual offices providing accountancy services to local customers. In the latter half of the 1980s, a merger boom occurred in the industry, and the creation of KPMG (1987), Ernst and Young International (1989), and Deloitte Touche Tohmatsu (1990) set the stage for the so-called "Big Six." With the merger of Price Waterhouse and Coopers & Lybrand in 1998, the current "Big Five" regime started. As shown in table 11.3, these five firms have a huge number of employees, including ample professionals, and operate worldwide. They believe that the source of competitive edge is offering clients a wide range of services with wide geographic coverage. Together with their technological superiority, these American-British accountancy firms have established their dominance in the world.

The form of foreign operation, however, is also peculiar. As pointed out by WTO (1998a), accountancy services have "the widespread nature of lo-

6. One of the references is United States Trade Representative (1998).

Table 11.3   **"Big Five" in the World**

Name of Company	Number of Employees	Operating Countries
PricewaterhouseCoopers	150,000	150+
KPMG	100,000	159+
Ernst and Young International	96,000	132+
Deloitte Touche Tohmatsu	82,000	130+
Arthur Andersen	72,000	84

Sources: [http://www/pwcglobal.com]; [http://www.oscaudit.or.jp]; [http://www.deloitte.com]; and [http://asahi.or.jp] (accessed on 10 June 2000).

cal qualification and licensing requirements, both in regard to individual practitioners and as conditions for the ownership and management of firms" (1). Therefore, foreign direct investment or the direct penetration of professional personnel is not a popular form of international operation. In many cases, the Big Five have business alliances or member firms in foreign countries and provide a franchise for them to use their brand names, often without substantial capital holdings. In many cases, even a profit-sharing contract does not exist with local partners.

In the case of Japan, there are currently four large auditing firms with more than 300 partner CPAs: ChuoAoyama Audit Corporation, Century Ota Showa & Co.,[7] Kansa Houjin Tohmatsu, and Asahi & Co. A brief profile of each firm is presented in table 11.4. Each firm has its own alliance relationship with the world Big Five. Chuo was originally with Coopers & Lybrand, and Aoyama was with Price Waterhouse. These two merged in April 2000 in response to the merger of Coopers & Lybrand and Price Waterhouse in 1998. Century was with KPMG, whereas Ota Showa was with Ernst & Young. However, they also merged in April 2000 to seek the advantage of economies of scale. The reformulation of corporate structure generates four large auditing firms of similar size in terms of the number of professionals, domestic branches, and audit clients. All four of these firms have related companies that conduct business consulting, tax advice, financial advisory, and other services. "One-stop service" is their sales strategy.

These auditing firms took great advantage of alliance relationships as a source of credibility and successfully expanded their operations. However, the current financial crisis in Japan revealed the existence of massive nonperforming loans and lenient financial management, and the quality of auditing services came under strong criticism. In the name of enhancing the quality of accountancy services, the world giants have recently placed more pressure on the management of Japanese accountancy firms. In particular, Japanese accountants were upset when the Big Five requested that Japanese firms add to their audit reports of March 1999 a line saying that the

---

7. Century Ota Showa & Co. was renamed Shin Nihon & Co. in July 2001.

Table 11.4   Profiles of Large Auditing Firms in Japan

	ChuoAoyama Audit Corporation (PricewaterhouseCoopers)	Century Ota Showa & Co. (KPMG; Ernst and Young International)	Kansa Houjin Tohmatsu (Deloitte Touche Tohmatsu)	Asahi & Co. (Arthur Andersen)
Amount of capital (millions of yen)	1,273	1,744	n.a.	1,242
Number of persons	2,544	2,736	2,154	2,311
Partner CPAs	340	331	312	329
Staff with CPA	961	1,217	766	862
Staff with junior CPA	674	666	659	604
Other	569	522	417	516
Number of domestic branches	26	36	30	37
Number of audit clients	3,846	4,987	2,422	3,846
Statutory audit for SEC and commercial laws	755	2,549	682	608
SEC law audit	87	(included above)	51	74
Commercial law audit	1,071	(included above)	815	1,032
School	226	284	122	295
Labor union	83	201	47	73
Other	1,299	1,953	705	1,764
Data as of:	04/01/2000	04/01/2000	03/31/1999 12/31/1999	03/31/2000

Sources: [http://www.chuoaoyama.or.jp]; [http://www.oscaudit.or.jp]; [http://www.tohmatsu.co.jp]; and [http://www.asahi.or.jp].

Note: Companies in parentheses indicate ally companies. n.a. = not available.

report was prepared along the Japanese standard and thus was not necessarily internationally accepted.[8]

Another wave of globalization has come from the effort of establishing the International Accounting Standards (IAS). The International Accounting Standards Committee (IASC) is an independent private-sector body, and its objective is to achieve uniformity in the accounting principles that are used by businesses and other organizations for financial reporting around the world.[9] It was established by private accountancy groups from nine countries (Australia, Canada, France, West Germany, Japan, Mexico, the Netherlands, the United States, and the United Kingdom) in 1973. Since 1983, the IASC's members have included all the professional accountancy bodies that are members of the International Federation of Accountants (IFAC). Table 11.5 presents the number of member bodies as of June 2000. The IASC board makes decisions on accounting principles and issues them in the form of IAS. Each board member country's delegation consists of two individuals as well as a technical advisor if desired. These individuals include accountants and persons from other business fields.

The support of the International Organization of Securities Commissions (IOSCO) has reinforced the activities of IASC. The IOSCO is an international organization for government agencies that supervise security markets. In 1987, the IOSCO joined the consultative group of the IASC and got involved in formulating the contents of IAS. In 1995, the IOSCO agreed that the IASC would complete "core standards" by 1999 and that on successful completion, the IOSCO would consider endorsing IAS for cross-border offerings. Voices demanding international convergence of accounting systems intensified, and the IASC accelerated the formation of IAS. The core standards were completed with the approval of IAS 39 in December 1998, and the IOSCO started reviewing IAS core standards in 1999. Table 11.6 presents the list of IASC standards. In May 2000, the IOSCO recommended that its members allow multinational issuers to use thirty IASC standards in cross-border offerings and listings. These cooperative movements by the IOSCO have encouraged member countries to conform their accounting systems to the IAS.

Table 11.7 was constructed from the list of companies that declare that their financial statements conform to the IAS without qualification. Nine hundred sixty one companies in seventy three countries (including international organizations) have applied the IAS. This indicates that the IAS is al-

8. In the Asian crisis, the Big Five were criticized for the quality of accountancy services provided by their business alliances in Asian countries. In October 1998, the World Bank issued a request to the Big Five to stop putting their names to accounts published in the Asian economies unless such accounts followed international financial reporting standards (WTO 1998a, 1). The Big Five's request to Japanese firms in March 1999 was made in this context. The criticism has actually provided a great opportunity for the Big Five to strengthen their control over their foreign alliances as well as expanding their consultancy operation abroad.

9. The following information is from [http://www.iasc.org.uk].

Table 11.5   The Organization of the International Accounting Standards Committee
                    (IASC)

Member bodies
   143 accounting organizations in 104 countries (including 5 associate members and 4
      affiliate members)
The IASC Board (two-and-a-half-year term ending 30 June 2000)
   Board members
      Chairman and Vice-Chairman
      Australia
      Canada
      France
      Germany
      India and Sri Lanka
      Japan
      Malaysia
      Mexico
      The Netherlands
      Nordic Federation of Public Accountants
      South Africa and Zimbabwe
      United Kingdom
      United States
      International Council of Investment Associations (ICIA)
      Federation of Swiss Industrial Holding Companies
      International Association of Financial Executives Institutes (IAFEI)
   Observers
      European Commission
      United States Financial Accounting Standards Board (FASB)
      International Organization of Securities Commissions (IOSCO)
      People's Republic of China

Source: [http://www.iasc.org.uk] (accessed on 10 June 2000).

ready visible in the world business community and is regarded as something useful with which many companies can add credibility to their business practices. Because detailed accounting methods are different across countries, many experts are not optimistic about the complete international convergence of accounting standards. However, it is now very likely that each country's accounting standards will be harmonized in the direction of IAS. The IAS are overall close to the U.S. accounting standard although not identical. Initially the United States was not very positive about the convergence of accounting standards. However, it gradually noticed the strategic importance of convergence and started trying to influence the contents of IAS. The European Commission, on the other hand, announced in June 2000 that all firms in the European stock market should apply the IAS by 2005 (see *Nihon Keizai Shinbun*, 28 June 2000). This announcement would accelerate the worldwide application of IAS as well as possibly intensifying a struggle between the United States and the European Union (EU) over the initiatives establishing the standards.

Table 11.6   List of Current International Accounting Standards (IAS) Committee Standards

IAS 1	Presentation of financial statements
IAS 2	Inventories
IAS 3	No longer effective. Replaced by IAS 27 and IAS 28.
IAS 4	Withdrawn. Replaced by IAS 16, 22, and 38.
IAS 5	No longer effective. Replaced by IAS 1.
IAS 6	No longer effective. Replaced by IAS 15.
IAS 7	Cash flow statements
IAS 8	Profit or loss for the period, fundamental errors, and changes in accounting policies
IAS 9	Research and development costs (will be superseded by IAS 38 effective 1/7/99)
IAS 10	Events after the balance sheet date
IAS 11	Construction contracts
IAS 12	Income taxes
IAS 13	No longer effective. Replaced by IAS 1.
IAS 14	Segment reporting
IAS 15	Information reflecting the effects of changing prices
IAS 16	Property, plant, and equipment
IAS 17	Leases
IAS 18	Revenue
IAS 19	Employee benefits
IAS 20	Accounting for government grants and disclosure of government assistance
IAS 21	The effects of changes in foreign exchange rates
IAS 22	Business combinations
IAS 23	Borrowing costs
IAS 24	Related party disclosures
IAS 25	Accounting for investments
IAS 26	Accounting and reporting by retirement benefit plans
IAS 27	Consolidated financial statements and accounting for investments in subsidiaries
IAS 28	Accounting for investments in associates
IAS 29	Financial reporting in hyperinflationary economies
IAS 30	Disclosures in the financial statements of banks and similar financial institutions
IAS 31	Financial reporting of interests in joint ventures
IAS 32	Financial instruments: Disclosures and presentation
IAS 33	Earnings per share
IAS 34	Interim financial reporting
IAS 35	Discontinuing operations (1/1/99)
IAS 36	Impairment of assets (1/7/99)
IAS 37	Provisions, contingent liabilities, and contingent assets
IAS 38	Intangible assets
IAS 39	Financial instruments: Recognition and measurement
IAS 40	Investment property

Source: [http://www.iasc.org.uk] (accessed on 11 June 2000).

Table 11.7   The Number of Companies using International Accounting Standards

Country	Number of Companies	Country	Number of Companies
Australia	6	Korea	2
Austria	10	Kuwait	49
Bahamas	4	Latvia	7
Bahrain	10	Lithuania	5
Barbados	4	Luxembourg	5
Belgium	5	Macedonia	2
Bermuda	2	Malaysia	5
Botswana	4	Malta	8
Brazil	2	Mexico	3
British Virgin Islands	2	The Netherlands	4
Bulgaria	2	Norway	5
Canada	45	Oman	8
Cayman Islands	1	Panama	61
China	121	Papua New Guinea	1
Croatia	8	Poland	7
Curacao	2	Portugal	2
Cyprus	11	Qatar	1
Czech Republic	16	Romania	8
Denmark	12	Russia	17
Egypt	1	Saudi Arabia	2
Estonia	2	Singapore	1
Fiji	1	Slovak Republic	4
Finland	15	Slovenia	6
France	34	South Africa	24
Georgia	1	Spain	1
Germany	161	Switzerland	106
Ghana	1	Tanzania	1
Greece	1	Turkey	19
Guyana	1	Turks and Caicos Islands	2
Hong Kong, China	11	Ukraine	4
Hungary	14	United Arab Emirates	17
India	1	United Kingdom	1
Italy	12	United States	2
Jamaica	1	Venezuela	1
Japan	10	Zimbabwe	15
Jordan	1	International	12
Kazakhstan	1	Total	961

Source: http://www.iasc.org.uk (accessed on 11 June 2000).

Notes: "International" includes African Development Bank, European Bank for Reconstruction and Development (EBRD), Eutelsat, International Accounting Standards Committee, International Bank for Reconstruction and Development (IBRD), International Federation of Accountants (IFAC), International Federation of Stock Exchanges (FIBV), International Finance Corporation (IFC), International Olympic Committee, International Organization of Securities Commissions (IOSCO), Organisation for Economic Cooperation and Development (OECD), and World Bank. "Japan" includes Dai-Ichi Kangyou Bank, Fujitsu Ltd., Kajima, Kansai, Kirin Brewery, Nissan Motor, Sakura Bank, Sanwa Bank, Sasebo, and Toray. "United States" includes American Institute of Certified Public Accountants and FMC.

11.5   The "Accounting Big Bang" in Japan

In the latter half of the 1990s, the Japanese accounting system became increasingly incompatible with modern corporate management, particularly on its usage of the book value of assets, the improper treatment of affiliates, and, as a consequence, the insufficient disclosure of corporate performance. A long-lasting recession in the 1990s gradually eroded public confidence in the Japanese economic system. Particularly since 1997, the bad performance of large construction companies has diluted the credibility of the traditional accounting practices. In addition, incompatibility with the international accounting system generated inconvenience for globalizing firms. Although institutions are interlocked with each other and resist being changed, even accountants with vested interests found themselves being forced to launch a drastic reform. Now the introduction of the new accounting system is ongoing. Because its impact is expected to be substantial, the series of reforms is called the "accounting Big Bang."

The Business Accounting Council under the Ministry of Finance has led the reform.[10] Council members have included academics, accountants, and other private-sector representatives, but the council has maintained the substantial direct involvement of the government. The active stance of the government has made the legislative process and other institutional modification easier. However, it has also reflected the nonexistence of an independent professional body to set up the accounting standard.

Table 11.8 is a list of notes the council submitted to the finance minister in the past few years. These will largely harmonize the Japanese business accounting system with the IAS. In particular, the introduction of the following three elements will generate a substantial impact:

1. Consolidated financial statements (which calculate economic substance by adding the parent company's and subsidiaries' accounts)
2. Market value calculation
3. Statements of cash flows (C/S) in addition to balance sheets (B/S) and income statements (P/L)

The introduction of consolidated financial statements is complemented by the amendment of the competition law in January 1998 allowing the formation of pure stock-holding companies. Major changes in tax effect accounting and corporate pension accounting will be also implemented. All of these will be completed in a few years.

The reform of the accounting system is expected to substantially change the basic structure of Japanese management system and corporate governance. For example, it was a commonly observed practice for Japanese

---

10. Since July 2000, the Business Accounting Council has been placed under the newly established Financial Services Agency (FSA).

Table 11.8   A Series of Notes by the Business Accounting Council

Date of Announcement	Name of Note	Date of Implementation
6/16/1997	Note on the revision of consolidated financial statements	From F/Y 1999
3/13/1998	Note on the accounting standard for research and development expenditures and others	From F/Y 1999
3/13/1998	Note on the standard of middle consolidated financial statements and others	From F/Y 2000
3/13/1998	Note on the standard of consolidated cash flow statements and others	From F/Y 1999
6/16/1998	Note on the accounting standard for corporate pension	From F/Y 2000 and after
10/30/1998	Note on the accounting standard for tax effect accounting	From F/Y 1999 (or earlier)
10/30/1998	Treatment of affiliates and related companies in consolidated financial statements	From F/Y 1999 (or earlier)
1/22/1999	Note on the accounting standard for financial commodities	From F/Y 2000 and after
2/19/1999	Treatment of the financial statement format	From F/Y 1999

Source: Miyata (1999).

firms to hide their losses by manipulating affiliates' accounts; such practice was called *tobashi* in Japanese. However, consolidated financial statements will make such nontransparent practice difficult to implement. The introduction of market value calculation will eradicate the old style of management based on *fukumi-eki* (unrealized gains or losses from the gap between the book value and market value of assets and liabilities a company holds); in principle, *fukumi-eki* must explicitly be recorded from now on. The market value calculation will also discourage the practice of cross-share holding (*kabushiki mochiai*).[11] In short, the reform of the accounting system will let the market have much clearer signals for corporate performances, which would in turn ignite fundamental corporate reform. Although the presence of foreign traders in Japanese stock exchanges became larger in the 1990s, the wave of cross-border mergers and acquisitions (M&As) has not arrived at Japan yet; although drastically increased in recent years, only 2 percent of world M&As came to Japan in 1999.[12] The modernization of the accounting system, however, may change the pattern of foreign investors.

How far the accounting Big Bang will have an impact is now the subject

11. The cross-share holding practice has steadily subsided since the beginning of the 1990s (see Ito 2000, 8–9 for an example). However, the reason why the introduction of market value calculation discourages cross-share holding is not entirely clear in the academic and nonacademic writing.

12. According to UNCTAD (2000, 108), cross-border M&A sales in Japan in 1999 were $15.9 billion, whereas world total M&A sales were $720.1 billion.

of great debate. Some scholars have a rather pessimistic view; they claim that changes in the accounting system may not necessarily reform the basic attitude of Japanese management. However, it is true that the Japanese accounting system will largely be in accordance with IAS in a few years, although it will not be in complete conformity with it. The author thinks that the overall enhancement of transparency in business accounting will encourage a substantial reform of the corporate governance of Japanese companies.[13]

A recent phenomenon of interest is a big boom in studying for American CPA qualification in Japan. One of the preparatory schools for foreign professional qualifications, Anjo International, is an illustration. Kotaro Anjo, a former worker in Nomura Securities, established it in 1995 with only twenty students. As of March 1999, it claimed to have 8,000 students in ten locations (Anjo 1999). The government also provides tuition subsidies for students in preparatory schools. Although the some professional accountants criticize the for U.S. CPA boom (see, e.g., Kojima 1999), it is now a common practice for auditing firms to treat a U.S. CPA as one of the eligibility qualifications for their recruits. At this moment, young people with U.S. CPAs work mainly as assistants because the Japanese license is required for official auditing services. However, along with the convergence of accounting systems, the qualification system of accountants would be reviewed in the future.

The JICPA has recently requested to the Ministry of Finance (MOF) a drastic increase in successful candidates in CPA qualifying exams (see Fukuda 2000 and *Nihon Keizai Shinbun*, 10 June 2000, evening edition). In the past, the JICPA was rather conservative in expanding the number of CPAs. Since the end of World War II, the Japanese have not had a strong inclination toward individual professionalism in general, but accountants and lawyers have been exceptions. Accountants have long been treated as highly prestigious professionals. Although audit fees have been strictly regulated, accountancy has also been regarded as a high-income profession.[14] The JICPA therefore claimed that a reckless increase in the number of accountants might degrade professional quality. However, the JICPA recently changed its stance in the opposite direction due to the recent drastic increase in the demand for CPAs as well as a sneaking penetration of U.S.

13. A questionnaire survey (Matsuo et al. 1999) reports that security market analysts particularly appreciate the introduction of consolidated financial statements and market value calculation and reveal support for the convergence toward the IAS or the U.S. accounting standard.

14. Despite the regulation on auditing fees, the cost of audits borne by audit firms has drastically increased recently. Chiyoda (1999) points out that Japanese audit firms tend to make up for the increase in auditing cost by providing consultant services to clients. This may be a trend against the current move in the United States, where the SEC announced a policy to separate audit services from consultant services in order to prevent undesirable collusions of audit firms and clients (*Nihon Keizai Shinbun,* 21 June 2000, evening edition).

CPA qualification. In June 2000, the Advisory Group to the Finance Minister on the CPA system made public a drastic reform plan to double the number of successful applicants in CPA exams. Interesting enough, the plan also includes several measures enhancing the quality of accountancy services, such as the three-year periodical update of CPA registration as well as the liberalization of audit fees and the length of audit periods (see *Nihon Keizai Shinbun*, 24 June 2000).

In addition, the JICPA and the MOF (and now the Financial Services Agency [FSA]) have started preparing the establishment of a private body to take care of business accounting standards.[15] So far, the MOF (FSA) has been directly in charge of Japanese business accounting standards. However, quick response to the wave of globalization requires a permanent group of private specialists. Moreover, to participate in the process of constructing IAS, Japan has to send a private group to the IASC. The IASC restructured its organization in January 2001, and Tatsumi Yamada of Japan was elected as one of the board members who will be in charge of constructing IAS. Such movements, however, should be accompanied with stronger monitoring by the public sector, as the SEC works in the United States. A critical private watch over the performance of accounting business must also be fostered in the Japanese business community.[16]

11.6   Liberalization Efforts in the World Trade Organization

Accountancy services were included in the framework of the General Agreement on Trade in Services (GATS) from the beginning. The GATS first requests member countries to obey the MFN principle (Article II) and transparency requirement (Article III) for all sorts of trade in services.[17] Then the market access obligation (Article XVI) and national treatment principle (Article XVII) are imposed on the basis of the Schedule of Specific Commitments (the positive list method). Whereas market access refers to quantitative measures to deter foreign entry, qualitative measures to possibly deter entry are taken care of by Article VI on domestic regulation,

15. A newspaper article reporting an interview with Hiroshi Nakachi, the president of the JICPA (*Nihon Keizai Shinbun*, 8 June 2000) and another article (*Nihon Keizai Shinbun*, 9 August 2000) disclosed the intention of JICPA that a private body for establishing new accounting standards would be organized in the first half of 2001.

16. In the United States, one of the major motivations for big mergers in the industry was to prepare for private suits against accountancy firms. In other words, the performance of accountancy firms is always under potential criticism. On the other hand, there have so far been very few lawsuits against accountancy firms in Japan.

17. The MFN principle is applied with some limited exceptions. In particular, Article II: 2 and the Annex on Article Exemptions admit that member countries can specify sectors temporarily exempted from the MFN obligation (the negative list method). In fact, seven countries—Costa Rica, the Dominican Republic, Honduras, Panama, Thailand, Turkey, and Venezuela—directly specify either accountancy services or professional services in general for MFN exemptions (WTO 1998a, 13).

which loosely states that "unnecessary barriers to trade in services" must be removed.

Because policies related to trade in services are often strongly domestic in nature, the borderline issue between international policy discipline and domestic policies often comes up, particularly for market access obligation and domestic regulation. The WTO often uses the word "impediments" or "barriers" for regulations not conforming to international policy discipline. However, please note that some government regulations not following the market access requirement, for example, may have a good domestic reason for existence.

Table 11.9 is an illustrative list of "impediments" to trade in accountancy services prepared by the WTO. Following a number of general impediments, nine specific impediments are listed. These impediments restrict trade in services conducted in four modes: cross-border (mode 1), consumption abroad (mode 2), commercial presence (mode 3), and natural persons (mode 4). Notably, "residence/establishment requirements" are common barriers for mode-1 service trade. "Professional certification/entry requirements," "restrictions on business structures," and "differences in accounting, auditing, and other standards" can be significant impediments for mode-3 and mode-4 service trade.

Table 11.10 presents the percentage of member countries that declare (a) full commitment, (b) partial commitment, or (c) no commitment for each type of services and for each mode of services trade. In the case of accounting, auditing, and bookkeeping services, a number of countries declare full commitment for mode-2 services trade (for both market access

Table 11.9   Impediments to Trade in Accountancy Services

A. General impediments
   Restrictions on international payments
   Restrictions on the mobility of personnel
   Impediments to technology and information transfer
   "Buy National" public procurement practices
   Differential taxation treatment/double taxation
   Monopolies
   Subsidies
B. Specific impediments
   Nationality requirements
   Residence/establishment requirements
   Professional certification/entry requirements
   Compartmentalization/scope of practice limitations/incompatibilities
   Restrictions on advertising, solicitation, and fee-setting
   Quantitative restrictions on the provision of services
   Differences in accounting, auditing, and other standards
   Restrictions on business structures
   Restrictions on international relationships/use of firm name

Source: WTO (1998a, table 4).

Table 11.10   Commitment Percentage, by Sector and Mode of Supply (professional services)

	Cross-border			Consumption Abroad			Commercial Presence			Natural Persons		
	Full	Partial	No	Full	Partial	No	Full	Partial	No	Full	Partial	No
A. Market Access												
Legal services	18	67	16	24	67	9	4	87	9	2	91	7
Accounting, auditing, and bookkeeping services	29	41	30	41	45	14	9	89	2	2	86	13
Taxation services	44	44	12	53	44	3	15	82	3	0	88	12
Architectural services	52	26	22	68	20	12	24	72	4	0	92	8
Engineering services	50	28	22	55	28	17	24	72	3	0	85	5
Integrated engineering services	59	22	19	66	22	13	31	59	9	0	94	6
Urban planning and landscape architectural services	45	36	18	52	36	12	24	73	3	0	97	3
Medical and dental services	34	29	37	61	34	5	21	68	11	0	87	13
Veterinary services	54	19	27	69	23	8	31	58	12	4	81	15
Services provided by midwives, nurses, physiotherapists	33	33	33	47	53	0	20	80	0	0	93	7
Other	33	67	0	33	67	0	0	100	0	0	100	0
B. National Treatment												
Legal services	22	60	18	31	58	11	16	76	9	2	91	7
Accounting, auditing, and bookkeeping services	34	36	30	50	36	14	32	64	4	4	80	16
Taxation services	41	41	18	56	35	9	35	56	9	12	71	18
Architectural services	52	30	18	64	22	14	56	38	6	8	80	12
Engineering services	45	31	24	60	21	19	52	43	5	9	79	12
Integrated engineering services	63	19	19	72	13	16	72	13	16	9	78	13
Urban planning and landscape architectural services	52	30	18	61	24	15	58	33	9	9	85	6
Medical and dental services	47	18	34	66	24	11	45	45	11	3	87	11
Veterinary services	62	12	27	81	8	12	58	35	8	8	77	15
Services provided by midwives, nurses, physiotherapists	40	27	33	53	47	0	53	47	0	0	93	7
Other	33	50	17	33	50	17	33	67	0	17	67	17

Source: WTO (1998a, table 6).

Notes: Full = Full commitment (indicated by "None" in the market access or national treatment column of the Schedule). Partial = Partial commitment (limitations are inscribed in the market access or national treatment column of the Schedule). No = No commitment (indicated by "Unbound" in the market access or national treatment column of the Schedule). Percentages may not add up to 100 due to rounding. Basis of total is listed sectors.

Table 11.11   **Analysis of Types of Measures (number of measures in accountancy services)**

Type of Limitation	Mode			
	1	2	3	4
A. Market Access				
Number of suppliers	—	—	2	—
Value of transactions or assets	—	—	12	1
Number of operations	—	—	—	—
Number of natural persons	1	1	3	31
Type of legal entity	4	2	22	1
Participation of foreign capital	—	—	17	1
Other measures, n.e.c.	4	4	19	2
B. National Treatment				
Tax measures, subsidies, grants and other financial measures	—	—	—	—
Nationality requirements	1	1	7	6
Residency requirements	4	2	11	8
Licensing, standards, qualifications	6	4	16	12
Registration requirements	2	2	6	2
Authorization requirement	3	1	5	2
Performance requirements	—	—	—	—
Technology transfer requirements	—	—	1	—
Local content, training requirements	—	—	1	—
Other measures, n.e.c.	1	—	4	3

Source: WTO (1998a, table 8).

Note: The number of "Other measures, n.e.c." in part A is large because a number of entries in the Schedules could not be classified into one or the other of the distinct categories of limitations. In some cases, this was due to a lack of specificity in the description of the measure, whereas in others it was because the measure itself did not correspond to any of the categories.

and national treatment), although this mode is not very important for accountancy services. For mode-1 service trade, about one-third of the countries provides full commitment, whereas another one-third gives no commitment. As for mode-3 and mode-4 services trade, most of the countries provide partial commitments. Table 11.11 presents the measures for the cases of partial commitments.

Table 11.12 summarizes the commitment to market access and national treatment made by Japan. For market access, a key impediment for modes 1, 2, and 3 is the specification of accountancy service providers. In addition, the commercial presence requirement inhibits mode-1 service trade. As for national treatment, services trade of mode 1, 2, and 3 does not have any barriers. For mode 4, Japan does not announce any commitment for market access or national treatment. We cannot say that all of these "impediments" must be removed. In fact, some of them—the license requirement, for example—would have good domestic reasons for keeping it. However, the

Table 11.12   Commitment in Japan by Mode of Supply in Accountancy Services

Cross-Border	Consumption Abroad	Commercial Presence	Natural Persons
A. Market Access			
Partial commitment (services must be provided by natural persons or auditing firms; auditing firms must have commercial presence)	Partial commitment (services must be provided by natural persons or auditing firms; auditing firms must have commercial presence)	Partial commitment (services must be provided by natural persons or auditing firms)	No commitment
B. National Treatment			
Full commitment	Full commitment	Full commitment (except as indicated in the horizontal section)	No commitment (except as indicated in the horizontal section)

Source: MOFA (1998, 274–275).

commercial presence requirement, for example, may have a much shakier justification in the present globalizing world. The boundary of international policy discipline is again fuzzy in some aspects.

Colecchia (2000) recently tried to quantify the barrier to market access for services, using accountancy services as a pilot case. She constructed a flow chart to check the existence of various types of barriers and provide a formula for calculating aggregate levels of protection. In her calculation, the overall protection index for the United Kingdom, France, Australia, and the United States was 0.0, 0.3, 0.85, and 1.25, respectively. The index for Japan would probably fall between those for Australia and the United States, which indicates that Japan is not at least a very protective country compared with other developed countries. However, direct access to the Japanese market by foreign accounting firms or foreign accountants is very limited so far.

From 1995, the WTO's Council for Trade in Services had the Working Party on Professional Services and sought a way of facilitating international trade in accountancy services as a forerunning case among professional services. In May 1997, the council adopted the working party's report titled "Guidelines for Mutual Recognition Agreements or Arrangements in the Accountancy Sector" (WTO 1997). The purpose of this guideline is to provide "practical guidance for governments, negotiating entities or other entities entering into mutual recognition negotiations on accountancy services." It encourages bilateral agreements for international harmonization of institutions related to accountancy services and provides a guideline for them to convey the information to the WTO for transparency purposes. This guideline is nonbinding in nature.

In December 1998, the council endorsed another report of the working party, titled "Disciplines on Domestic Regulation in the Accountancy Sector" (WTO 1998b). The press release states: "Most professional services, and many others, are heavily regulated, and for good reasons: but it is also true that regulations can be an unnecessary, and usually unintended, barrier to trade in services" (WTO 1998b).

From this view, the disciplines are adopted to ensure that "measures relating to qualification requirements and procedures, technical standards and licensing requirements and procedures, technical standards and licensing requirements do not constitute such barriers." The report explicitly states that the disciplines do not address measures subject to scheduling under Articles XVI (market access) and XVII (national treatment) of the GATS. Rather, they ensure that domestic regulations meet the requirements of Article VI: 4 (removal of unnecessary barriers to trade) of the GATS. The text of the report contains provisions on transparency, licensing requirements, licensing procedures, qualification requirements, qualification procedures, and technical standards. The disciplines did not have an immediate legal effect but would be included in the new round of service negotiations started in 2000. The council, however, decided on a "standstill provision," effective immediately. Then, in April 1999, the Working Party on Domestic Regulation was launched as a replacement for the Working Party on Professional Services.

The formation of GATS as well as additional work by the working party in the Council for Trade in Services has successfully specified the framework of liberalization. The Disciplines on Domestic Regulations may provide some liberalization pressure on member countries. However, the hardcore negotiation for liberalization, particularly on market access and national treatment, has not yet taken place. The WTO so far takes a rather conservative stance and respects country-specific regulations. All the difficult negotiation issues are sent to the new round, where we will start discussing the contents of liberalization commitments.

11.7   Conclusion

The story of the accountancy sector provides a profound opportunity for us to consider the nature of liberalization in the globalization era. Once we go beyond the liberalization of merchandise trade, it is inevitable for us to confront a delicate issue on the borderline between "pure" domestic policies and international commercial policies under international policy discipline. The borderline is not at all clear in many cases. It depends on our feeling of how important the logic of domestic regulations with institutional divergence is and how far globalization provides incentives for accepting international policy discipline. This borderline issue emerges when we step forward the liberalizing of various fields such as services, foreign direct investment, intellectual property rights, environment, labor, and so on.

As for the accountancy services in Japan, we could have treated them as a sector under a "pure" domestic policy twenty years before. The accountancy services were based on a country-specific regulatory framework and were backed up by a pure domestic system of professional qualification. However, the wave of globalization has come since the latter half of the 1980s, and the effort to establish an international accounting standard started bearing fruit in the 1990s. At the same time, the slump in the Japanese economy since the beginning of the 1990s has revealed various structural problems in the Japanese economic system. As a result, a drastic convergence toward IAS has started in Japan as an "accounting Big Bang." Since the accounting system is deeply rooted in the fundamental institutional structure, many experts do not predict complete convergence with the international standards. However, the movement in the direction of international harmonization would change the borderline of "pure" domestic policies, and the scope of international policy discipline would be enlarged.

In Japan, major impediments to accountancy services trade remain in the form of the service provider qualification and residence requirements. The penetration of IAS, however, would erode the justification for having a country-specific institutional structure. The pressure of globalization would also come from the effort of mutual recognition among European countries and others. At least from a not-legally-obliged portion of accountancy services, the liberalization is likely to proceed in the near future. That would in turn encourage the fundamental reform of the legal structure and accountant qualification system. The position of Japanese CPAs would become a delicate one in the liberalization process.

References

Anjo, Kotaro. 1999. *Kokusai Kaikei Kijun to Beikoku Kounin Kaikei-shi* (IAS and U.S. CPA) Tokyo: Keizai Hourei Kenkyuu-kai.

Arai, Kiyomitsu. 1999. *Nihon no Kigyou Kaikei Seido: Keisei to Tenkai* (*Corporate accounting system: Formation and development*). Tokyo: Chuo Keizai Sha.

Chiyoda, Kunio. 1999. Kounin Kaikei-shi to Masukomi (On the CPAs and mass media). *Kigyou Kaikei* 51 (3): 4–12.

Colecchia, Alessandra. 2000. Measuring barriers to market access for services: A pilot study on accountancy services. In *Impediments to trade in services: Measurement and policy implications,* ed. Christopher Findlay and Tony Warren, 245–66. London: Routledge.

Fukuda, Shinya. 2000. Kounin Kaikei-shi no Kazu no Juujitsu ni Tsuite (On the increase in Japanese CPAs). *JICPA Journal* 12 (6): 4–5.

Ito, Kunio. 2000. *Zeminaaru Gendai Kaikei Nyuumon* (*Seminar: The introductory contemporary accounting*), 3rd ed. Tokyo: Nihon Keizai Shinbun-sha.

Kojima, Yoshiteru. 1999. Sakkon no Beikoku CPA Buumu wo Kiru (Criticism on the recent boom of U.S. CPAs). *Kigyou Kaikei* 51 (3): 2–3.

Matsuo, Nobumasa. 1999. Anarisuto kara Mita Nohon no Kaikei Seido Kaikaku: Kaikei Kijun no Kokusai-ka ni Mukete (The reform of Japanese accounting systems: Toward the internationalization of accounting standards)." *Kigyou Kaikei* 51 (6): 93–105; 51 (7): 99–112.

Ministry of Foreign Affairs (MOFA). Government of Japan. 1998. *1998 Nen Ban WTO Saabisu Boueki Ippan Kyoutei (1998 F/Y WTO General Agreement on Trade in Services)*. Tokyo: Japan Institute of International Affairs.

Miyata, Keiichi. 1999. Wagakuni Kigyou Kaikei Seido Kaikaku wo Meguru Genjou to Tenbou (Current progress and prospects of the reform of business accounting systems in Japan). December. Available from [http://www.boj.or.jp/wakaru/etc/kaikei.htm]. 10 June 2000.

United Nations Conference on Trade and Development (UNCTAD). 2000. *World investment report 2000: Cross-border mergers and acquisitions and development.* New York: United Nations.

United States Trade Representative (USTR). 1998. *1998 National Trade Estimate Report on Foreign Trade Barriers*. Washington, D.C.: GPO.

World Trade Organization (WTO). 1997. "Guidelines for Mutual Recognition Agreements or Arrangements in the Accountancy Sector." May 29. Available from [http://www.wto.org]. 10 June 2000.

———. 1998a. "Accountancy services: Background Note by the Secretariat." December 4. Available from [http://www.wto.org]. 10 June 2000.

———. WTO. 1998b. "Disciplines on Domestic Regulation in the Accountancy Sector." December 14. Available from [http://www.wto.org]. 10 June 2000.

Comment   Edwin L.-C. Lai

This is an interesting paper that describes in detail the historical background of the regulation of accounting services in Japan as well as the current situation and future prospects of trade liberalization in the sector. It is very informative, and it helps people who are unfamiliar with the sector to understand the issues that are involved in liberalization of trade in this area. It highlights how Japanese regulation in accounting services impedes, either intentionally or unintentionally, foreign market access and trade in the services.

One point the author emphasizes is that to allow for free trade in accounting services, Japan needs to adopt the modern accountancy system. The introduction of the modern accountancy system would necessitate the introduction of other institutions of modern capitalism, such as corporate culture and corporate governance. I am not so clear on this point. The big question is to what extent liberalization in accounting services would affect the corporate governance of Japanese firms. Corporate culture and corporate governance are very history- and culture-dependent, and it is hard to imagine a change in the accounting system can change all that. Perhaps it is more likely that, accompanied by other fundamental changes in Japanese capitalism, a reform in the accounting system can set in motion some other

Edwin L.-C. Lai is associate professor of economics at the City University of Hong Kong.

changes that can have long-term impacts. I suspect that in the near future, Japanese firms can find innovative ways to satisfy accounting rules yet essentially maintain the corporate governance. For example, it is probably that Japanese firms can hide their performance in a more sophisticated way under the new system. As the saying goes, "where there is a will, there is a way." The process of convergence in corporate governance would therefore probably be a long one. Take the case of intellectual property rights protection in less developed countries: the laws can be adopted, yet enforcement is a key problem.

To put the point in another way, if Japanese firms find that they have to change fundamentally their way of doing business, it seems likely that they would resist trade liberalization in accounting services or in other services that impose similar effects, such as legal services. They can then lobby for the use of administrative measures, such as qualification and residency requirements, to block free trade of such services.

This makes me think of a more general question. If countries need to change their systems in a fundamental way so as to allow trade in services, would they be willing to do that? Is this exactly why it is so much harder to liberalize trade in services than trade in goods? If these are important concerns of the countries, should the rest of the world ask the countries to change their system so as to allow freer trade? I think doing this is going beyond the scope of the World Trade Organization's mandate. The role of the World Trade Organization (WTO) is limited to promoting freer trade. The best that free-traders can do in this case is to focus on trade liberalization and try to minimize the impact of trade on other aspects of the country, such as institutional change that is unrelated to trade per se.

It is beyond the power of the WTO to ask Japan to change its corporate systems. It is probably Japan that wants to change its own system. Therefore, in contrast to the author, I believe the direction of causation is just the opposite: globalization, the need for harmonization, and the disillusion in Japan with Japanese corporate governance can help to change the Japanese accounting system to a more internationally accepted standard. Such a movement can give the incentive to the Japanese business community to accept freer trade in areas such as accounting and legal services. It would then help trade liberalization in these sectors.

The author emphasizes that the major impediments to trade in accountancy services remain on service provider qualification and residence requirements. This shows a major difference between trade in goods and trade in services. In the case of trade in goods, the two pillars of liberalization are the "national treatment" and "most-favored nation" principle. In the case of trade in services, these two principles are not enough. As table 11.12 of the paper shows, although Japan has full commitment to national treatment in "cross-border trade," "consumption abroad," and "commercial presence," there is only partial commitment to market access on these aspects. Apparently, major access impediments come from professional certification and entry requirements and differences in accounting and auditing standards. As Richard H. Snape points out (chap. 3 in this volume), a generic principle that governs this aspect of trade liberalization in services should be established by the WTO.

Finally, some international comparison would be enlightening. For example, one could compare the number of foreign certified public accountants and foreign accounting establishments in countries such as Germany, France, and China with those of Japan. Of course, the availability of data could be a problem.

Comment   Aaditya Mattoo

The paper contains a wealth of information and numerous valuable insights. In fact, my comments are based mostly on what I have learned from the paper. The main suggestion I have is to step back a little and provide a clearer conceptual structure and a fuller analysis of the policy options. Some of my suggestions are probably beyond the scope of the paper and really subjects for future research. Nevertheless, it may be helpful to focus on three sets of issues.

The Relationship between Accounting Standards and Qualification and Licensing Requirements for Accounting Professionals

It may be useful to clarify the link between the different layers of regulation affecting accountancy services. At the top, and analogous to output standards, are the accounting standards themselves.[1] The precise nature of these standards depends, as the paper nicely shows, on the purpose for which the accounts are prepared—the requirements of securities and exchange law, commercial law, or corporation tax law. At the layer below, and analogous to input standards, are the qualification and licensing requirements for accounting professionals, governed by the law on certified public accountants (CPAs) and the registration requirements of the Japanese Institute of CPAs.

The demand for the services of accountants is a demand derived from the existing business transparency standards (which may be mandatory or vol-

Aaditya Mattoo is a senior economist in the Development Economics Research Group at the World Bank in Washington, D.C.

The author has benefited greatly from discussions with Fukunari Kimura and the comments of Claude Trolliet.

1. Accounting standards are not the only output standards in the sector: there are also auditing standards, management accounting standards or guidelines, and so on.

untary). Hence, the qualification and licensing requirements imposed on accountants depend, in principle, on the nature of these "output" standards. The current segmentation in the market for accountancy services is attributable in part to the heterogeneity of business transparency standards, both within countries and across countries.[2]

Is this heterogeneity inevitable, since transparency objectives necessarily differ—from shareholders to tax authorities, and from country to country? Or can we conceive of a core set of universal transparency requirements that would satisfy everybody? The paper rightly emphasizes the strong link between national standards and the historical and legal context of each country, but it also shows how, in this globalizing world, there is bound to be convergence on business transparency standards. Evidence of this phenomenon are the increasingly successful attempts by the International Accounting Standards Committee (IASC) to develop a set of international accounting standards (IAS) that could constitute such a universal core.

A Conceptual and Empirical Analysis of the Barriers to Trade

The barriers to trade in accountancy services are both explicit (i.e., targeted at foreigners alone) and implicit (i.e., targeted ostensibly at both nationals and foreigners). In order to understand the consequences of the different barriers, a first step would be to identify the stage they affect: qualification, licensing, establishment, or operation. Domestic competition is evidently influenced by the manner in which the CPA qualification examination is conducted. One approach would be to test for a certain essential level of skills that accountants must have and pass everybody who achieves that level. The number of accountants would then be determined by the market. Another approach, and this seems to me closer to actual practice, is to decide first on the number of accountants that are to be qualified and determine the skill threshold accordingly. The exhortations by industry to the Ministry of Finance, which runs the examination, to increase the number of accountants do raise the question: are they suggesting that standards be lowered to an unacceptable level or simply that a quota be relaxed without any serious affect on required standards?[3]

A fundamental restriction affecting the ability of foreigners to enter the sector is the residency requirement, which undoubtedly has a de facto dis-

2. Of course, the segmentation of the input market is not due only to differences in output standards. Input standards can have a life of their own. For instance, in some countries, qualification and licensing requirements for accountants were developed before the accounting standards themselves.

3. Of course, the level of the standards may have little relation to the size of the quotas, because the direction of causality may well be from the degree of competition to the quality of service. Thus, protected markets often have poor standards.

criminatory effect. What, if any, is the rationale for these residency requirements—enforcement? A second restriction, which has an impact at the establishment stage, is the restriction on the use of foreign business names.
This could serve to protect national providers from competition by reputable foreigners in a market where reputation provides a crucial competitive edge.

In assessing the impact of these different restrictions, it is important first
to identify the binding constraints. Some of the more obvious restrictions
may be circumvented: for instance, restrictions on foreign investment have
been sidestepped by a system of alliances. In identifying the constraints at
different stages, it may be useful to consider the flowcharts developed by
Alessandra Colecchia at the Organization for Economic Cooperation and
Development (OECD).

Empirical work is bound to be worthwhile but difficult—and beyond the
scope of this paper. An obvious approach is to begin with some simple price
or cost comparisons. How much do standard accounting activities cost in
Japan relative to other markets? How do the earnings of accountants in
Japan compare with those elsewhere? Alternatively, it may be possible to
look at the costs of the barriers themselves: how much would it cost a foreign accountant to qualify in Japan—in terms of both actual expenses and
lost earnings? Some of the costs of barriers are less easily quantifiable: for
example, what is the cost to Japanese firms (e.g., in terms of raising finance)
of not having access to the most reputable accountancy firms? Then the
most difficult question of all: in analyzing the estimated differences in costs
between countries, how is one to distinguish between legitimate differences
and protection?

The Policy Options in Dealing with Barriers to Trade

There could have been more discussion of how the barriers to trade are
best addressed. With regard to the explicit, which include residency requirements and restrictions on the use of foreign names, the solution would
seem straightforward: They should be eliminated unless the authorities can
demonstrate that some legitimate objective is served by maintaining them.

The implicit barriers arising from the heterogeneity of regulations, both
in the form of accounting standards and in the form of qualification and licensing requirements, are more difficult to eliminate. There are in principle
three related routes.

Harmonization

As noted above, the work of the IASC in developing international accounting standards would do much to further the liberalization of trade in
accountancy services. Once the basic "output" standards are harmonized,

there is less basis for segmentation of the market through heterogeneous "input" standards in terms of country-specific qualification and licensing requirements.[4] Efforts are apparently also being made by the International Federation of Accountants (IFAC) to develop international norms for the requirements imposed on professionals. The WTO has little role in this context except to encourage such harmonization by creating a legal presumption in favor of international standards (as in Article VI:5[b] of the General Agreement on Trade in Services [GATS]).

Mutual Recognition Agreements

Such agreements can in principle have a powerful liberalizing impact, but they have proved difficult to conclude where they would matter most.[5] There are, of course, sound policy criteria to conclude such agreements—sufficient similarity in the basic norms and sufficient confidence in trading partners' conformity assessment systems—but these considerations are often secondary to political-economic considerations. Because such agreements must necessarily be concluded at the sectoral level (with the detailed involvement of the regulators and domestic professionals), there is a limited incentive to conclude such an agreement even between countries with similar costs—since the end of market segmentation necessarily means lower rents for the industry in at least one country—and much less of an incentive between countries with significantly different costs. Here, again, the WTO has a limited role. A set of non-binding guidelines for recognition agreements in accountancy has been developed. The main objective (as in GATS Article VII on recognition) is to allow such agreements because of their liberalizing impact but to prevent their being used as a means of discrimination against excluded countries.

Multilateral Disciplines and the Necessity Test

I have discussed the meaning of this test and how it can be given a sound economic interpretation in my paper in this volume (chap. 2). The new WTO disciplines on domestic regulations for the accountancy sector already incorporate such a test. However, the elaboration of the disciplines is somewhat disappointing; for instance, regulators are only required to "take account" of qualifications obtained in other countries. Nevertheless, the scope of application of the test to sift the legitimate from the protectionist

---

4. Full harmonization may be slow to emerge. Initially, international standards may have an impact only on companies that need international comparability for their financial information. The corner shop, which has only basic accounting needs, may remain unaffected. This output market segmentation may be reflected in a continued segmentation in the accountancy profession, between those providing services to small national businesses and those servicing clients with international needs.

5. Apparently, mutual recognition agreements have had limited impact even within the European Union.

depends on the differences between countries in their basic standards. The more the harmonization and the greater the universal content of the basic standards, the less reason there is for imposing elaborate requalification and relicensing requirements. It is in this sense that the work of IASC and IFAC must be seen as the font of all liberalization efforts in this sector.

Services Trade in East Asia

Shujiro Urata and Kozo Kiyota

12.1 Introduction

Services trade has been attracting the increasingly greater attention and interest of many people, including policy makers, businesspeople, researchers, and others. There are various reasons for such developments. First, services trade has been increasing faster than goods trade in the last few decades, and it has become an important means of conducting international economic activities, besides goods trade and foreign direct investment (FDI). Between 1980 and 1997 world service exports grew at the annual average rate of 7.8 percent, faster than the corresponding rate for goods exports of 6.7 percent (figure 12.1).

Another reason for the increased interest in services trade, which is closely related to the reason noted above, is a rising share of services in domestic economic activities. The share of services in gross domestic product (GDP) in the world economy increased from 55.4 percent in 1980 to 61.9 percent in 1997 (World Bank, *World Development Indicators 2000,* CD version). The increasing share of services in economic activities is attributable not only to the increase in demand for services but also to the changes in the supply side of service sectors. In addition to the rise in income resulting from economic growth, rapid technological progress in service sectors such as communications services contributed to the increase in demand for such services by reducing their prices. Moreover, technological progress has given rise to new forms of business such as e-commerce, which has resulted in promoting

Shujiro Urata is professor of economics at Waseda University. Kozo Kiyota is an assistant professor of business administration at Yokohama National University.

The authors are grateful for useful comments from Ponciano S. Intal, Jr., Richard Snape, Takatoshi Ito, and other seminar participants.

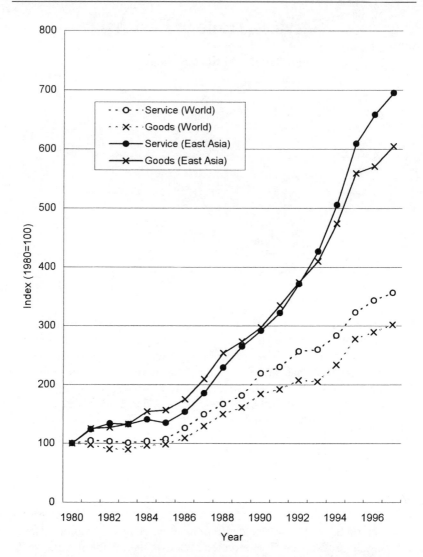

Fig. 12.1   Exports of goods and services in the world and East Asian countries
Sources: Services trade: IMF *Balance of Payments Statistics,* 1999. Goods (merchandise) trade: World Bank *World Development Indicators,* 1999.

telecommunications and distribution and other services. Indeed, there is an increasing recognition on the part of businesspeople as well as policy makers that improving the competitiveness of the service sector is an important factor for the realization of a strong company and economy.

It is also important to note that deregulation in the service sector and liberalization in services trade have contributed significantly to the growth of service activities and services trade. Indeed, the multilateral rules on services trade were established in the Uruguay Round negotiations in the form of the General Agreement on Trade in Services (GATS), and the GATS became effective under the newly created World Trade Organization (WTO) in 1995. The impacts of the GATS do not seem to have been realized much yet, due to the limited time since its enactment, but its impacts on the promotion of services trade will be substantial by providing a freer and more stable environment for conducting services trade.

In light of the increasing importance of services trade, this paper examines the patterns and the determinants of services trade in East Asia. Because the important characteristics of services are intangibility and non-storability, services trade typically requires simultaneity in production and consumption, or physical contact. However, mainly because of technical progress in telecommunication services, some types of services trade can be conducted without physical contact.

Services trade takes four different modes, cross-border supply, consumption abroad, commercial presence, and the presence of natural persons. In this paper we examine services trade from two different perspectives. One examines services trade that is recorded in the balance-of-payments statistics. Although the classification of the items recorded in the balance-of-payments statistics into the four modes of services trade indicated above is not straightforward, most items can be classified into services trade in the forms of cross-border supply and consumption abroad.[1] Services trade in the forms of commercial presence and the presence of natural persons has increased notably in recent years as a result of rapid increase in foreign direct investment (FDI) and movements of professionals, but we do not analyze these types of services trade because of a lack of necessary data for the analysis. The other perspective that we adopt examines trade in services that are embodied in goods trade. Such analysis may be warranted, because a significant portion of services are nontradable and thus traded indirectly through goods trade.

The structure of the paper is as follows. Section 12.2 provides a brief overview of services trade in East Asia to set the stage for the following analyses. Section 12.3 examines the determinants of services trade by applying a regression analysis, and section 12.4 estimates trade in services that are embodied in goods trade by using input-output tables. Finally, section 12.5 presents some concluding comments.

12.2 Recent Trends of Services Trade in East Asia

East Asian economies registered substantial increases in services trade (figure 12.2 and table 12.1).[2] Among them, China recorded a particularly

---

1. It should be noted that royalty receipts and payments, which are recorded in the balance-of-payments statistics, are an exception in that they do not represent cross-border services but are derived from particular contractual arrangements involving factor services.

2. Due to the lack of appropriate price deflators, services trade in this paper is analyzed in terms of nominal prices.

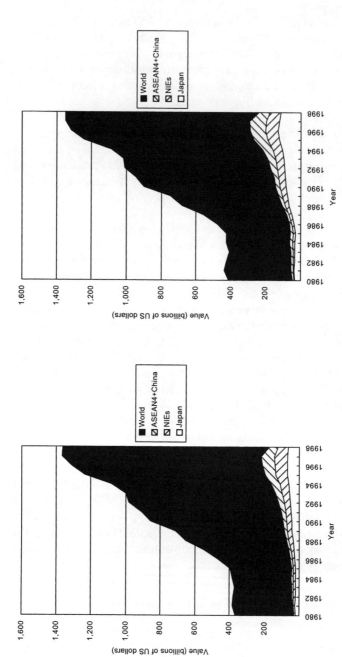

Fig. 12.2   Trends of services trade between 1980 and 1998: services total; *A*, Services total: credit; *B*, Services total: debit

Sources: International Monetary Fund *Balance of Payments Statistics on CD-ROM*, 1999. For Taiwan, Economic Research Department of the Central Bank of China, *Balance of Payments Quarterly; Taiwan District, the Republic of China* (various years).

Notes: World for 1988–98 world total, and for 1980–87, sum of the 175 countries since world total is not available; NIEs: Korea, Taiwan, and Singapore; ASEAN-4: Indonesia, Malaysia, the Philippines, and Thailand.

Table 12.1   Sectoral Trends of Services Trade in East Asian Economies (US$ millions, current prices)

	Total				Transportation				Travel				Communications			
	Credit	Debit	Net	NXR	Credit	Debit	Net	NXR	Credit	Debit	Net	NXR	Credit	Debit	Net	NXR
Japan																
1985	21648.0	31252.4	−9604.4	−0.18	22506.3	35924.3	−13418.0	−0.23	3224.3	36764.1	−33539.9	−0.84	500.2	840.7	−340.5	−0.25
1990	41384.1	84281.4	−42897.3	−0.34	21270.3	28384.5	−7114.2	−0.14	3743.0	28806.1	−25063.1	−0.77	1162.8	1594.2	−431.4	−0.16
1995	65274.0	122626.0	−57352.0	−0.31												
1998	62412.0	111833.0	−49421.0	−0.28												
Korea																
1985	3822.8	3364.2	458.6	0.06	1857.6	1564.1	293.5	0.09	784.3	605.9	178.4	0.13	58.9	11.6	47.3	0.67
1990	9636.9	10251.8	−614.9	−0.03	3179.1	3998.1	−819.0	−0.11	3161.1	2768.0	393.1	0.07	394.6	161.8	232.8	0.42
1995	22827.3	25806.1	−2978.8	−0.06	9272.1	9645.0	−372.9	−0.02	5150.4	6340.7	−1190.3	−0.10	560.7	641.6	−80.9	−0.07
1998	24579.7	23951.4	628.3	0.01	10204.0	8982.9	1221.1	0.06	5933.2	2898.2	3035.0	0.34	655.6	1133.1	−477.5	−0.27
Singapore																
1985	4687.9	3554.3	1133.6	0.14	966.3	1393.5	−427.2	−0.18	1701.7	613.1	1088.6	0.47				
1990	12810.8	8641.6	4169.3	0.19	2225.3	3513.1	−1287.8	−0.22	4649.5	1799.9	2849.6	0.44				
1995	29817.3	17760.5	12056.8	0.25	5125.7	5927.4	−801.7	−0.07	7744.3	5024.4	2720.0	0.21				
1998	18327.0	17996.7	330.3	0.01	4450.8	5981.0	−1530.2	−0.15	4590.3	5035.0	−444.7	−0.05				
Taiwan																
1985	2559.0	5433.0	−2874.0	−0.36	883.0	1737.0	−854.0	−0.33	962.0	1999.0	−1037.0	−0.35	46.0	17.0	29.0	0.46
1990	7008.0	14658.0	−7650.0	−0.35	2323.0	3753.0	−1430.0	−0.24	1741.0	4984.0	−3243.0	−0.48	315.0	264.0	51.0	0.09
1995	15016.0	24053.0	−9037.0	−0.23	4548.0	6333.0	−1785.0	−0.16	3287.0	8458.0	−5171.0	−0.44	563.0	493.0	70.0	0.07
1998	16768.0	24169.0	−7401.0	−0.18	3656.0	5774.0	−2118.0	−0.22	3372.0	7331.0	−3959.0	−0.37	629.0	519.0	110.0	0.10
China																
1985	3055.0	2524.0	531.0	0.10	1302.0	1524.0	−222.0	−0.08	979.0	314.0	665.0	0.51	13.0	7.0	6.0	0.30
1990	5855.0	4352.0	1503.0	0.15	2706.0	3245.0	−539.0	−0.09	1738.0	470.0	1268.0	0.57	159.0	13.0	146.0	0.85
1995	19130.3	25222.8	−6092.5	−0.14	3352.1	9526.1	−6174.0	−0.48	8730.0	3687.6	5042.4	0.41	755.7	217.4	538.3	0.55
1998	24057.0	28980.0	−4923.0	−0.09	2461.0	9071.0	−6610.0	−0.57	12602.0	9205.0	3397.0	0.16	819.0	207.0	612.0	0.60
Indonesia																
1985	844.0	5135.0	−4291.0	−0.72	42.0	1576.8	−1534.8	−0.95	548.0	591.0	−43.0	−0.04	113.0	98.0	15.0	0.07
1990	2488.0	6056.0	−3568.0	−0.42	70.0	2794.7	−2724.7	−0.95	2153.0	836.0	1317.0	0.44	85.0	40.0	45.0	0.36
1995	5469.0	13540.0	−8071.0	−0.42		4861.0			5229.0	2172.0	3057.0	0.41				
1998	4479.0	11813.0	−7334.0	−0.45		3731.0			4255.0	2102.0	2153.0	0.34				

(continued)

Table 12.1 (continued)

	Total				Transportation				Travel				Communications			
	Credit	Debit	Net	NXR	Credit	Debit	Net	NXR	Credit	Debit	Net	NXR	Credit	Debit	Net	NXR
Malaysia																
1985	1934.3	3926.6	−1992.3	−0.34	690.3	1408.0	−717.7	−0.34	621.8	1158.3	−536.4	−0.30				
1990	3859.0	5484.5	−1625.6	−0.17	1197.8	2531.4	−1333.5	−0.36	1684.0	1450.3	233.7	0.07				
1995	11601.6	14980.8	−3379.2	−0.13	2465.7	5609.3	−3143.7	−0.39	3968.8	2314.3	1654.3	0.26				
1998																
The Philippines																
1985	2235.0	867.0	1368.0	0.44	200.0	372.0	−172.0	−0.30	506.0	37.0	469.0	0.86				
1990	3244.0	1761.0	1483.0	0.30	246.0	980.0	−734.0	−0.60	466.0	111.0	355.0	0.62				
1995	9348.0	6926.0	2422.0	0.15	274.0	2051.0	−1777.0	−0.76	1136.0	422.0	714.0	0.46				
1998	7477.0	10107.0	−2630.0	−0.15	324.0	1983.0	−1659.0	−0.72	1418.0	1950.0	−532.0	−0.16				
Thailand																
1985	2041.3	1814.6	226.7	0.06	462.6	1126.3	−663.8	−0.42	1169.2	280.7	888.5	0.61				
1990	6419.0	6309.2	109.8	0.01	1327.0	3575.8	−2248.8	−0.46	4324.5	1432.3	2892.3	0.50				
1995	14845.2	18803.8	−3958.6	−0.12	2454.6	7779.8	−5325.1	−0.52	8035.0	4271.2	3763.9	0.31	198.9	120.7	78.2	0.24
1998	13155.6	11998.1	1157.5	0.05	2670.8	4603.6	−1932.7	−0.27	6173.6	1959.6	4214.1	0.52	158.9	54.5	104.5	0.49

	Construction				Insurance				Finance				Computer and Information			
	Credit	Debit	Net	NXR	Credit	Debit	Net	NXR	Credit	Debit	Net	NXR	Credit	Debit	Net	NXR
Japan																
1985																
1990																
1995	6559.3	3205.0	3354.3	0.34	295.6	2508.4	−2212.8	−0.79	305.5	456.9	−151.4	−0.20	1338.3	3532.0	−2193.8	−0.45
1998	7736.1	5526.9	2209.2	0.17	42.3	2368.8	−2326.4	−0.96	1607.8	2151.7	−543.9	−0.14				
Korea																
1985					−19.5	56.5	−76.0	−2.05	2.4	46.7	−44.3	−0.90	1.4	12.1	−10.7	−0.79
1990					3.9	−20.4	24.3	−1.47	0.7	11.0	−10.3	−0.88	3.2	50.1	−46.9	−0.88
1995					−19.9	254.7	−274.6	−1.17	105.0	129.7	−24.7	−0.11	4.9	93.3	−88.4	−0.90
1998					51.5	142.8	−91.3	−0.47	145.1	108.9	36.2	0.14	4.5	90.4	−85.9	−0.91

	(1)	(2)	(3)	(4)	(5)	(6)	(7)	(8)	(9)	(10)	(11)	(12)	(13)	(14)
Singapore														
1985					47.7	115.0	−67.3	−0.41						
1990					88.4	779.0	−690.5	−0.80						
1995					354.5	973.8	−619.4	−0.47	30.0					
1998					417.7	732.9	−315.2	−0.27	712.0					
Taiwan														
1985	12.0	5.0	7.0	0.41	41.0	138.0	−97.0	−0.54	13.0			15.0		
1990	31.0	0.0	31.0	1.00	146.0	210.0	−64.0	−0.18	465.0			22.0		
1995	111.0	275.0	−164.0	−0.42	418.0	508.0	−90.0	−0.10	7.0	17.0	0.40	45.0	23.0	
1998	160.0	342.0	−182.0	−0.36	699.0	526.0	173.0	0.14	900.0	−188.0	−0.12	98.0	−75.0	−0.62
China														
1985					196.0	69.0	127.0	0.48						
1990					227.0	94.0	133.0	0.41						
1995					1852.1	4273.3	−2421.2	−0.40	27.0			134.0		
1998	594.0	1120.0	−526.0	−0.31	384.0	1758.0	−1374.0	−0.64	163.0	−136.0	−0.72	333.0	−199.0	−0.43
Indonesia														
1985						139.2								
1990						234.3								
1995						451.0								
1998						334.0								
Malaysia														
1985					1.6									
1990					2.6									
1995					7.2									
1998														
The Philippines														
1985					4.0	6.0	−2.0	−0.20						
1990	3.0	5.0	−2.0	−0.25	14.0	59.0	−45.0	−0.62						
1995	10.0	58.0	−48.0	−0.71	62.0	109.0	−47.0	−0.27						
1998	37.0	218.0	−181.0	−0.71	24.0	43.0	−19.0	−0.28						
Thailand														
1985					9.7	96.9	−87.1	−0.82						
1990					10.3	336.0	−325.7	−0.94						
1995	19.2	162.8	−143.6	−0.79	98.7	961.2	−862.5	−0.81						
1998	94.0	124.0	−30.1	−0.14	51.2	591.8	−540.6	−0.84						

(continued)

Table 12.1 (continued)

	Royalties and License Fees				Other Business Services				Personal, Cultural, and Recreational				Government, n.i.e.			
	Credit	Debit	Net	NXR	Credit	Debit	Net	NXR	Credit	Debit	Net	NXR	Credit	Debit	Net	NXR
Japan																
1985																
1990																
1995	6005.0	9417.2	-3412.1	-0.22	24436.8	31870.7	-7433.9	-0.13	133.3	560.2	-426.8	-0.62	1307.7	1078.5	229.2	0.10
1998	7388.0	8947.3	-1559.3	-0.10	17077.5	28131.7	-11054.2	-0.24	428.8	1261.2	-832.4	-0.49	617.0	1128.2	-511.2	-0.29
Korea																
1985	3.2	322.8	-319.6	-0.98	861.6	612.7	248.9	0.17					272.9	127.2	145.7	0.36
1990	37.1	1364.4	-1327.3	-0.95	2375.6	1696.8	678.8	0.17					481.6	202.2	279.4	0.41
1995	299.2	2384.8	-2085.6	-0.78	6760.8	5806.5	954.3	0.08					694.1	411.9	282.2	0.26
1998	260.1	2369.3	-2109.2	-0.80	6575.0	7705.1	-1130.1	-0.08	14.1	91.8	-77.7	-0.73	736.6	428.9	307.7	0.26
Singapore																
1985					1881.7	1390.8	490.9	0.15					90.4	41.8	48.6	0.37
1990					5755.5	2483.4	3272.1	0.40					92.0	66.2	25.9	0.16
1995					16499.8	5716.6	10783.2	0.49					93.0	118.2	-25.3	-0.12
1998					8784.7	6135.5	2649.1	0.18					83.5	112.3	-28.7	-0.15
Taiwan																
1985	2.0	150.0	-148.0	-0.97	562.0	773.0	-211.0	-0.16					21.0	533.0	-512.0	-0.92
1990	121.0	582.0	-461.0	-0.66	2260.0	3567.0	-1307.0	-0.22					71.0	735.0	-664.0	-0.82
1995	241.0	937.0	-696.0	-0.59	5759.0	5775.0	-16.0	0.00					89.0	1071.0	-982.0	-0.85
1998	317.0	1419.0	-1102.0	-0.63	7069.0	6172.0	897.0	0.07	23.0	165.0	-142.0	-0.76	108.0	923.0	-815.0	-0.79

China												
1985	63.0	420.0	−357.0	−0.74	435.0	347.0	88.0	0.11	130.0	263.0	−133.0	−0.34
1990					918.0	291.0	627.0	0.52	107.0	239.0	−132.0	−0.38
1995					3740.0	6930.1	−3190.1	−0.30	700.3	588.2	112.1	0.09
1998	15.0	39.0	−24.0	−0.44	6941.0	6459.0	482.0	0.04	17.0	205.0	−188.0	−0.85
Indonesia												
1985					254.0	2703.0	−2449.0	−0.83	125.0			
1990					265.0	2033.0	−1768.0	−0.77	158.0			
1995						5648.0			127.0	310.0	−183.0	−0.42
1998						5389.0			139.0	217.0	−78.0	−0.22
Malaysia												
1985					520.7	1248.1	−727.3	−0.41	99.9	112.4	−12.5	−0.06
1990					885.1	1412.3	−527.2	−0.23	89.5	90.6	−1.1	−0.01
1995					4996.8	6897.9	−1901.1	−0.16	163.3	159.3	4.0	0.01
1998												
The Philippines												
1985		17.0			1152.0	413.0	739.0	0.47	373.0	22.0	351.0	0.89
1990	1.0	38.0	−37.0	−0.95	2167.0	528.0	1639.0	0.61	347.0	40.0	307.0	0.79
1995	2.0	99.0	−97.0	−0.96	7839.0	4167.0	3672.0	0.31	25.0	20.0	5.0	0.11
1998		70.0			5662.0	5823.0	−161.0	−0.01	12.0	20.0	−8.0	−0.25
Thailand												
1985	0.0	45.5	−45.5	−1.00	256.2	192.9	63.3	0.14	143.5	72.3	71.3	0.33
1990		170.5			629.7	645.7	−16.1	−0.1	127.5	148.9	−21.4	−0.08
1995	0.6	629.8	−629.2	−1.00	3844.4	4675.2	−830.7	−0.10	193.7	203.1	−9.4	−0.02
1998	7.1	513.9	−506.8	−0.97	3918.6	4026.5	−107.9	−0.01	81.4	124.4	−42.9	−0.21

Sources: International Monetary Fund (1999a); Economic Research Department, Central Bank of China (various years).

Notes: NXR (net export ratio) = (exports − imports)/(exports + imports). n.i.e. = not included elsewhere. Blank cells indicate data not available.

high growth of services trade. Between 1985 and 1998 the credit and debit of total services for China increased eight- and tenfold respectively. The rates of expansion for other economies were less spectacular, but they were still very high. Some of the economies that exhibited high growth in total services trade include Korea and Thailand, whose credit (debit) in total services trade expanded by six (seven) and six (seven) times, respectively, during the 1985–98 period. Among the East Asian economies that recorded rapid expansion in services trade Japan is an exception, because its value of total services trade increased much more slowly. This pattern is particularly notable for credit, which increased less than threefold from 1985 to 1998, whereas its debit increased slightly more than threefold. In spite of the relatively lower growth of services trade, the value of services trade for Japan is still significantly greater when compared to other East Asian economies. In 1998 the credit in total services trade for Japan is 2.5 times as large as that of the country with the second largest value, Korea, whereas the debit for Japan is more than three times as large as the economy with the second largest value, China.

Rapid growth of services trade by East Asian economies resulted in an increase in their shares in world services trade. Specifically, the share of East Asia in world total service credit increased sharply, from 12.2 percent in 1985 to 20.3 percent in 1997, before declining to 15.4 percent in 1998. The corresponding share in world total debit was higher but exhibited a similar pattern for the period under study, because it increased from 15.8 percent in 1985 to 29.5 percent in 1996 before declining to 23.3 percent in 1998. For service credit, Japan and the newly industrialized economies (NIEs) accounted for similar shares, around 5.5 percent each of the world service credit, whereas the corresponding share for the Association of Southeast Asian Nations (ASEAN) and China combined was somewhat smaller at 4.4 percent. As to the service debit, Japan had a notably high share of world service debit at 10.8 percent, whereas the shares for the NIEs and the ASEAN and China combined were similar at 6.4 and 6.1 percent, respectively.[3]

An examination of the services trade for the East Asian economies at the sectoral level reveals a number of interesting developments (tables 12.1, 12.2, and 12.3).[4] Most economies' transportation services account for a substantial part of their service credits. In 1998 more than 20 percent of service credits came from transportation services for Japan, Korea, Singapore,

3. For individual services trade categories, the East Asian economies account for a very small share in world services trade except for a few cases. The categories in which the shares exceed 10 percent of world total are only found for Japan. Japan's shares of world credit in construction and in royalties and license fees in 1998 were 20.8 and 11.5 percent, respectively, whereas its shares of world debit in construction, computer and information, and royalties and license fees were 22.6, 20.9, and 14.3 percent, respectively.

4. The differences in the availability of data among the East Asian economies make it difficult for us to make accurate comparisons.

Table 12.2   Composition of Services Trade for East Asian Economies: Credit (%)

	Total	Transportation	Travel	Communications	Construction	Insurance	Finance	Computer and Information	Royalties License and Fees	Other Business Services	Personal, Cultural, and Recreational	Government, n.i.e.
Japan												
1985	100.0											
1990	100.0											
1995	100.0	34.5	4.9	0.8	10.0	0.5	0.5		9.2	37.4	0.2	2.0
1998	100.0	34.1	6.0	1.9	12.4	0.1	2.6	2.1	11.8	27.4	0.7	1.0
Korea												
1985	100.0	48.6	20.5	1.5		-0.5	0.1	0.0	0.1	22.5		7.1
1990	100.0	33.0	32.8	4.1		0.0	0.0	0.0	0.4	24.7		5.0
1995	100.0	40.6	22.6	2.5		-0.1	0.5	0.0	1.3	29.6		3.0
1998	100.0	41.5	24.1	2.7		0.2	0.6	0.0	1.1	26.7	0.1	3.0
Singapore												
1985	100.0	20.6	36.3			1.0				40.1		1.9
1990	100.0	17.4	36.3			0.7				44.9		0.7
1995	100.0	17.2	26.0			1.2				55.3		0.3
1998	100.0	24.3	25.0			2.3				47.9		0.5
Taiwan												
1985	100.0	34.5	37.6	1.8	0.5	1.6	1.2		0.1	22.0		0.8
1990	100.0	33.1	24.8	4.5	0.4	2.1			1.7	32.2		1.0
1995	100.0	30.3	21.9	3.7	0.7	2.8			1.6	38.4		0.6
1998	100.0	21.8	20.1	3.8	1.0	4.2	4.2	0.1	1.9	42.2	0.1	0.6
China												
1985	100.0	42.6	32.0	0.4		6.4				14.2		4.3
1990	100.0	46.2	29.7	2.7		3.9				15.7		1.8
1995	100.0	17.5	45.6	4.0		9.7				19.6		3.7
1998	100.0	10.2	52.4	3.4	2.5	1.6	0.1	0.6	0.3	28.9	0.1	0.1

(continued)

Table 12.2 (continued)

	Total	Transportation	Travel	Communications	Construction	Insurance	Finance	Computer and Information	Royalties License and Fees	Other Business Services	Personal, Cultural, and Recreational	Government, n.i.e.
Indonesia												
1985	100.0	5.0	64.9							30.1		
1990	100.0	2.8	86.5							10.7		
1995	100.0		95.6	2.1								2.3
1998	100.0		95.0	1.9								3.1
Malaysia												
1985	100.0	35.7	32.1			0.1				26.9		5.2
1990	100.0	31.0	43.6			0.1				22.9		2.3
1995	100.0	21.3	34.2			0.1				43.1		1.4
1998	100.0											
The Philippines												
1985	100.0	8.9	22.6			0.2				51.5		16.7
1990	100.0	7.6	14.4		0.1	0.4			0.0	66.8		10.7
1995	100.0	2.9	12.2		0.1	0.7			0.0	83.9		0.3
1998	100.0	4.3	19.0		0.5	0.3				75.7		0.2
Thailand												
1985	100.0	22.7	57.3			0.5			0.0	12.5		7.0
1990	100.0	20.7	67.4			0.2				9.8		2.0
1995	100.0	16.5	54.1	1.3	0.1	0.7			0.0	25.9		1.3
1998	100.0	20.3	46.9	1.2	0.7	0.4			0.1	29.8		0.6

Sources and Notes: See table 12.1

Table 12.3   **Composition of Services Trade for East Asian Economies: Debit (%)**

	Total	Transportation	Travel	Communications	Construction	Insurance	Finance	Computer and Information	Royalties and License Fees	Other Business Services	Personal, Cultural, and Recreational	Government, n.i.e.
Japan												
1985	100.0											
1990	100.0											
1995	100.0	29.3	30.0	0.7	2.6	2.0	0.4		7.7	26.0	0.5	0.9
1998	100.0	25.4	25.8	1.4	4.9	2.1	1.9	3.2	8.0	25.2	1.1	1.0
Korea												
1985	100.0	46.5	18.0	0.3		1.7	1.4	0.4	9.6	18.2	0.1	3.8
1990	100.0	39.0	27.0	1.6		-0.2	0.1	0.5	13.3	16.6	0.2	2.0
1995	100.0	37.4	24.6	2.5		1.0	0.5	0.4	9.2	22.5	0.4	1.6
1998	100.0	37.5	12.1	4.7		0.6	0.5	0.4	9.9	32.2	0.4	1.8
Singapore												
1985	100.0	39.2	17.3			3.2				39.1		1.2
1990	100.0	40.7	20.8			9.0				28.7		0.8
1995	100.0	33.4	28.3			5.5				32.2		0.7
1998	100.0	33.2	28.0			4.1				34.1		0.6
Taiwan												
1985	100.0	32.0	36.8	0.3	0.1	2.5	0.2	0.3	2.8	14.2	1.0	9.8
1990	100.0	25.6	34.0	1.8	0.0	1.4	3.2	0.2	4.0	24.3	0.5	5.0
1995	100.0	26.3	35.2	2.0	1.1	2.1	0.0	0.2	3.9	24.0	0.6	4.5
1998	100.0	23.9	30.3	2.1	1.4	2.2	3.7	0.4	5.9	25.5	0.7	3.8
China												
1985	100.0	60.4	12.4	0.3		2.7				13.7		10.4
1990	100.0	74.6	10.8	0.3		2.2				6.7		5.5
1995	100.0	37.8	14.6	0.9		16.9				27.5		2.3
1998	100.0	31.3	31.8	0.7	3.9	6.1	0.6	1.1	1.4	22.3	0.1	0.7

(*continued*)

Table 12.3   (continued)

	Total	Transportation	Travel	Communications	Construction	Insurance	Finance	Computer and Information	Royalties and License Fees	Other Business Services	Personal, Cultural, and Recreational	Government, n.i.e.
Indonesia												
1985	100.0	30.7	11.5			2.7				52.6		2.4
1990	100.0	46.1	13.8			3.9				33.6		2.6
1995	100.0	35.9	16.0	0.7		3.3				41.7		2.3
1998	100.0	31.6	17.8	0.3		2.8				45.6		1.8
Malaysia												
1985	100.0	35.9	29.5							31.8		2.9
1990	100.0	46.2	26.4							25.7		1.7
1995	100.0	37.4	15.4							46.0		1.1
1998												
The Philippines												
1985	100.0	42.9	4.3		0.3	0.7			2.0	47.6		2.5
1990	100.0	55.7	6.3		0.8	3.4			2.2	30.0		2.3
1995	100.0	29.6	6.1		2.2	1.6			1.4	60.2		0.3
1998	100.0	19.6	19.3			0.4			0.7	57.6		0.2
Thailand												
1985	100.0	62.1	15.5			5.3			2.5	10.6		4.0
1990	100.0	56.7	22.7			5.3			2.7	10.2		2.4
1995	100.0	41.4	22.7	0.6	0.9	5.1			3.3	24.9		1.1
1998	100.0	38.4	16.3	0.5	1.0	4.9			4.3	33.6		1.0

Sources and Notes: See table 12.1.

Taiwan, Malaysia, and Thailand.[5] The figure is remarkably high for Korea at 41.5 percent. Travel is an important source of service credit for many East Asian economies. Travel has a particularly large share for Indonesia, China, and Thailand, where the respective shares in total service credits were 95, 52, and 47 percent, reflecting abundance in attractive tourist spots. Another similarity among the East Asian economies regarding sectoral services trade patterns is a high share of other business services, such as consulting and accountant services, as for many economies its share exceeded 20 percent. There are other interesting observations to be made. Japan showed relatively high shares of construction and royalties in its total service credit, reflecting competitiveness in construction services and innovative activities. It should also be noted that Japan's large credit in construction services appears to be attributable to its large official development assistance (ODA) in infrastructure, which generates exports of construction services from Japan. China recorded a relatively high share for insurance from 1985 to 1995, around 4–10 percent of total, but the share declined sharply to 1.6 percent in 1998.

Turning to the debits of services trade by sectors, one also finds a relatively large share of transportation services: for virtually all economies its share is greater than 20 percent in 1998. Coupled with the observation on sectoral shares in service credit examined earlier, this finding appears to indicate a substantial amount of intraindustry trade in transportation services.[6] The shares of travel in total service debit vary widely among the East Asian economies. Relatively rich economies such as Japan, Singapore, and Taiwan registered high values of around 20–30 percent, whereas lower-income economies, including Indonesia, Malaysia, the Philippines, and Thailand, recorded relatively low values of around 15 percent. These observations are consistent with our expectation that because overseas travel is expensive rich economies can afford it but poor countries cannot. Korea and China appear to be exceptions from the groups discussed above. Although Korea is a relatively rich country in East Asia, its share of travel in total service debit is smaller compared with other high-income countries. One important reason is the economic crisis in 1998, which made it difficult for Koreans to travel abroad. The corresponding share for China in 1998 is quite high at 31.8 percent. This high share may be attributable to the appreciation of Chinese yuan vis-à-vis other Asian currencies, which promoted overseas travel by the Chinese.

5. A somewhat peculiar treatment of transportation services in services trade in the balance-of-payments statistics should be noted. Use of national carriers by nationals for overseas transport is not recorded as international trade in services, whereas use of foreign carriers is recorded as international trade in services. Such treatment underestimates the magnitude of international trade in transportation services.

6. Intraindustry trade in the discussion refers to cases in which both credits and debits are recorded in comparable magnitude for the same services trade category.

"Other business services" is another service category for which the shares in total service debit are quite high for all the East Asian economies. Similar to the finding on transportation services, a significant amount of intra-industry trade appears to be conducted. Royalty payments and insurance recorded debits for many East Asian economies. It is worth noting that a significant amount of intraindustry trade in royalties is undertaken in Japan, whereas for Korea and Taiwan royalty payments are significantly greater than royalty receipts. These differences in the patterns of royalty trade between Japan on the one hand and Korea and Taiwan on the other hand appear to be attributable to the differences in their innovative capabilities.

So far we have examined services trade for the East Asian economies by looking at credit and debit separately. The balance (credits-debits) of services trade gives us useful information on the competitiveness of the service sector concerned. One may divide the East Asian economies into three groups according to the patterns of overall service trade balance. One group, consisting of Japan, Taiwan, Indonesia, and Malaysia, registers a deficit for all the years shown in table 12.1. The second group, consisting of Singapore, the Philippines, and Thailand, exhibits a surplus for all or most years. The last group, consisting of Korea and China, does not show consistent patterns for the period under study.

The balance of services trade at the sectoral level reveals wide variations among the East Asian economies, with one notable exception: insurance. For insurance all economies recorded an import surplus, reflecting a lack of competitiveness of the insurance industry in East Asia vis-à-vis those in the United States and United Kingdom. Japan registered deficit in all the categories except construction, and huge deficits were recorded in travel and other business services, whereas the Philippines had deficits in all the categories.[7] For Korea a sizeable surplus was recorded for travel, but notable deficit was shown in royalties and license fees. As noted earlier, the depressed economic situation resulting from the economic crisis contributed significantly to a surplus in the travel account for Korea by reducing the demand for overseas travel. Singapore's overall surplus mainly comes from a huge surplus in other business services, whereas Taiwan's overall deficit is mainly due to the deficits in transportation services and travel. China, Malaysia, and Thailand had a similar pattern in that they recorded quite a big surplus in the travel account but a huge deficit in transportation services. Similar to the case of Korea, sizable surpluses in the travel account in 1998 for both Thailand and Malaysia are due to the economic crisis. In addition, for Malaysia a large deficit was recorded for other business services. The information on Indonesia is incomplete, but it indicates a large surplus in travel.

As was indicated earlier, the information on trade balance may give use-

---

7. Unless otherwise noted, the discussions in this section refer to the data for 1998.

Table 12.4   **Foreign Direct Investment Restrictiveness for Services in East Asia and Other Countries**

Region/Country	Average	Business	Communications	Distribution	Finance	Transport
East Asia	59	51	70	46	72	57
China	47	36	82	28	45	46
Indonesia	56	56	64	53	55	53
Korea	67	57	69	63	88	57
Malaysia	31	32	42	8	61	12
The Philippines	73	48	76	48	95	98
Thailand	81	78	84	78	88	78
Developed countries						
Australia	29	18	44	18	45	20
Canada	31	23	51	20	38	24
Japan	19	6	35	5	36	11
New Zealand	19	9	43	8	20	13
United States	17	1	35	0	20	3

Source: World Bank (2000).
Note: Index value of 100 represents maximum restrictiveness.

ful information on the competitiveness of the sectors concerned. However, one should note that services trade, especially debits, for many East Asian economies is likely to be distorted because of restrictions on services trade imposed by these economies. This point may be seen clearly from the restrictiveness of the FDI regime in service sectors for the East Asian economies in table 12.4. Among the East Asian economies Thailand, the Philippines, and Korea had particularly restrictive regimes before the crisis, whereas Japan and Malaysia had relatively open regimes. Without these restrictions, the debits in services trade would have been greater, contributing to the deterioration of the trade balance. One example of government regulations in services trade is the governments' bias for national carriers in the allocation of landing rights in airports and berthing space in seaports, limiting the imports of transportation services. It should also be noted that bilateral aviation agreements between countries effectively strangle competition on a great number of international routes.

12.3   The Determinants of Services Trade

The previous section discussed the changing patterns of services trade for East Asian economies. In this section we investigate the determinants of the patterns of trade in services, or the determinants of the patterns of comparative advantage in services. We use the figures on trade balance, which we examined in the previous section, as an indicator of comparative advantage. Those sectors with trade surplus are interpreted to possess a comparative advantage and those with trade deficit a comparative disadvantage.

12.3.1   Previous Studies

As for merchandise trade, a number of empirical studies have been conducted to identify the determinants of the patterns of comparative advantage based on the cross-commodity regression framework. Most of these studies confirmed the validity of the Heckscher-Ohlin theorem to explain the patterns of comparative advantage.[8] The pioneering work is by Baldwin (1971), who conducted the cross-commodity regression analysis to examine the relationship between factor endowment and the pattern of comparative advantage for the United States. Regressing cross-sectoral trade performance on cross-sectoral factor intensity for production, he found that the U.S. manufacturing sector had a comparative advantage in the production of human capital–intensive products. This finding is consistent with the Heckscher-Ohlin theorem, because the United States is relatively well endowed with human capital compared to its trading partners. Following Baldwin (1971), Stern and Maskus (1981) and Urata (1983) investigated the determinants of comparative advantage for the United States in 1958–1976 and for Japan in 1967 and 1975, respectively. Stern and Maskus also found that the United States had a comparative advantage in human capital–intensive products, supporting the validity of the Heckscher-Ohlin theorem. The findings by Urata were different in that Japan was shown to have a comparative advantage in human capital–intensive products in its trade with developing countries, whereas it was shown to have a comparative advantage in capital-intensive products in its trade with developed countries. Realizing that Japan was placed somewhere in the middle of developed and developing countries in terms of the level of economic development, he argued that the findings were consistent with the Heckscher-Ohlin theorem. Although these analyses contributed to the empirical investigation of the determinants of foreign trade and comparative advantage, the empirical framework used in these analyses has been subject to criticisms, indicating that the cross-commodity regression analysis is rather weak on theoretical foundation.[9]

Bowen (1983) proposed an alternative approach—a cross-country regression approach. He collected factor endowment data for thirty-four countries for five years from 1963 to 1975 and investigated the relationship between factor abundance and comparative advantage. The estimation re-

8. Deardorff (1984) and Leamer and Levinsohn (1995) perform the detailed survey work of the empirical international trade analyses.

9. See, for example, Leamer and Bowen (1981) and Bowen, Hollander, and Viaene (1998). Aw (1983) derived the sufficient condition that the sign of cross-commodity regression coefficients coincide with the theoretical implication. An alternative approach to support the cross-commodity regression is suggested by Petri (1991). He constructed the estimation model by relaxing the assumption of factor price equalization.

sults revealed the availability of human capital, or skilled labor, as an important determinant of U.S. comparative advantage. His results are consistent with the results from the cross-commodity regression analyses. Leamer (1984) argued that the cross-country regression framework has stronger theoretical foundations than the cross-commodity approach.

Concerning services trade, however, there have been few attempts to investigate the determinants of comparative advantage. One of the first rigorous attempts to identify the determinants of the pattern of comparative advantage in services trade was conducted by Sapir and Lutz (1981). They applied the cross-country regression analysis to services trade, focusing on freight transportation, other transportation, and insurance for 1971. Their results showed that the Heckscher-Ohlin theorem could go a long way in explaining trade patterns in services and that economies abundant with physical and human capital had a comparative advantage in services.

Sazanami and Urata (1990), following Sapir and Lutz (1981), performed a detailed analysis of the determinants of services trade. They also applied the cross-country regression framework, expanding the coverage to include travel, intellectual property rights, and engineering services, in addition to freight transportation, other transportation, and insurance services, which were examined by Sapir and Lutz. Their results indicated that both physical and human capitals were important determinants of services trade.

We extend the earlier studies by expanding the coverage of the services trade and the countries for the analysis. In our analysis, there are eleven services trade categories: transportation; travel; communications; construction; insurance; financial services; computer and information; royalties and license fees; other business services; personal, cultural, and recreational services; and government services. Our country coverage includes 108 countries at maximum. We apply the cross-country regression framework to analyze the determinants of the patterns of comparative advantage in services trade and investigate whether the Heckscher-Ohlin theorem performs well in explaining services trade.

12.3.2 The Empirical Framework for the Analysis of the Determinants of Services Trade

This section briefly explains the methodology, or the cross-country regression analysis framework, that is applied in this paper.[10]

Suppose that there are $n(= 1, \cdots, N)$ countries, $i(= 1, \cdots, I)$ commodities, and $j(= 1, \cdots, J)$ factors. Let country n's net export (exports minus imports) vector be $\mathbf{T}_n(I \times 1$ vector), its output vector be $\mathbf{Q}_n(I \times 1$ vector), and its final demand vector be $\mathbf{C}_n(I \times 1$ vector). Assume that country

---

10. For more details, see Bowen and Sveikauskas (1992) and Bowen, Hollander, and Viaene (1998).

n's factor input matrix is $\mathbf{A}_n (J \times I$ matrix), whose element is a_{ij}, and the number of commodities is equal to or greater than the number of factors $(I \geq J)$. The net export of factor in country n is provided by

$$(1) \qquad \mathbf{F}_n \equiv \mathbf{A}_n \mathbf{T}_n \equiv \mathbf{A}_n \mathbf{Q}_n - \mathbf{A}_n \mathbf{C}_n.$$

Suppose that each country satisfies traditional Heckscher-Ohlin-Vanek (hereafter, HOV) assumptions—full employment ($\mathbf{E}_n = \mathbf{A}_n \mathbf{Q}_n$, where $\mathbf{E}_n$ is the factor endowment vector $(J \times 1)$ in country n), identical technologies ($\mathbf{A}_n = \mathbf{A} \forall n$), and identical and homothetic preferences ($\mathbf{C}_n = \mu_n \mathbf{C}_W$, where μ_n is country n's share of world expenditure and $\mathbf{C}_W$ is world final demand). Assume that the world final demand equals to the world output ($\mathbf{C}_W = \sum_{n=1}^{N} \mathbf{C}_n = \sum_{n=1}^{N} \mathbf{Q}_n = \mathbf{Q}_W$). Let world income be Y_W, country n's income be Y_n and its balance of trade be b_n.[11] With these assumptions, equation (1) is rewritten as

$$(2) \qquad \mathbf{F}_n \equiv \mathbf{A} \mathbf{T}_n = \mathbf{E}_n - \mu_n \mathbf{A} \mathbf{Q}_W = \mathbf{E}_n - (\alpha_n - \lambda_n) \mathbf{E}_W,$$

where $\mathbf{E}_W = \sum_{n=1}^{N} \mathbf{E}_n$ is the factor endowment vector $(J \times 1)$ in the world, α_n is the ratio of country n's income to the world income (Y_n / Y_W), and λ_n is the ratio of trade imbalance to the world income (b_n / Y_W).[12] Assume that there is an equal number of commodities and factors (this implies that $\mathbf{A}$ is square and can be inverted). Then we can rewrite equation (2) as

$$(3) \qquad \mathbf{T}_n = (\mathbf{A}^t)^{-1} [\mathbf{E}_n - (\alpha_n - \lambda_n) \mathbf{E}_W],$$

where superscript t means *transpose*. Therefore, the trade balance of commodity i in the country n, t_{ni}, is

$$(4) \qquad t_{ni} = (\mathbf{A}_i^t)^{-1} [\mathbf{E}_n - (\alpha_n - \lambda_n) \mathbf{E}_W],$$

where $(\mathbf{A}_i^t)^{-1}$ is a $1 \times J(=K)$ vector whose elements a_{nij}^{-1} are the row elements of the inverse of the factor requirements matrix corresponding to commodity j.

Suppose that trade is balanced ($\lambda_n = 0$) and let $\mathbf{p}$ and $\mathbf{w}$ be the vector of world output prices $(I \times 1)$ and world factor prices $(J \times 1)$, respectively. The expenditure share of country n, α_n, and long-run zero profit condition, $\mathbf{w}^t = \mathbf{p}^t \mathbf{A}^{-1}$, imply

$$(5) \qquad Y_W \alpha_n = \mathbf{p}^t \mathbf{Q}_n = \mathbf{p}^t (\mathbf{A}^{-1} \mathbf{E}_n) = \mathbf{w}^t \mathbf{E}_n.$$

The last term in equation (5) means national expenditure in terms of factor prices. From equation (5), the following equation can be derived:

$$(6) \qquad t_{ni} = \mathbf{R}_i^t \mathbf{E}_n,$$

---

11. The relationship between b_n and μ_n is as follows. Supposing that $\mathbf{p}$ is the vector of commodity prices, and using $\mathbf{T}_n = \mathbf{Q}_n - \mu_n \mathbf{C}_W$, $\mathbf{p}^t \mathbf{T}_n = \mathbf{p}^t \mathbf{Q}_n - \mu_n \mathbf{p}^t \mathbf{C}_W$, where superscript t means *transpose*, we obtain $b_n = Y_n - \mu_n Y_W$.

12. $\mathbf{A} \mathbf{Q}_W = \mathbf{A} \sum_{n=1}^{N} \mathbf{Q}_n = \sum_{n=1}^{N} \mathbf{A} \mathbf{Q}_n = \sum_{n=1}^{N} \mathbf{E}_n = \mathbf{E}_W$ and $\mu_n = \alpha_n - \lambda_n$.

where $\mathbf{R}_i^t$ is a vector of factor requirement for net exports with elements r_{nij} = $[a_{nij}^{-1} - (y_{Wi}w_j / Y_W)]$.[13] Equation (6) is the equation for cross-country regression analysis.

From equation (6), we derive the specification of the equation to be estimated as (7).

$$(7) \qquad T_{ni} = \beta_{0i} + \beta_{1i}K_{ni} + \beta_{2i}L_{ni} + \beta_{3i}H_{ni} + \varepsilon_{ni},$$

where T is net export, K is physical capital, L is labor, H is human capital, and ε is an error term.

12.3.3 The Hypotheses

Following the findings from the earlier studies, and given the information on factor requirements for the provision of services, we constructed the following hypotheses concerning the determinants of the patterns of comparative advantage in services trade.

We classify eleven service categories into three categories and establish the hypotheses to be tested by the regression analyses. The first category includes physical capital– and human capital–intensive services. Communications, insurance, financial, computer and information, royalty and license fee, and government services are classified into this category. The second category is physical capital–intensive services. Transportation and construction services are included in this group because these services require a large scale of physical capital, such as ships and construction machines. Finally, the third category includes travel, other business services, and personal, cultural, and recreational services, the provision of which requires labor and human capital services. Table 12.5 summarizes the expected signs from the regression analysis, based on the Heckscher-Ohlin theorem.

12.3.4 The Data

The data for the dependent variables (net export of services) are taken from the International Monetary Fund (IMF; 1999a). The data for the independent variables (labor, physical and human capital) are taken from the World Bank (1999) for all the countries except for Taiwan, for which the Council for Economic Planning and Development, Republic of China (1999) is used.

Because most countries do not provide physical capital data, we constructed the proxy for physical capital by using the information on gross domestic fixed investment. Specifically, we accumulated the value of gross fixed investment from 1960 by using the perpetual inventory method with a

---

13. The world output of commodity i is y_{wi}. Derivation of equation (6) is described in appendix A.

Table 12.5   **Expected Signs of Regression Coefficients**

	Regression Equations					
	Total (1)	Transportation (2)	Travel (3)	Communications (4)	Construction (5)	Insurance (6)
K	?	+	−	+	+	+
L	?	−	+	−	−	−
H	?	−	+	+	−	+

	Regression Equations					
	Finance (7)	Computer and Information (8)	Royalties and License Fees (9)	Other Business Services (10)	Personal, Cultural, and Recreational (11)	Government, n.i.e. (12)
K	+	+	+	−	−	+
L	−	−	−	+	+	−
H	+	+	+	+	+	+

Notes: K = capital stock; L = labor force; H = human capital. + and − indicate expected signs of regression coefficients. ? indicates undetermined. Section 3 provides detailed descriptions of the hypotheses.

10 percent depreciation rate.[14] Labor is defined as the labor force (population aged fifteen to sixty-four). As for human capital, we use the postsecondary school enrollment ratio as a proxy in the absence of better indicators, such as the number of researchers and scientists.

12.3.5   The Results

We estimated the coefficients for the variables in equation (7) by using ordinary least squares (OLS), and the results are shown in tables 12.6–12.8. Table 12.6 presents the results of the estimation for equation (7), where the dependent variable is net export. The results indicate that only a few estimated coefficients are consistent with our expectation and statistically significant. Specifically, labor (L) is shown to affect negatively the competitiveness of transportation and insurance services. Physical capital (K) has a positive impact on construction and computer and information services.

Recognizing that services trade tends to be distorted by border measures and regulatory restrictions, we examined the determinants of the patterns of service exports only, and the results are shown in table 12.7. As expected, the results turn out to be more favorable. According to our results, physical capital has a significantly positive impact on the exports of transportation, construction, and computer and information services, and human capital (H) has a significantly positive impact on the exports of travel, communi-

14. For some countries, 1960 data are not available. In this case, we regard the first available year as the initial year.

Table 12.6 Estimation Results: Net export, 1995 (US$ millions, 1990 prices)

Regression Equations

	Total (1)	Transportation (2)	Travel (3)	Communications (4)	Construction (5)	Insurance (6)	Finance (7)	Computer and Information (8)	Royalties and License Fees (9)	Other Business Services (10)	Personal, Cultural, and Recreational (11)	Government, n.i.e. (12)
Constant	−1147.98	−335.781*	−170.718	124.970	178.095*	20.336	−443.613	−68.788	−1165.85	−65.232	−364.316	−117.128
	(−0.937)	(−1.865)	(−0.290)	(1.481)	(1.767)	(0.433)	(−1.554)	(−0.696)	(−1.065)	(−0.287)	(−0.895)	(−1.608)
K	−0.153	−0.546	−0.869	−0.195	0.301**	−0.228**	−0.028	0.268*	0.645	0.080	−0.031	0.283
	(−0.027)	(−0.791)	(−0.337)	(−0.975)	(7.995)	(−2.399)	(−0.104)	(1.845)	(0.502)	(0.094)	(−0.233)	(0.912)
L	−0.295	−5.852**	9.264	0.406	−2.478	−2.044**	17.089	−3.605	5.598	−2.104	8.199	0.133
	(−0.022)	(−3.911)	(1.590)	(0.221)	(−0.695)	(−2.139)	(1.293)	(−1.092)	(0.313)	(−0.740)	(0.808)	(0.183)
H	41.071	5.582	18.232	−2.469	−1.154	0.841	10.067	0.129	16.076	8.943	2.694	2.515
	(1.049)	(1.030)	(0.894)	(−0.989)	(−0.457)	(0.439)	(1.088)	(0.041)	(0.514)	(1.187)	(0.409)	(1.002)
R^2	0.013	0.320	0.049	0.454	0.581	0.414	0.224	0.612	0.229	0.026	0.178	0.294
Adj. R^2	−0.015	0.300	0.020	0.428	0.543	0.392	0.166	0.563	0.177	−0.004	0.075	0.272
F-statistic	0.463	16.029***	1.723	17.483***	15.269***	19.295***	3.855**	12.612***	4.366***	0.865	1.732	13.311***
N	109	106	105	67	37	86	44	28	48	102	28	100

Sources: IMF (1999a); World Bank (1999). For Taiwan, Economic Research Department, Central Bank of China (various years); Council for Economic Planning and Development, Republic of China (1999).

Notes: K = capital stock; L = labor force; H = human capital. The estimation method is OLS; t-statistics are in parentheses; standard deviation is based on White's heteroskedasticity consistent estimator.

***Significant at the 1 percent level.

**Significant at the 5 percent level.

*Significant at the 10 percent level.

Table 12.7   **Estimation Results: Export, 1995 (US$ millions, 1990 prices)**

						Regression Equations						
	Total (1)	Transportation (2)	Travel (3)	Communications (4)	Construction (5)	Insurance (6)	Finance (7)	Computer and Information (8)	Royalties and License Fees (9)	Other Business Services (10)	Personal, Cultural, and Recreational (11)	Government, n.i.e. (12)
Constant	814.355	282.735	165.660	10.468	534.496	-10.069	-834.056**	-166.280*	-1019.46	612.89**	-256.369	-152.019
	(0.534)	(1.012)	(0.232)	(0.156)	(1.579)	(-0.216)	(-2.208)	(-1.843)	(-1.085)	(2.274)	(-1.002)	(-0.797)
K	13.229	3.244**	3.384	0.151	0.543***	0.078	-0.063	0.196*	1.423	2.756***	-0.032	0.827
	(1.519)	(2.167)	(0.921)	(0.967)	(6.653)	(0.877)	(-0.180)	(1.958)	(1.125)	(3.472)	(-0.254)	(0.961)
L	13.263	1.247	9.119	1.046	-5.803	1.632*	25.701*	1.109	4.771	0.877	10.689	0.929
	(0.583)	(0.296)	(0.967)	(0.616)	(-0.477)	(1.946)	(1.746)	(0.223)	(0.267)	(0.381)	(1.511)	(0.420)
H	127.433*	29.055**	55.140c	5.264**	-0.431	6.340**	32.584***	8.088**	26.214	27.300**	7.571	6.717
	(1.892)	(2.078)	(1.744)	(2.022)	(-0.072)	(2.264)	(2.288)	(2.720)	(0.899)	(2.401)	(1.436)	(1.013)
R^2	0.614	0.680	0.425	0.454	0.496	0.218	0.479	0.721	0.570	0.588	0.390	0.447
Adj. R^2	0.603	0.670	0.407	0.429	0.452	0.189	0.440	0.688	0.542	0.575	0.317	0.430
F-statistic	55.604***	72.135***	24.893***	18.543***	11.462***	7.696***	12.542***	21.553***	20.367***	46.635***	5.335***	26.097
N	109	106	105	71	39	87	45	29	50	102	29	101

Sources and Notes: See table 12.6.

cations, insurance, financial, computer and information, and other business services. The estimated results on labor show that labor has a statistically significantly positive impact on insurance and finance, which is not consistent with our expectation.

Table 12.8 reports the results of the estimation, which uses imports as the dependent variable. The results are mixed in the sense that many of the estimated coefficients are either with unexpected signs or with expected signs but not statistically significant. These results are expected, because service imports are subject to various restrictions. It is to be noted that the estimated coefficients on per capita GDP (PGDP) are mostly positive and in many cases statistically significant. The introduction of this variable is based on the recognition that demand patterns for services are not identical among countries with different income levels. Indeed, our findings appear to indicate that countries with high income levels tend to demand greater amounts of service imports, reflecting their preference for a variety. This also explains the significant amount of intraindustry trade in services.

Our results of the analysis of the determinants of services trade indicate that the Heckscher-Ohlin theorem does explain the patterns of trade of some services, but its applicability is shown to be quite limited. These results may be due to the presence of restrictions and other barriers to services trade and also to product differentiation in services, giving rise to intraindustry trade, for which the Heckscher-Ohlin model does not have much validity. To discern the determinants of the patterns of services trade, analyses using more detailed data have to be conducted.

12.4 Services Trade Embodied in Goods Trade

One of the special characteristics of services in general is that production and consumption take place simultaneously, as discussed in section 12.1. This characteristic makes it difficult for producers and consumers of services located in different places, let alone different countries, to trade services. Furthermore, various restrictions on services trade limit and distort services trade, as discussed in the previous section. Noting the increasing share of services in production of goods, one realizes that services are "traded" internationally in the form of goods trade. One may describe such trade as trade in embodied services. For example, production of cars requires service inputs such as distribution and communication services, which enable producers to purchase parts and components. If a car is exported, such an export results in export of distribution and communication services indirectly. Based on the recognition of this point, we compute indirect trade of services for a selected number of East Asian economies.

12.4.1 Previous Studies

Despite the importance of trade in services embodied in goods trade, there have been few studies that analyzed such a type of services trade

Table 12.8   Estimation Results: Import, 1995 (US$ millions, 1990 prices)

	Regression Equations											
	Total (1)	Transportation (2)	Travel (3)	Communications (4)	Construction (5)	Insurance (6)	Finance (7)	Computer and Information (8)	Royalties and License Fees (9)	Other Business Services (10)	Personal, Cultural, and Recreational (11)	Government, n.i.e. (12)
Constant	366.960	277.711	−154.386	−118.597	100.661	−36.792	−192.257	−89.133	−53.999	288.399	17.772	2.539
	(0.501)	(1.321)	(−0.699)	(−0.984)	(0.362)	(−0.732)	(−1.594)	(−1.515)	(−0.953)	(1.277)	(0.144)	(0.029)
K	11.141***	3.264***	3.494***	0.339	0.044	0.276	−0.071	−0.123	0.652***	2.076***	−0.054	0.581
	(3.376)	(3.687)	(2.804)	(0.881)	(0.235)	(1.653)	(−0.466)	(−0.824)	(8.531)	(4.921)	(−0.917)	(0.983)
L	20.573*	8.957***	2.297	0.636	2.332	3.767**	12.179**	6.238	0.087	5.304***	4.845	0.653
	(1.758)	(2.631)	(0.510)	(0.182)	(0.196)	(2.162)	(2.221)	(1.021)	(0.056)	(4.004)	(1.144)	(0.450)
H	10.038	3.483	4.496	6.541	−7.234	2.710	3.550	−1.708	−0.323	−6.756	−0.347	5.211
	(0.329)	(0.424)	(0.348)	(1.136)	(−1.278)	(1.228)	(0.762)	(−0.416)	(−0.155)	(−0.908)	(−0.063)	(1.029)
PGDP	687.244***	163.516***	243.299**	3.705	53.932*	11.990	24.702	29.564*	53.734**	190.856***	17.877	−10.994
	(2.670)	(3.101)	(2.301)	(0.223)	(1.689)	(1.629)	(1.368)	(1.817)	(2.320)	(2.875)	(1.195)	(−0.640)
R^2	0.781	0.848	0.688	0.481	0.345	0.726	0.359	0.489	0.824	0.716	0.214	0.472
Adj. R^2	0.773	0.842	0.676	0.448	0.273	0.715	0.301	0.413	0.812	0.705	0.116	0.452
F-statistic	92.817***	143.41***	56.237***	14.604***	4.750***	62.339***	6.164***	6.460***	72.393***	64.313***	2.814***	21.866***
N	109	108	107	68	41	99	49	32	67	107	37	102

Sources: See table 12.6.

Notes: K = capital stock; L = labor force; H = human capital; PGDP = per capita GDP. The estimation method is OLS; t-statistics are in parentheses; standard deviation is based on White's heteroskedasticity consistent estimator.

***Significant at the 1 percent level.

**Significant at the 5 percent level.

*Significant at the 10 percent level.

empirically. Early attempts are Tucker and Sundberg (1988) and Grubel (1988). The study by Tucker and Sundberg investigated the embodied services for Australia and Thailand in 1975 and for Singapore in 1973. Their results indicated that Australia exported approximately one-half of its services in the form of embodied services in manufactured goods, whereas for Thailand the share was somewhat smaller at one-third. The share of embodied service exports in total service exports was found to be significantly smaller for Singapore at 18.5 percent. Grubel (1988) analyzed trade in embodied services for Canada in 1973 and 1983. He found that the value of embodied service exports increased rapidly from 1973 to 1983, to result in the situation in 1983, when the surplus in the balance of embodied services trade was greater than the deficit in the balance of services trade.

Sazanami and Urata (1990) examined the patterns of disembodied and embodied services trade for Japan in 1975 and 1985 and for the United States in 1982. They found that embodied services trade was significantly greater than disembodied services trade for both Japan and the United States. Specifically, the ratios of embodied service exports (imports) to total exports (imports) for Japan in 1975 and 1985 were 0.78 (0.82) and 0.79 (0.80), whereas the corresponding ratio for the United States in 1982 was 0.78 (0.90). Urata (1994) extended the earlier analysis to study embodied and disembodied services trade for Japan in 1990. His findings were similar to the earlier ones, in that he also found the proportions of embodied service exports and imports in total exports and imports to be as high as 0.78 and 0.76. Urata also examined the patterns of embodied and disembodied services trade for six service categories: electricity, gas, and water; commercial services; financial services; real estate; transportation and communications services; and other services. He found the proportion of embodied service exports to total exports to be extremely high for electricity, gas, and water, and for real estate at 0.99. Because we know that these services are more or less completely nontradable, his findings are understandable. Financial services and other services are found to have relatively high proportions, around 0.9, whereas the corresponding values for commercial services and for transportation and communications services are significantly lower, both at 0.7. Commercial services and transportation and communications services are relatively easily traded compared to other services.

The previous studies on developed countries showed the importance of services trade that is embodied in goods trade. Below we examine the patterns of embodied and disembodied services trade for selected East Asian economies.

12.4.2 The Methodology for Computing Services Trade Embodied in Goods Trade

Suppose that there are $n(= 1, \cdots, N)$ countries and $i(= 1, \cdots, I)$ commodities. Let country n's export vector be $\mathbf{EX}_n$ ($I \times 1$ vector), its import vector be $\mathbf{IM}_n$ ($I \times 1$ vector), and $\mathbf{D}_n$ ($I \times 1$ vector) be domestic final demand. Output vector and input-output coefficient matrices are $\mathbf{Q}_n$ ($I \times 1$ vector) and $\mathbf{A}_n$ ($I \times 1$ matrix), respectively. Let us drop country index n for simplicity. Incorporating these definitions, we can write the balance of domestic production and consumption as

(8) $$\mathbf{Q} \equiv \mathbf{AQ} + \mathbf{D} + \mathbf{EX} - \mathbf{IM}.$$

With an assumption that $\mathbf{IM}_n$ is exogenously determined and some manipulations, we obtain the magnitude of domestic production induced by exports and imports (import substitution) as

(9) $$\mathbf{Q}_{EX} \equiv (\mathbf{I} - \mathbf{A})^{-1}\mathbf{EX}$$

and

(10) $$\mathbf{Q}_{IM} = (\mathbf{I} - \mathbf{A})^{-1}\mathbf{IM}.$$

Let us turn to the computation of services embodied in goods trade. Suppose that service industries are labeled from $h + 1$ to I in i. In other words, if $i \leq h$, sector i is agriculture or manufacture (merchandises), whereas if $i > h$, sector i is services. Let $\mathbf{EX}'[= (ex_1', ex_2', \cdots, ex_h', 0, \cdots, 0)]$ and $\mathbf{IM}'[= (im_1', im_2', \cdots, im_h', 0, \cdots, 0)]$ be export and import vectors of goods. Domestic production induced by exports and imports of goods can be obtained as $\mathbf{Q}_{EX'}$ and $\mathbf{Q}_{IM'}$ below:

(11) $$\mathbf{Q}_{EX'} \equiv (\mathbf{I} - \mathbf{A})^{-1}\mathbf{EX}'$$

and

(12) $$\mathbf{Q}_{IM'} = (\mathbf{I} - \mathbf{A})^{-1}\mathbf{IM}'.$$

Using equations (11) and (12) and taking the computed values for service sectors, we obtain trade in services that are embodied in goods trade.

A few words of caution are in order. It should be noted that computed services are those required for the production of import-competing goods and not imports. To calculate service content of imports, one needs to use an input-output table of an exporting country. However, such an exercise is difficult because it requires input-output tables of all the export source countries, and thus it is not attempted. This does not cause a problem if the production technologies, or input-output relations, are identical between countries, as assumed in the Heckscher-Ohlin model. However, in reality they are different, and thus one has to be careful in interpreting the estimated results presented and discussed below.

Another problem is the use of national input-output tables for the computation of service content of exports. This is because the production technologies used for the production of exports tend to be different from those for the production of goods sold in the domestic market. This problem is likely to appear in the case of transportation and distribution services. Generally export production does not use much transportation services, retail services, or multilayered wholesale services when compared with production for the domestic market.

One should also note that the differences in the treatment of services among the countries make it difficult to conduct an international comparison. In countries where markets are not well developed, services are provided in-house, and thus service transactions do not appear in the statistics. By contrast, in countries where markets are more developed, services are traded between firms, and thus service transactions appear in the statistics. These differences in the treatment of services do influence the estimated service contents of goods trade.

12.4.3   The Data

We computed services trade embodied in goods trade for China, Malaysia, the Philippines, Singapore, and Taiwan in 1990, and for Japan in 1990 and 1995. The choice of the sample depended on the availability of comparable input-output tables, which are required for the computation. Input-output tables of the selected East Asian economies are obtained from the Statistical Research Department of the Institute of Developing Economies (1995, 1996a, 1996b, 1997a, 1997b). These input-output tables are compiled by the Statistical Research Department of the Institute of Developing Economies as a part of the project for the construction of the international input-output table of ten countries that include Japan, the United States, and eight East Asian economies. Japan's input-output tables are obtained from the Management and Coordination Agency (1994, 1999).

The definition of services in an input-output table is different from that of the balance-of-payments (BOP) statistics, which were used in the earlier section. For instance, an input-output table in general excludes travel, which accounts for a large portion of services trade in the BOP statistics. On the other hand, an input-output table includes electricity, gas, and water, which are not included in the BOP statistics. These differences in the definition and treatment of services preclude one from making direct comparisons of the figures derived from these different sources. The description of the input-output tables used in the analysis is given in appendix B.

12.4.4   The Results

The results of the computation of embodied services trade, which are obtained by applying equations (11) and (12) to the selected East Asian economies for 1990, are shown with the statistics on disembodied service trade in table 12.9. As was the case for developed countries, which were reviewed earlier, embodied services trade accounts for a large portion of services trade. Indeed, the proportions of embodied service exports/imports in total (disembodied and embodied) service exports/imports for the sample

Table 12.9   Services Trade Embodied in Merchandise Trade in East Asian Economies in 1990 (US$ millions, 1990 prices)

Industry	Disembodied Services Trade			Embodied Services Trade		
	Export	Import	Net Export	Export	Import	Net Export
Japan						
1 Electricity, gas, and water supply	165,723	26,942	138,780	13,149,399	12,468,012	691,387
2 Construction	0	0	0	3,274,452	2,581,194	693,258
3 Wholesale and retail trade	14,352,331	2,261,530	12,090,800	26,474,230	18,054,904	8,419,326
4 Transportation	24,072,422	17,087,009	6,985,413	9,001,284	7,648,865	1,352,419
5 Telephone and telecommunication	269,563	373,776	-104,213	3,040,356	2,372,713	667,643
6 Finance and insurance	2,979,764	5,217,370	-2,237,606	13,420,837	13,813,276	-392,439
7 Education and research	71,103	122,743	-51,640	1,113,012	917,207	195,805
8 Other services	6,327,060	25,204,409	-18,877,348	33,251,728	23,759,020	9,492,709
9 Public administration	0	0	0	209,223	184,211	25,012
Services total	48,237,966	50,293,779	-2,055,814	102,934,521	81,789,402	21,145,119
Singapore						
1 Electricity, gas, and water supply	78,029	2,824	75,204	1,988,459	2,608,706	-620,247
2 Construction	899	20,155	-19,256	239,933	314,245	-74,312
3 Wholesale and retail trade	3,927,314	0	3,927,314	3,254,863	4,036,984	-782,121
4 Transportation	5,462,566	2,213,387	3,249,179	2,931,646	3,547,289	-615,644
5 Telephone and telecommunication	62,961	90,785	-27,824	709,438	839,429	-129,992
6 Finance and insurance	1,354,385	375,146	979,239	7,379,651	8,084,740	-705,090
7 Education and research	4,907	0	4,907	29,358	33,530	-4,172
8 Other services	700,561	-2,487,769	3,188,330	5,446,530	6,661,873	-1,215,343
9 Public administration	16,090	0	16,090	175,445	252,494	-77,049
Services total	11,607,712	214,529	11,393,183	22,155,322	26,379,292	-4,223,969
Taiwan						
1 Electricity, gas, and water supply	7,304	4,168	3,135	4,899,250	3,972,096	927,154
2 Construction	332	1,439	-1,107	404,026	347,759	56,267
3 Wholesale and retail trade	3,113,611	0	3,113,611	6,642,919	4,809,402	1,833,518
4 Transportation	4,588,750	2,264,257	2,324,493	2,931,658	2,262,502	689,156
5 Telephone and telecommunication	219,366	371,155	-151,789	662,708	478,703	184,006
6 Finance and insurance	195,684	204,426	-8,742	6,492,693	5,132,105	1,380,588
7 Education and research	10,586	139,026	-128,440	180,673	143,739	36,923
8 Other services	2,272,077	5,069,753	-2,979,676	6,388,856	5,566,621	822,235
9 Public administration	0	0	0	0	0	0

Malaysia

1 Electricity, gas, and water supply	0	8,069	-8,069	1,519,973	2,409,893	-889,915
2 Construction	441	8,223	-7,782	156,779	195,347	-38,568
3 Wholesale and retail trade	1,280,256	141,839	1,138,417	2,473,014	3,537,798	-1,064,784
4 Transportation	277,090	1,149,384	-872,294	1,783,233	2,038,432	-255,199
5 Telephone and telecommunication	178,304	26,863	151,442	323,051	432,795	-109,744
6 Finance and insurance	124,496	216,343	-91,846	778,184	1,049,251	-271,068
7 Education and research	0	0	0	14,549	22,722	-8,173
8 Other services	232,327	1,685,331	-1,453,004	3,985,993	5,212,395	-1,226,402
9 Public administration	0	0	0	33,583	55,217	-21,634
Services total	2,092,915	3,236,052	-1,143,137	11,068,362	14,953,850	-3,885,487

The Philippines

1 Electricity, gas, and water supply	0	0	0	274,539	640,444	-365,905
2 Construction	37,349	6,410	30,939	13,894	30,041	-16,147
3 Wholesale and retail trade	2,440,361	0	2,440,361	812,378	1,819,521	-1,007,143
4 Transportation	656,354	613,853	42,501	270,114	648,859	-378,745
5 Telephone and telecommunication	315,292	13,473	301,820	20,646	60,275	-39,628
6 Finance and insurance	638,617	447,545	191,073	213,624	618,785	-405,161
7 Education and research	4,441	0	4,441	1,860	3,543	-1,684
8 Other services	1,834,353	482,363	1,351,990	256,418	666,457	-410,039
9 Public administration	0	0	0	47	104	-57
Services total	5,926,768	1,563,644	4,363,124	1,863,520	4,488,029	-2,824,509

China

1 Electricity, gas, and water supply	8,731	147,153	-138,422	3,230,780	3,945,761	-714,982
2 Construction	0	0	0	0	0	0
3 Wholesale and retail trade	-3,289,131	0	-3,289,131	4,931,889	4,413,710	518,179
4 Transportation	3,548,063	156,444	3,391,620	4,027,245	3,708,750	318,495
5 Telephone and telecommunication	219,032	31,910	187,122	169,442	159,815	9,627
6 Finance and insurance	192,421	5,734	186,687	3,160,741	3,023,251	137,490
7 Education and research	5,280	29,017	-23,736	160,720	164,375	-3,655
8 Other services	952,663	515,132	437,531	1,527,370	1,429,882	97,489
9 Public administration	0	0	0	0	0	0
Services total	1,637,059	885,390	751,669	17,208,186	16,845,544	362,643

Sources: Statistical Department, Institute of Developing Economies (1995, 1996a, 1996b, 1997a, 1997b); Management and Coordination Agency (1994). Sectoral classification is based on the Statistical Department, Institute of Developing Economies (1998).

Notes: Disembodied services trade is defined as services trade directly traded. Embodied services trade is defined as services trade embodied in goods trade. For the estimation method and data sources, see appendix.

countries range between 0.24 (exports for the Philippines) and 0.95 (imports for China).[15] The extremely low figure for the Philippines is attributable to a large share of semiconductors and electrical machinery in its goods exports that require only assembling operation and not much service content. Among the service sectors the proportion of embodied services in total services is found to be generally high for most sectors, reflecting the non-tradable nature of services. However, the corresponding proportions are lower for transportation services for most countries and relatively low for telecommunications and finance and insurance for some countries, reflecting the relatively high tradability of these services.

It is of interest to find that for some countries the direction of the balance of embodied services trade differs from disembodied services trade. Specifically, for Japan the balance of disembodied services trade is in deficit, whereas that of embodied services trade is found to be in surplus. Indeed, the balance of embodied services trade is so large that it offsets the deficit in disembodied services trade, to result in a net surplus in total (disembodied and embodied services trade) services trade. The situation is the opposite for Singapore and the Philippines, for which the balance of disembodied services trade is in surplus, whereas that of embodied services trade is in deficit. For these countries, the absolute magnitude of the balance of embodied services trade is smaller than that of disembodied services trade, and therefore the overall services trade balance is in surplus. Unlike the countries examined so far, Taiwan and China recorded a surplus in both disembodied and embodied services trade, whereas Malaysia recorded a deficit in both types of services trade.

We saw above the magnitude of disembodied and embodied services trade for the selected East Asian countries. It would be of interest to examine the impact of the trade and production of goods on embodied services trade. We computed the magnitude of embodied services trade resulting from the production of 1 million dollars' worth of goods exports and imports. Such normalization would provide us with useful information for the international comparison of the patterns of trade and production.

The results of the computation, shown in table 12.10, indicate that the magnitude of services embodied in goods trade, or service content of goods trade, is similar for Japan, Singapore, Taiwan, and Malaysia, whereas the corresponding values for the Philippines and China are significantly smaller. These differences in service content of goods trade reflect the differences in the patterns of trade and the structure of production for these two groups of countries. The countries in the former group tend to trade goods that embody a large amount of services or tend to have overall production

---

15. The figures referred here exclude the sectors that record negative values in disembodied services trade, that is, other services in Singapore and wholesale and retail trade in China. We need to investigate the meanings of these figures before we make any interpretations.

Industry	Japan				Singapore				Taiwan			
	Exports	Imports	Intensity	Rank	Exports	Imports	Intensity	Rank	Exports	Imports	Intensity	Rank
1 Electricity, gas, and water supply	49	53	0.913	8	56	60	0.934	8	77	72	1.070	7
2 Construction	12	11	1.098	4	7	7	0.935	7	6	6	1.008	8
3 Wholesale and retail trade	98	77	1.269	1	92	93	0.988	6	104	87	1.198	3
4 Transportation	33	33	1.018	6	82	81	1.012	4	46	41	1.124	4
5 Telephone and telecommunication	11	10	1.109	3	20	19	1.035	3	10	9	1.201	2
6 Finance and insurance	50	59	0.841	9	208	186	1.118	1	102	93	1.097	5
7 Education and research	4	4	1.050	5	1	1	1.073	2	3	3	1.090	6
8 Other services	123	102	1.211	2	153	153	1.002	5	100	101	0.995	9
9 Public administration	1	1	0.983	7	5	6	0.851	9	0	0	1.711	1
Services total	381	350	1.089		623	606	1.029		449	411	1.092	

Industry	Malaysia				The Philippines				China			
	Exports	Imports	Intensity	Rank	Exports	Imports	Intensity	Rank	Exports	Imports	Intensity	Rank
1 Electricity, gas, and water supply	57	80	0.712	8	43	43	1.011	5	51	70	0.734	8
2 Construction	6	6	0.906	2	2	2	1.091	2	0	0	0.753	7
3 Wholesale and retail trade	93	117	0.789	6	127	121	1.053	4	78	78	1.002	2
4 Transportation	67	68	0.988	1	42	43	0.982	6	64	66	0.974	
5 Telephone and telecommunication	12	14	0.843	4	3	4	0.808	9	3	3	0.951	4
6 Finance and insurance	29	35	0.837	5	33	41	0.815	8	50	54	0.938	5
7 Education and research	1	1	0.723	7	0	0	1.238	1	3	3	0.877	6
8 Other services	149	173	0.863	3	40	44	0.908	7	24	25	0.958	3
9 Public administration	1	2	0.687	9	0	0	1.055	3	0	0		
Services total	415	497	0.836		292	298	0.980		273	298	0.916	

Sources: See table 12.9.

Notes: Exports and imports are calculated from embodied services trade in table 8, assuming that total goods trade is 1 million dollars. Intensity is defined as the ratio of export service contents to import ones. For the estimation methods and data sources, see the main text and the appendix. Ranking is based on the intensity.

structure with a significant amount of services being used as inputs, compared to the countries in the latter group.

To compare service contents of goods exports and imports, we computed the ratio between them, or service content ratio. Using unity for the service content ratio as a cutoff value, we may divide the sample into two groups. One group, whose service content ratio is greater than unity, consists of Singapore and Taiwan, and the other group, whose service content ratio is less than unity, includes Japan, Malaysia, the Philippines, and China. These observations indicate that the countries in the first group export services through goods trade, whereas the countries in the second group import services through goods trade. If the Heckscher-Ohlin theorem can be applied to this analysis, one may interpret these results to indicate that the countries in the first group, Singapore and Taiwan, are well endowed with services in comparison with other factors of production vis-à-vis their trading partners. By contrast, the countries in the second group, Japan, Malaysia, the Philippines, and China, are likely to be poorly endowed with services. Recognizing that the share of services in production tends to increase with the level of economic development, we see that our findings are consistent with the Heckscher-Ohlin theorem except for Japan, because Japan, being a well-developed country, is expected to be a net exporter of services, not a net importer.

One should note here that there are several reasons that our results may deviate from the predictions from the Heckscher-Ohlin theorem.[16] First, the patterns of demand for services are likely to differ between countries, thus violating the assumption of the Heckscher-Ohlin theorem. Second, the patterns of demand for goods are likely to differ between countries as well, affecting the computed service content embodied in goods trade. Third, related to the second point, trade policy and discriminatory business practices distort demand patterns for goods, thus aggravating the problem just noted. Indeed, the unexpected result for the case of Japan may be due to these problems. The Japanese may have a greater preference for the consumption of services compared to their trading partners, leading to the deficit in services trade. One may also argue that Japan, with its poor natural resource endowments, imports a lot of raw materials, which embody a substantial amount of transportation services.[17] Furthermore, demand for raw materials in Japan may be upwardly biased, because manufactured imports are restricted by import restrictions and by discriminatory business practices against imports.

16. One should note that similar points have been made for the validity of the Heckscher-Ohlin theorem in explaining the pattern of goods trade. See, for example, Markusen et al. (1995).

17. This point can be confirmed by the following statistics. The 1995 ratio of transportation services to output for the primary products, which include raw materials, is 0.056, significantly higher compared with the corresponding ratio for the manufactured goods at 0.020.

The computed results of the service content of goods trade at the sectoral level show interesting patterns. It is found that the service contents at the sectoral level are very similar among East Asian countries. Electricity, gas and water supply, wholesale and retail trade, finance and insurance, and transportation services have relatively higher values compared to other services, reflecting their importance in goods production.

A comparison of the service contents of goods exports and imports gives interesting information concerning the patterns of services trade for the East Asian economies. Let us point out some notable observations. Japan is a huge net importer of electricity, gas, and water supplies, public administration, transportation services, and education and research, while being a net exporter of wholesale and retail services. Because of the fact that Japan is a highly educated society, the finding on education and research contents does not seem to reflect the reality, and thus it needs further examination. We will come back to this issue later, when we discuss the results obtained by using a more disaggregated input-output table. The finding on wholesale and retail services may counter our expectation, because the Japanese distribution sector is generally regarded as inefficient. Indeed, inefficiency may be a reason for our unexpected result. Specifically, the values used for our analysis are not adjusted for their quality or price, and, therefore, service inputs for Japan are overvalued in comparison with the value based on international prices, making Japan a large exporter of distribution services.[18]

Singapore is shown to be a large net exporter of finance and insurance services and education and research. These findings appear to be consistent with the characteristics of Singapore. Taiwan is found to be a notable net exporter of public administration services, telecommunications, and wholesale and retail services. Malaysia and the Philippines are large net exporters of transportation services and education and research, respectively, whereas China is a net importer for all the categories except wholesale and retail trade, for which China is a small net exporter.

The availability of detailed and more recent input-output tables for Japan enables us to investigate the patterns of services trade, both disembodied and embodied in goods, in more detail. Such analysis would shed light on the important patterns, which would be masked by the analysis at the aggregated level. The results of the computation are shown in tables 12.11 and 12.12. Some of the interesting observations follow.

The basic patterns of disembodied and embodied services trade remained more or less the same for 1990 and 1995. Japan registered a deficit

18. The overblown or inefficient nature of the distribution sector in Japan may be shown by the relatively large share of wholesale services in total cost of production in Japan at 0.030, significantly higher compared with the case for Taiwan at 0.012; both figures are taken from respective input-output tables.

Table 12.11   **Changes of Services Trade Embodied in Merchandise Trade in Japan between 1990 and 1995 (US$ millions, current prices)**

	Disembodied Services Trade			Embodied Services Trade		
	Exports	Imports	Net Exports	Exports	Imports	Net Exports
1990						
Construction	0	0	0	3,253	2,581	672
Electric power, gas, stream, and hot water supply	142	20	122	12,082	11,207	875
Water supply and waste disposal	26	7	19	1,761	1,955	–195
Commerce	14,352	2,262	12,091	26,366	18,055	8,311
Finance and Insurance	2,980	5,217	–2,238	13,293	13,813	–520
Finance	2,553	4,841	–2,288	11,880	11,762	118
Insurance	426	376	50	1,413	2,051	–638
Real estate	28	53	–25	6,240	5,174	1,066
Transport	26,926	17,087	9,839	17,079	17,838	–759
Communication and broadcasting	270	375	–105	4,211	3,182	1,029
Public administration	0	0	0	208	184	24
Education and research institute	71	123	–52	18,306	8,619	9,688
Education	0	1	–1	300	167	133
Research institutes (including research and development)	71	122	–51	810	750	60
Research and development (intra-enterprise)	0	0	0	17,196	7,701	9,494
Medical service, health, and social security	1	7	–5	2	1	0
Other public service	275	194	81	660	665	–5
Business services	3,245	7,228	–3,983	31,120	23,127	7,993
Advertising agencies	469	2,002	–1,533	4,149	2,760	1,389
Inquiry and information services	663	1,511	–848	4,695	3,159	1,536
Personal services	2,776	17,724	–14,948	981	610	371
Amusement and recreational services	236	1,840	–1,604	693	404	289
Services total	51,093	50,296	796	135,562	107,012	28,550
1995						
Construction	0	0	0	5,547	4,425	1,122
Electric power, gas, stream, and hot water supply	263	13	250	19,073	16,746	2,327
Water supply and waste disposal	43	7	36	2,541	2,347	193
Commerce	32,953	1,663	31,290	42,637	34,072	8,565
Finance and insurance	6,135	10,915	–4,780	22,565	22,149	416
Finance	5,251	8,809	–3,559	20,759	19,808	951
Insurance	885	2,106	–1,221	1,806	2,341	–535
Real estate	55	48	7	7,906	7,132	774

Table 12.11   (continued)

	Disembodied Services Trade			Embodied Services Trade		
	Exports	Imports	Net Exports	Exports	Imports	Net Exports
Transport	32,302	24,034	8,267	25,130	26,313	−1,183
Communication and broadcasting	7,966	3,426	4,539	8,434	7,951	483
Public administration	0	0	0	578	511	67
Education and research institute	229	349	−119	29,088	14,467	14,621
Education	0	2	−1	437	289	148
Research institutes (including research and development)	229	347	−118	1,690	1,138	552
Research and development (intra-enterprise)	0	0	0	26,961	13,040	13,921
Medical service, health, and social security	1	8	−7	2	2	0
Other public service	501	418	83	961	878	83
Business services	8,602	16,583	−7,981	43,411	35,765	7,646
Advertising agencies	1,088	3,584	−2,496	6,531	4,933	1,598
Inquiry and information services	1,483	3,051	−1,568	3,765	2,718	1,047
Personal services	4,671	29,898	−25,227	1,133	871	262
Amusement and recreational services	284	2,327	−2,044	735	561	175
Services total	93,720	87,362	6,358	209,006	173,629	35,377

Sources: Sector classification is based on Management and Coordination Agency (1994, 1999).
Note: For the estimation method and data sources, see the main text and the appendix.

in disembodied services trade but a huge surplus in embodied services trade, indicating that Japan exported services through goods trade for both years under study. As to disembodied trade, personal services registered a sizable deficit, and the magnitude of the deficit increased from 1990 to 1995. Turning to the results on embodied services trade, one finds that an analysis at disaggregated level yields a larger figure concerning embodied service total compared to the results obtained from the analysis at the aggregated level, indicating that an analysis using disaggregated data captures the complexity of input-output relations better. It is of interest to observe that in-house research and development (R&D) in the private firms contributes significantly to the export of education and research services, which reflects that goods embodying private research activities tend to be exported at large scale. This in turn appears to indicate that Japan has a comparative advantage in the sectors that conduct in-house R&D actively. An examination of the results on business services reveals that the categories that are

Table 12.12   **Changes of Service Contents in Merchandise Trade in Japan between 1990 and 1995 (US$ millions, current prices)**

	1990				1995			
	Export	Import	Intensity	Rank	Export	Import	Intensity	Rank
Construction	12	11	1.102	6	14	13	1.076	3
Electric power, gas, stream, and hot water supply	45	48	0.943	10	48	49	0.978	7
Water supply and waste disposal	7	8	0.787	14	6	7	0.929	11
Commerce	99	77	1.277	3	107	100	1.075	4
Finance and Insurance	50	59	0.842	12	57	65	0.875	13
Finance	44	50	0.883		52	58	0.900	
Insurance	5	9	0.602		5	7	0.662	
Real estate	23	22	1.055	7	20	21	0.952	9
Transport	64	76	0.837	13	63	77	0.820	14
Communication and broadcasting	16	14	1.157	5	21	23	0.911	12
Public administration	1	1	0.989	9	1	1	0.971	8
Education and research institute	69	37	1.858	1	73	42	1.727	1
Education	1	1	1.572		1	1	1.300	
Research institutes (including research and development)	3	3	0.945		4	3	1.275	
Research and development (intra-enterprise)	64	33	1.953		68	38	1.775	
Medical service, health, and social security	0	0	1.045	8	0	0	0.994	6
Other public service	2	3	0.868	11	2	3	0.940	10
Business services	116	99	1.177	4	109	105	1.042	5
Advertising agencies	16	12	1.315		16	14	1.137	
Inquiry and information services	18	14	1.300		9	8	1.189	
Personal services	4	3	1.406	2	3	3	1.117	2
Amusement and recreational services	3	2	1.501		2	2	1.126	
Services total	507	458	1.108		526	509	1.034	

Sources: See table 12.11.

Notes: Service contents are calculated from disembodied trade in table 10, assuming that total goods trade is 1 million dollars. See also notes to table 12.10. Blank cells indicate data not available.

separately analyzed (advertising, and inquiry and information services) account for only a small portion of business services. A closer examination of this sector is needed. The results of the normalization exercise, which are presented in table 12.12, show that Japan is a huge net exporter of private R&D, whereas it is a net importer of many other services, particularly transportation services and finance and insurance services.

12.5   Conclusions

We found in this paper that services trade in East Asia has been increasing in the recent decades. However, the share of East Asia in world services trade is smaller than the corresponding share for the goods trade. We also found that, unlike in goods trade, many East Asian economies register a deficit in services trade. Our analysis of trade in services that are embodied in goods trade reveals that a large magnitude of services is traded via goods trade. Indeed, for many countries the overall balance in services trade (disembodied and embodied) turns out to be surplus, because the trade surplus in embodied services trade is greater than the trade deficit in disembodied services trade.

An examination of the sectoral distribution of service exports and imports for the East Asian economies shows that many economies are heavily engaged in services trade (exports and imports) in transportation, travel, and other services, with some variations among the economies. Our analysis of the determinants of services trade indicates that the Heckscher-Ohlin model can explain the patterns of trade in a few types of services trade, such as computer and information, but generally its validity cannot be confirmed. One important factor that may reduce the validity of the trade models such as the Heckscher-Ohlin model in explaining the pattern of services trade is the presence of various barriers, such as government regulations.

Further research on trade in services is acutely needed for several reasons. First, the importance of services trade appears likely to increase in the future, for the following reasons. Deregulation in the service sector and liberalization in services trade are likely to proceed, not only because of the multilateral and regional arrangements in services, but also because of the realization on the part of policy makers and business circles that such policy changes are required to increase the competitiveness of the service sector for improving the competitiveness of the overall economy. Indeed, the World Bank (2000) reports the results of a study that show that liberalization of trade in goods and services would lead to a median GDP increase of 3.9 percent among major East Asian countries.

Technical progress in telecommunications, which is likely to take place, would promote services trade by lowering the cost of conducting trade in services and by developing new means of cross-border service transactions. Expected increase in FDI would also promote trade in services, especially through the mode of commercial presence.

This paper analyzed the patterns and determinants of two types of services trade, cross-border supply and consumption abroad, and did not examine other types of services trade, namely, services trade conducted through commercial presence and the presence of natural persons. As noted earlier, services trade conducted through these forms is likely to increase rapidly in the future. Therefore, among various areas for possible future research in services trade, we think that services trade conducted through commercial presence and the presence of natural persons are of particular importance. To conduct such analysis and other types of analyses on services trade, it has to be emphasized that the quality and quantity of data on services trade have to be improved and expanded.

Appendix A

Derivation of Regression Equation

Appendix A derives the regression equation. From equations (4) and (5), we have

(A1)
$$t_{ni} = (\mathbf{A}_i^t)^{-1}\left(\mathbf{E}_n - \frac{\alpha_n \mathbf{E}_W}{Y_W}\right)$$

$$= (\mathbf{A}_i^t)^{-1}\mathbf{E}_n - \frac{(\mathbf{A}_i^t)^{-1}\mathbf{E}_W(\mathbf{w}^t \mathbf{E}_n)}{Y_W}$$

$$= \left[(\mathbf{A}_i^t)^{-1} - \frac{(\mathbf{A}_i^t)^{-1}\mathbf{E}_W \mathbf{w}^t}{Y_W}\right]\mathbf{E}_n .$$

Let y_{wi} be the world output of commodity j. Substituting the world full employment condition, $(\mathbf{A}_i^t)^{-1}\mathbf{E}_W = y_{Wi}$, into equation (A1), we obtain

(A2)
$$t_{ni} = \left[(\mathbf{A}_i^t)^{-1} - \frac{y_{wi}\mathbf{w}^t}{Y_W}\right]\mathbf{E}_n$$

(6)
$$= \mathbf{R}_j^t \mathbf{E}_n$$

where $\mathbf{R}_j^t$ is a vector of factor requirement for country n's net export of commodity i with elements $r_{nij} = (a_{nij}^{-1} - y_{Wi}w_j/Y_W)$.

Appendix B

Data Description: Input-Output Tables for East Asian Countries

The Input-Output Tables for East Asian Economies

The data used in this paper are taken from the Statistical Research Department of the Institute of Developing Economies (1995, 1996a, 1996b,

1997a, 1997b). The input-output tables of East Asian countries are compiled by the Statistical Research Department of the Institute of Developing Economies as a part of its project on the construction of the international input-output table to include Japan, the United States, and East Asian countries. The number of sectors for the input-output tables used in our analysis is as follows: 106 for China (25 service sectors), 187 for Japan (54 service sectors), 96 for Malaysia (37 services), 177 for Philippines (38 services), 174 for Singapore (45 services), and 150 for Taiwan (45 services). Since the sector classification is different among these tables, we aggregated the service sectors into nine categories by using the conversion method developed by the Statistical Department of the Institute of Developing Economies (1998). Table 12B.1 presents these nine categories. This classification is also used in the construction of the international input-output

Table 12B.1   Classification Code of Services Trade

No.	Code	Industry
	International Input-Output Table	
1	51	Electricity, gas, and water supply
2	52	Construction
3	53A	Wholesale and retail trade
4	53B	Transportation
5	54A	Telephone and telecommunication
6	54B	Finance and insurance
7	54C	Education and research
8	54D	Other services
9	55	Public administration
	Japanese Input-Output Table	
1	17	Construction
2	18	Electric power, gas, stream, and hot water supply
3	19	Water supply and waste disposal
4	20	Commerce
5	21	Finance and insurance
6	22	Real estate
7	23	Transport
8	24	Communication and broadcasting
9	25	Public administration
10	26	Education and research institute
11	27	Medical service, health, and social security
12	28	Other public service
13	29	Business services
14	30	Personal services

Sources: Management and Coordination Agency (1994, 1999); Statistical Department, Institute of Developing Economies (1998).

Notes: The definition of "Education and research institute" in the Japanese input-output table is slightly different from the international input-output table. The Japanese input-output table includes private research and development activity but the international input-output table does not include it.

table. Because the sample countries' input-output tables are reported in the local currency, we converted these tables in local currency into those in the U.S. dollar. The exchange rate used for the conversion is the annual average exchange rate (rf) from IMF (1999b) and from the Council for Economic Planning and Agency (1999) for Taiwan.

The Japanese Input-Output Tables

For the detailed analysis of Japan's services trade embodied in goods trade, we used a 187-sector input-output table for 1990 and a 186-sector table for 1995. The sector classification used in the international input-output table is different from that of the Japanese input-output table. Some sectors in the Japanese input-output table are excluded from the international input-output table. For instance, research and development at private companies and repair service are separately recorded in the Japanese input-output table, but they are not included in the international input-output table.

When we examined the changes in embodied services in detail, we used the Japanese classification. This is because Japanese input-output tables provide detailed sector classification, which enabled us to perform more detailed analyses than the international input-output table. Therefore, the estimated results of embodied services for Japan differ between those based on the international input-output table and those based on the detailed Japanese input-output tables.

The calculated results are aggregated to the thirty-two-sector level (fourteen services), whose sector classification is shown in table 12B.1. The sector classification for the 1995 input-output table differed from that for the 1990 input-output table. However, the differences disappeared when we aggregated these two tables into thirty-two sectors.

References

Aw, Bee Y. 1983. The interpretation of cross-section regression tests of the Heckscher-Ohlin theorem with many goods and factors. *Journal of International Economics* 14 (1–2): 163–67.

Baldwin, Robert E. 1971. Determinants of the commodity structure of U.S. Trade. *American Economic Review* 61:126–46.

Bowen, Harry P. 1983. Changes in the international distribution of resources and their impact on U.S. comparative advantage. *Review of Economics and Statistics* 65 (3): 402–14.

Bowen, Harry P., Abraham Hollander, and Jean-Marie Viaene. 1998. *Applied international trade analysis.* Basingstoke, England: Macmillan.

Bowen, Harry P., and Leo Sveikauskas. 1992. Judging factor abundance. *Quarterly Journal of Economics* 107 (2): 599–620.

Council for Economic Planning and Development, Republic of China. 1999. *Taiwan statistical databook*. Taipei: Council for Economic Planning and Development.

Deardorff, Alan V. 1984. Testing trade theories and predicting trade flows. In *Handbook of international economics*, Vol. 1, ed. R. W. Kenen and P. B. Jones, 467–518. Amsterdam: North-Holland.

Economic Research Department, Central Bank of China. 1999. *Balance of payments quarterly, Taiwan District, Republic of China*. Taipei: Economic Research Department, Central Bank of China.

Grubel, Herbert G. 1988. Direct and embodied trade in services. In *Trade and investment in services*, ed. Chung H. Lee and Naya Seiji, 53–76. Boulder, Colo.: Westview Press.

International Monetary Fund (IMF). 1999a. *Balance of payments statistics* [CD-ROM]. Washington, D.C.: IMF.

———. 1999b. *International financial statistics* [CD-ROM]. Washington, D.C.: IMF.

Leamer, Edward E., and Harry P. Bowen. 1981. Cross-section tests of the Heckscher-Ohlin theorem: Comment. *American Economic Review* 71 (4): 1040–43.

Leamer, Edward E., and James Levinsohn. 1995. International trade theory and the evidence. In *Handbook of international economics*, ed. Gene M. Grossman and Kenneth Rogoff, 1341–94. Amsterdam: Elsevier.

Management and Coordination Agency. 1994. *1990 input-output table for Japan*. Tokyo: Management and Coordination Agency.

———. 1999. *1995 input-output table for Japan*. Tokyo: Management and Coordination Agency.

Markusen, James R., James R. Melvin, William H. Kaimpfer, and Keith E. Maskus. 1995. *International trade: Theory and evidence*. New York: McGraw-Hill.

Petri, Peter A. 1991. Market structure, comparative advantage, and Japanese trade under the strong yen. In *Trade with Japan: Has the door opened wider?* ed. Paul Krugman, 51–82. Chicago: University of Chicago Press.

Sapir, Andre, and Ernst Lutz. 1981. Trade in services: Economic determinants and development related issues. World Bank Staff Working Paper no. 480. Washington, D.C.: World Bank, August.

Sazanami, Yoko, and Shujiro Urata. 1990. *Services trade: Theory, present and future topics* (in Japanese). Tokyo: Toyokeizai Shinpo-sha.

Statistical Department, Institute of Developing Economies. 1995. *Philippines Input-Output Table, 1990*. Tokyo: Institute of Developing Economies.

———. 1996a. *China Input-Output Table, 1990*. Tokyo: Institute of Developing Economies.

———. 1996b. *Taiwan Input-Output Table, 1990*. Tokyo: Institute of Developing Economies.

———. 1997a. *Malaysia Input-Output Table, 1990*. Tokyo: Institute of Developing Economies.

———. 1997b. *Singapore Input-Output Table, 1990*. Tokyo: Institute of Developing Economies.

———. 1998. *Asian international input-output table, 1990*. Tokyo: Institute of Developing Economies.

Stern, Robert M., and Keith F. Maskus. 1981. Determinants of the structure of U.S. foreign trade. *Journal of International Economics* 11 (2): 207–24.

Tucker, Ken, and Mark Sundberg. 1998. *International trade in services*. London: Routledge.

Urata, Shujiro. 1983. Factor inputs and Japanese manufacturing trade structure. *Review of Economics and Statistics* 65 (4): 678–84.
———. 1994. Service trade embodied in merchandise trade: The case of Japan (in Japanese). *World Economic Critiques* 38 (5): 50–56.
World Bank. 1999. *World development indicators* [CD-ROM]. Washington, D.C.: World Bank.
———. 2000. *East Asia: Recovery and beyond.* Washington, D.C.: World Bank.

Comment   Ponciano S. Intal, Jr.

This is a very good and well-written paper. There are a number of "golden nuggets" of information on trade in services in East Asia that I find very interesting. I enjoyed reading the paper. My comments focus on a number of the interesting findings of the authors as well as one puzzling result.

1. I find the information that Korea's and Taiwan's royalty payments are much higher than royalty receipts very interesting because of its policy implications for developing countries like the Philippines. The usual policy focus in developing countries tends to be on foreign investment; the technology market is barely given emphasis. The finding on Korea and Taiwan, like Japan in the 1950s and 1960s, suggests that developing countries wishing to play technology catch-up must be willing to pay for the technology and patent and be net royalty payers in the meantime as their research and development (R&D) capabilities are being upgraded.

2. The significant intraindustry trade in transportation services in East Asia is not quite surprising. With foreign trade growing so fast in the region during the past decade, exporters and importers have to be concerned with the reliability of shipping bottoms and airline cargo space. Transport service exports and imports in the region expanded apace with the growth of interregional trade in East Asia considering the governments' bias for national carriers in the allocation of landing rights in airports and berthing spaces in seaports, the importance of long-term relationships in transport services, and the nature of bilateral agreements in air transport. Nonetheless, on a *net* basis, the capital-poor but fast-growing countries (e.g., Thailand, Malaysia, China) can be expected to be net deficit countries in transportation services because ships and airplanes are capital intensive. The authors' econometric analysis of the determinants of services trade bears this out.

3. That Japan has a surplus position in construction services is very interesting considering the very high wage rate in Japan. I think this boils

Ponciano S. Intal, Jr. is professor of economics and executive director of the Angelo King Institute for Economic and Business Studies, De La Salle University, Manila, Philippines.

down to a large extent to official development assistance (ODA). Japan's ODA through the Overseas Economic Cooperation Fund or the Japan Bank for International Cooperation (JBIC) has been primarily in infrastructure, and the politics of bilateral ODA means that a substantial portion of that money goes back to the donor country through supply or construction contracts.

4. It is worth noting that virtually all the East Asian countries in the sample have a negative balance in insurance. This is probably the result of the growing importance of nonlife insurance with the fast growth of the economies, investment, and interregional trade. (For example, factory buildings may need to be insured against fire.) Risk pooling may need the growing demand and payment for reinsurance business, especially if the domestic insurance companies are comparatively small and have limited equity to cushion all the attendant risks. The reinsurance centers are primarily London and New York: hence, the negative balances in insurance for much of East Asia.

5. It is also interesting to note that there is some specialization in services trade in East Asia. Generally, each East Asian country has positive balances in one or two sectors and negative balances in the rest of the sectors. Thus, for example, Association of Southeast Asian Nations (ASEAN) countries and China, with cheaper labor and location-specific assets (e.g., the tropical climate for ASEAN, cultural assets for China) have a lock on travel, with Singapore and the Philippines also competitive in the sector of "other business services." As Urata and Kiyota have shown, this is generally consistent with the insights from the factor proportions theory. Nevertheless, the travel balances for 1998 are likely to have been bloated by the combination of outward travel restrictions in some ASEAN countries hit by the East Asian crisis (e.g., Malaysia) and by the favorable impact of the currency depreciation of ASEAN countries arising from the East Asian crisis.

The analysis of the determinants of services trade is well done. The authors seem to have succeeded in expanding and deepening previous analyses. There is one minor technical point that calls for clarification. Specifically, "government services, not included elsewhere" is included in the physical and human capital intensive services. The authors could give examples of this particular service industry in order for the readers to have better a understanding of why this is included under physical and human capital–intensive service industries.

The results of the analyses on embodied service exports are very interesting. I will start with one minor technical detail about Philippine service exports and then focus on one puzzling result related primarily to wholesale and retail trade. The authors found that Philippine exports have the lowest ratio of embodied service export to total service exports in the region. This stems from the heavy dependence of Philippine goods exports on semiconductors and electrical machinery, which are primarily assembled in a few export zones and industrial estates in the country and which rely a lot on imported inputs. As a result, the service content is small.

The authors highlight that, using a normalized embodied services trade, Japan is a net exporter of wholesale, retail, and transport services; China is a net exporter of wholesale and retail trade services; and Malaysia is a large net exporter of transportation services. Findings such as the above are puzzling. For example, it is surprising indeed that Japan is a net exporter of wholesale and retail services when the sector is not generally considered efficient and has one of the highest distribution margins in the industrialized world. The probable reason for the surprising result is that the authors used the input-output table of Japan to estimate the embodied wholesale and retail services exports of Japan. Given the multilayered nature of the Japanese distribution system (especially the wholesaling subsector) and the concomitant high distribution margins, Japan ended by becoming a net exporter of embodied and disembodied wholesale and retail services. Thus, it pays to have an inefficient wholesale and retail services subsector!

What can possibly be wrong here? The answer is that the authors used the internal distribution system of Japan (indicated in the input-output estimates) to get the estimates of the embodied services trade. Yet what may be more relevant is the distribution system of the export market in order to get the embodied services trade. Thus, for example, to bring a Toyota car to a customer in the Philippines means providing distribution network and possibly financing arrangements in the Philippines, and not the distribution network in Japan. Only in the case of the Japanese inputs into the Toyota would the distribution network in Japan become relevant. Even then, the whole distribution margin is not relevant because all that matters is the producer-to-Toyota link, rather than the whole gamut that includes the retail sector. In short, the use of the internal distribution system of Japan *overestimates* the embodied services exports of Japan.

The same overestimation of embodied exports could explain why China is a net exporter of wholesale and retail services and Malaysia a net exporter of transport services. In the case of China, national data must have been used in the estimation of the input-output table of China. However, China's major production bases for exports are concentrated in China's East Coast provinces (e.g., Guandong). Using national data, which probably show higher distribution margins, given the inefficiency of the distribution systems in certain parts of the country, effectively overestimates the embodied services exports of China. Similarly, Malaysia is a large net exporter of transport services, probably because the relatively high cost of transporting products between Peninsular Malaysia and Eastern Malaysia (e.g., Sabah) raises the national average transport margins in Malaysia's input-output estimates. Nonetheless, the relevant transport cost embodied in Malaysia's exports is likely just the cost of a brief trip from an electronics plant in Penang to Penang's airport or a somewhat longer trip from Petaling Jaya to the seaport. In either case, the embodied transport cost is much lower than the national estimate used in the estimation of input-output tables.

It is clear that the methodology of estimating embodied service exports needs to be improved. It is not clear, however, what can be done to address this problem of overestimation of the embodied services. In the meantime, we need to be more cautious in the interpretation of the results of embodied services trade.

Comment   Richard H. Snape

This paper focuses on services trade in the forms of cross-border trade and consumption abroad for a number of East Asian economies. It is an interesting paper, in particular in the manner in which it incorporates embodied trade. The omission of establishment abroad as a mode of trade, while understandable for reasons of data, may mean that some important causal relations are omitted, particularly where human capital is a major factor. Similarly, the omission of the movement of natural persons may miss some important labor-intensive effects. Are data relating to payments back home by foreign workers not available?

After describing the trends of services trade, the paper reports tests for the determinants of the trade in a Heckscher-Ohlin framework. The tests do not yield a great deal that supports the hypotheses where actual services trade is concerned, but they are more encouraging where embodied services trade is considered also.

My first comment picks up a point early in the paper where it is stated that rapid technological progress will reduce the price of services and increase the demand for them, and it is implied that this will increase the share of services in gross domestic product (GDP). Of course, there may be an increased share, particularly if the statisticians are very clever with their deflators. On the other hand, with the rapidly declining prices of communication and of related services, the *measured* share of GDP may fall, not rise.

The paper points out that barriers to trade may be important reasons that trade patterns may not be as expected from the Heckscher-Ohlin framework. I think that more attention could be given to transport. First is the simple point that international transport does not occur when, say, Japanese airlines or ships carry Japanese people or produce abroad. It is important to know how such transport is treated in the statistics and to consider

Richard H. Snape is deputy chairman of the Australian Productivity Commission and emeritus professor of economics, Monash University, Australia.

how one would wish to treat it in relation to the theories being tested. Second, the bilateral aviation agreements between countries effectively strangle competition on a great number of international routes. Travel between partner countries on the routes will not tell us much about comparative advantage.

The carriage between two foreign countries depends greatly on location. For example, Singapore, for carriage between Australia and the United Kingdom, can match its aviation agreements between Australia and the United Kingdom, and the carriage between Australia and the United Kingdom will depend on these agreements and location as much as on its factor endowments.

I turn now to embodied services and the authors' experiment of changing exports and imports of goods of each of the countries by $1 million and seeing what that implies for the embodied trade in services. We should note that for, say, Japan, it is the Japanese input-output relations for exports *and imports* that are being used. Thus, when we are looking at services embodied in imports of Japan, it is really the embodiment ratio for Japanese import-competing goods that is being used, not the embodiment ratio in the country that is actually exporting the goods to Japan. This causes me some concern, for (say) the domestic transport input to goods produced in Japan may be quite different from the domestic transport input of the same goods produced abroad. This may be particularly so for the bulk commodities that loom large in Japanese imports. Can one infer the domestic (Australian) transport component of iron ore produced in Australia for export to Japan from the domestic Japanese transport of iron ore produced in Japan for use in Japan?

Further, the fact that Japan is importing products—say, raw materials—that may have a large foreign domestic transport component in them is not likely to be because Japan has a high preference for the consumption of embodied services (as suggested in the paper), but simply because Japan does not have the raw materials or has a comparative disadvantage in them. In other words, we cannot really divorce the demand for and supply of the embodied services from the demand for and supply of the goods and services in which they are embodied. Heckscher-Ohlin is difficult to apply to embodied services for that reason. The embodied service is but one component of the goods—Heckscher-Ohlin should be applied at the frontier, to the product valued at that point.

We might also note that domestic transport costs are likely to be a higher proportion of the cost of raw materials than of the cost of more processed goods. Thus, import policies that favor the import of raw materials and discriminate against more processed goods will lead to a higher proportion of embodied services in imports than a more neutral import policy. The high embodied service component of imports would reflect that trade policy, not comparative advantage in services, nor preference for embodied services.

Another point is that we are at the mercy of statisticians with respect to embodied services. In cross-country comparisons, is the treatment of services supplied within the enterprise the same in each country in their input-output tables? If some countries treat in-house research and development in manufacturing as manufacturing while others treat it as a service input to manufacturing, the cross-country studies are damaged. Of course, there is also the well-known problem that a change from in-house to bought-in services will show changes in service inputs unless the statisticians are very careful. I note in table 12.11 that the Japanese statisticians are indeed careful insofar as in-house research and development (R&D) is concerned, but of course there are many other services for which the same point applies.

In conclusion, I found the paper very interesting, and one particular point of interest was in table 12.11, where the embodied intraenterprise R&D was shown to be a major component of Japan's exports.

Foreign Direct Investment and Services Trade
The Case of Japan

Kyoji Fukao and Keiko Ito

13.1 Introduction

Because many services are either untradable or at least difficult to trade, a substantial part of the international delivery of services is conducted through affiliates established within other countries. For this reason, it has been argued that the compilation of statistics on international sales of services must include information not only on *cross-border transactions,* as recorded in the balance-of-payment statistics, but also on services delivered through *establishment transactions* (Kravis and Lipsey 1988; Ascher and Whichard 1991). Being aware of this issue, the U.S. government has made efforts to improve official statistics, so that in the case of the United States, relatively reliable statistics on these two types of international transactions of services are available from the 1980s onward (U.S. Congress 1986; U.S. Department of Commerce 1995c, 1999). In contrast, although Japan has the second largest market for services in the world, Japan's official statistics on establishment transactions of services have many drawbacks in comparison with U.S. statistics.

In this paper, we estimate the sales and employment of Japanese affiliates of foreign firms (JAFFs) and foreign affiliates of Japanese firms (FAJFs) in the service sector at the three-digit industry level for the year 1995. Our estimation is based mainly on data provided by Toyo Keizai and the results of the Establishment and Enterprise Census of Japan, which is conducted by the Japan Management and Coordination Agency. Using our estimates, we compare Japan's establishment transactions with Japan's cross-border

Kyoji Fukao is professor of economics at Hitotsubashi University, Tokyo. Keiko Ito is a research assistant professor at the International Centre for the Study of East Asian Development in Kitakyushu, Japan.

transactions at the three-digit industry level. We also compare Japan's purchases of services from foreigners with U.S. purchases from foreigners. Although our new estimates possibly contain large estimation errors due to statistical deficiencies, we think that our results are more comprehensive and balanced than existing statistics on this issue.

According to our new statistics, actual foreign activities in Japan are much greater than those reported in Japan's Ministry of International Trade and Industry's (MITI, which is now the Ministry of Economy, Trade, and Industry, METI) survey, *Gaishi-kei Kigyo Doko Chosa* (*Survey on Trends of Business Activities by Japanese Subsidiaries of Foreign Firms*).

Probably the most commonly cited statistics on Japan's inward direct investment are those provided by the Ministry of Finance (MOF; 1999; the data are also available in Organization for Economic Cooperation and Development [OECD] 1999a). According to these data, Japan's outward direct investment stock in the service sector is nine times greater than the corresponding inward direct investment stock (table 13.1). Since no other OECD country has an imbalance of this magnitude, it has been argued that the imbalance indicates the closedness of the Japanese economy to inward direct investment in the service industries (General Agreement on Tariffs and Trade [GATT] 1995; MITI 1998b; Stern 2000).

However, since the MOF data only record cross-border capital flows, they do not necessarily correspond to the extent of affiliates' actual activities. For example, because of Japanese regulations, many foreign banks and insurance companies entered the Japanese market by setting up branches rather than founding subsidiary companies. This fact makes their investment flows relatively small compared with the actual magnitude of their affiliates' activities measured by sales or employment. According to our new statistics, imbalances between the activities of JAFFs and those of FAJFs are smaller than those indicated by the MOF's foreign direct investment (FDI) statistics. In terms of employment, the JAFF-FAJF ratio is 0.22.

Although our new estimates of foreign activities in Japan are larger than existing estimates, we found that foreign activities in Japan are substantially smaller than foreign activities in the United States. Japan's ratio of number of workers employed by majority-owned foreign affiliates to total number of workers is 0.4 percent, which is one-seventh of the corresponding U.S. ratio of 2.8 percent. We also found that, compared with the United States, Japan's purchases from foreigners are concentrated in a limited number of industries. Four industries—financial intermediary services, wholesale trade, air transportation, and hotels and lodging places—account for about 54 percent of Japan's total purchases of services from foreigners.

Because our data are compiled at the three-digit industry level, we can use them for cross-industry regression. We estimated an empirical model explaining the determinants of Japan's inward FDI penetration. We found that inward FDI penetration is closely related to several characteristics of

Table 13.1   **Japan's Inward and Outward FDI: Position at the End of March 2001 (billions of yen)**

Industry	Inward FDI Stock
Construction	21
Real estate	339
Commerce	2,028
Business and personal services	1,526
Transportation services	48
Communication services	1,155
Finance and insurance	2,595
Others	168
Nonmanufacturing total	7,880
Manufacturing	5,324
Total amount	13,203

	Outward FDI Stock
Agriculture and forestry	424
Fishery	257
Mining	5,193
Construction	821
Commerce	11,016
Finance and insurance	20,347
Business and personal services	11,398
Transportation services	7,862
Real estate	12,524
Others	1,824
Nonmanufacturing total	71,665
Manufacturing	34,187
Branches	1,656
Total amount	107,669

Sources: MOF (1999); [http://www.mof.go.jp].
Note: Cumulated value of FDI flows approved or notified from 1950 onwards.

industries. Japan's inward FDI penetration is relatively high in industries that have higher advertisement intensity, a lower presence of government activities, and a lower presence of official restrictions on inward FDI. We found that the presence of *keiretsu* does not have significant negative effects on FDI penetration.

The paper is organized as follows: In the succeeding section, we discuss existing data on Japan's international transactions of services through affiliates. In section 13.3, we explain how we estimated sales and employment by JAFFs and FAJFs in the service sector. In section 13.4, we provide a general overview of Japan's international transactions of services using our new statistics. In section 13.5, we undertake an econometric investigation of the determinants of Japan's FDI penetration in the service sector at the three-digit industry level. Section 13.6 concludes.

13.2   Existing Data on Japan's International Transactions of Services through Affiliates

In the case of inward direct investment in nonmanufacturing industries, MITI's survey *Gaishi-kei Kigyo Doko Chosa* (*Survey on Trends of Business Activities by Japanese Subsidiaries of Foreign Firms*) is the only official source on the sales and employment of foreign firms' Japanese subsidiaries.[1] According to this survey, foreign firms' Japanese subsidiaries employed only 63,000 workers in nonmanufacturing industries at the end of March 1998. The survey is loosely based on the U.S. Department of Commerce's survey of FDI in the United States, but MITI's survey has the following serious drawbacks for the purpose of studies on inward direct investment in the service sector.

1. It is not mandatory and suffers from a low response ratio. In the case of the survey for the 1997 fiscal year, only 49.5 percent of the questionnaires sent out were returned to MITI. Moreover, usually not all the questions in the returned questionnaires are answered.

2. The survey does not cover subsidiaries in real estate, finance, and insurance.

3. The survey covers only Japanese companies that are more than one-third foreign-owned and does not cover branches and other establishments directly owned by foreign firms.

4. In MITI's report on inward FDI, all the data on nonmanufacturing subsidiaries are aggregated into three industries only: commerce, services, and others (agriculture, construction, etc.). In the case of outward FDI, the data on nonmanufacturing subsidiaries are aggregated into six industries: agriculture, mining, construction, commerce, services, and others. No data at a more detailed industry level are published.

Because of the low response ratio and the exclusion of real estate, finance, and insurance, the number of subsidiaries covered by MITI's survey is substantially smaller than that of other surveys on foreign subsidiaries conducted by private companies. For example, the number of nonmanufacturing subsidiaries covered by the MITI survey for 1997 was only 983.[2]

---

1. MITI's other survey, *Kigyo Katsudo Kihon Chosa* (*Basic Survey on Business Activities by Enterprises*), also collects data on JAFFs as a part of information obtained on Japanese firms. However, this survey covers only the manufacturing and commerce sectors. Moreover, the response ratio of this survey is also low. In 1999, the Japan Management and Coordination Agency added questions on whether firms were majority-owned by foreigners to their survey *Service-gyo Kihon Chosa* (*Basic Survey on Service Sector*), which covers several service industries. A coming report of this survey probably includes some information on JAFFs.

2. Mainly focusing on manufacturing sectors, Kimura and Baldwin (1996) estimated sales and procurements by JAFFs and FAJFs using the results of MITI's surveys. They did not make adjustments to account for these problems.

The results of this survey on Japanese companies majority-owned by foreign firms are reproduced in OECD (1999b). In the case of inward direct investment in Japan's service sector, the formats of tables in the OECD publication are quite misleading. According to the publication, Japanese subsidiaries in finance, insurance, real estate, and business services that were majority-owned by foreign firms employed only 3,800 workers in 1996. However, this number is in fact only for business service subsidiaries, because MITI's survey does not cover the other subsectors.

Concerning foreign subsidiaries of Japanese firms, MITI conducts the survey *Kaigai Jigyo Katsudo Doko Chosa* (*Survey on Trends of Japan's Business Activities Abroad*), which covers foreign subsidiaries with more than a 10 percent Japanese ownership. This survey has similar setbacks as the survey on inward direct investment. It suffers from a low response ratio and does not cover Japanese-owned subsidiaries in the finance and insurance sector. According to this survey, foreign subsidiaries of Japanese firms employed 487,000 workers in nonmanufacturing sectors, excluding agriculture, fishery, and mining, at the end of March 1998.

Compared with these surveys by MITI, Toyo Keizai's microdata, *Gaishi-kei Kigyo Soran: CD-ROM-ban* (*Directory of Japanese Subsidiaries Abroad: CD-ROM version*) and *Kaigai Shinshutsu Kigyo Soran: CD-ROM-ban* (*Directory of Japanese Subsidiaries Abroad: CD-ROM version*) have a substantially broader coverage of subsidiaries. Toyo Keizai conducts its own surveys for this database.[3] Toyo Keizai also uses additional data, such as financial reports, for nonresponding firms. The data cover all industries. In principle, the Toyo Keizai data on inward FDI cover subsidiaries with a 49 percent or higher foreign ownership. However, in the case of listed or large subsidiaries, the data cover those with a 20 percent or higher foreign ownership. The data on outward FDI primarily cover foreign subsidiaries with a 20 percent or higher Japanese ownership in principle. Judging by the number of subsidiaries and number of workers employed by subsidiaries, the coverage of the Toyo Keizai data is much broader than that of MITI. In the case of foreign firms' Japanese subsidiaries in nonmanufacturing sectors excluding the primary sector, the Toyo Keizai data for 1997 cover 2,456 subsidiaries, which employed 204,000 workers.[4] In the case of foreign subsidiaries of Japanese firms in nonmanufacturing sectors excluding the primary sec-

3. In the case of inward FDI, Toyo Keizai and Dun & Bradstreet Japan Ltd. jointly conduct their surveys for this database.

4. A private company, Teikoku Data Bank Ltd., provides a database, "Cosmos," which covers 1.1 million Japanese firms for 1999. In the case of the nonmanufacturing sector, the database contains information on 1,236 firms that were more than one-quarter foreign-owned. The database was too expensive for us to use for this research. Some statistics on these firms are available at http://www.tdb.co.jp.

tor, the data for 1995 cover 10,378 subsidiaries, which employed 865,000 workers.

13.3 Estimation of Sales and Employment by JAFFs and FAJFs in the Service Sector

We use Toyo Keizai's data as the basic statistics for our estimation. Sales and employment data for JAFFs and FAJFs in service sectors at the three-digit level are estimated for the year 1995. We chose 1995 because the most recent input-output (I-O) tables (Japanese Government 1999) are available for this year.

Although the coverage is broader, the Toyo Keizai data have several shortcomings. We revised the data using additional statistics in the following way. (For details regarding the estimation procedures, please see the appendix.)

13.3.1 Branches and Other Establishments Directly Owned by Foreign Firms

In the case of the banking and insurance sector, the Toyo Keizai data cover Japanese branches and other establishments directly owned by foreign firms. However, the data only partially cover such establishments in other sectors. The Statistics Bureau of the Japan Management and Coordination Agency (1998) records the number of workers employed by Japanese branches and other establishments directly owned by foreign firms at the four-digit industry level.[5] We used these data for estimations on Japanese branches and other establishments directly owned by foreign firms. In the case of outward investment, Toyo Keizai's database covers such establishments. According to the Toyo Keizai data, foreign establishments directly owned by Japanese firms employed 44,000 workers in 1995.

13.3.2 Estimation of Sales

Although for most subsidiaries the number of workers is reported in the Toyo Keizai data, information on sales is not available for many subsidiaries. In the case of Japanese subsidiaries of foreign firms, we calculated each industry's average value of sales per worker from data on subsidiaries, for which both the number of workers and the sales were available. We used these values in order to estimate the sales of subsidiaries for which data on

---

5. *Jigyosho Kigyo Tokei Chosa (Establishment and Enterprise Census of Japan)*, conducted by the Japan Management and Coordination Agency, is the most basic and important survey on Japanese establishments and covers all the industries. The survey collects both data on establishments and data on enterprises, and these two sets of data are linked. In the survey, companies are asked whether they are majority-owned by foreign firms. Therefore, the data collected in this survey are ideal for a compilation of statistics on the number of workers employed by all the JAFFs. However, such statistics are not included in the report on this survey, and we did not have enough time to get access to microdata of the survey.

sales were not available in the Toyo Keizai database and sales by Japanese branches and other establishments directly owned by foreign firms.[6] In the case of foreign subsidiaries of Japanese firms, we used both microdata of MITI's survey and Toyo Keizai's data to get average values of sales per worker for subsidiaries at the three-digit industry level. Using these values, we estimated the sales of subsidiaries for which information on sales were not available in the Toyo Keizai database. Since employment data is more reliable than sales data, we will mainly use employment data for international comparison and regression analysis.

For wholesale and retail trade and financial intermediary services, sales are not a suitable measure of activities. In the case of trade services, we estimated the distribution margins of JAFFs. Using 1995 I-O tables, we calculated the average values of distribution margins per worker in the wholesale and retail trade sectors. Multiplying the total number of workers of JAFFs by these average values, we derived our estimations for their distribution margins. In the case of subsidiaries in financial intermediary services, following Toyo Keizai, we use current incomes instead of sales as a measure of activities.

13.3.3 Industry Classification

Toyo Keizai's industry classification, which has thirty-one nonmanufacturing sectors, is not detailed enough for our analysis.[7] We therefore reclassified all subsidiaries into one of fifty-one sectors using information on the subsidiary's line of business, which is included in the Toyo Keizai data. Table 13.2 shows the correspondence between our own classification and several other standard classifications.[8,9] In our estimation, affiliates are classified according to their primary industry. Therefore, services supplied by JAFFs that are engaged in industries that are not classified as "services" are excluded from our estimation. For example, computer-related services provided by computer makers are not included. In the case of the United

6. We have also examined financial reports. Because the majority of foreign-owned firms are unlisted and the Toyo Keizai usually reports sales in the case of listed firms, this strategy did not help us substantially. We thought that the sales-employment ratio might be different for firms of different scale, and so we investigated whether this ratio depended on the scale of firm for several major industries, but we found no significant relationship.

7. Toyo Keizai's classification contains eleven wholesale trade sectors. For the other nonmanufacturing subsidiaries, it contains only twenty sectors.

8. We aimed at setting the target of our analysis as broad as possible. Our classification includes all the nonmanufacturing industries except agriculture, fishery, forestry, and mining. Our data cover electricity, gas, and water supply, which are not covered by the General Agreement on Trade in Services, and agricultural services and ship and aircraft repair, which are not classified in the service sector in Standard Industrial Classification for Japan (Statistics Bureau, Japan Management and Coordination Agency 1993).

9. For definitions of industries in Japan's, the United States', and the General Agreement on Tariffs and Trade (GATT) secretariat's classifications systems, see United Nations (1991), GATT (1991), Statistics Bureau of Japan Management and Coordination Agency (1993), MITI (1999c), Japanese Government (1999), and Nijhowne and Usher (1999).

Table 13.2   Correspondence Table

Fukao-Ito Industry Code	Definition	1995 Japan I-O Standard Classification	1992 U.S. I-O Standard Classification	1992 BEA Classification for FDI in the U.S. Establishment Data	GATT Secretariat Classification
1	Construction and civil engineering	4111-011, 4112-011, 4131-011, 4112-021, 4121-011, 4131-031, 4131-021, 4132-011, 4132-021, 4132-031, 4132-099	11, 12	15, 16, 17, 6522	3.A, 3.C, 3.E, 3.B, 3.D
2	Electricity	5111-001, 5111-041	680100, 780200, 790200	491, 4931	171, 1.F.j[a]
3	Gas supply	5121-011, 5122-011	680201, 680202	492, 4932	172, 1.F.j[a]
4	Steam and hot water supply	5211-011, 5211-021	680301[a]	494[a], 4953[a], 4959[a]	173, 1.F.j[a]
5	Water supply	5211-031	680302[a]	4952, 496[a]	180
6	Sewerage systems				6.A
7	Sanitary services	5212-011, 5212-021	680302[a]	4953[a], 4959[a], 496[a]	6.B, 4.B
8	Wholesale trade	6111-011	69A	50, 51	4.A
9	Retail trade	6112-011	69B	52-57, 59	4.C, 4.D[a]
10	Financial intermediary services	6211-011, 6211-012, 6211-013, 6211-014	70A	60, 61, 62	7.B.a-1, 6.B
11	Life insurance	6212-011	70B[a]	63[a], 64[a]	7.A.a, 7.A.c.d[a]
12	Casualty insurance	6212-021			7.A.b, 7.A.c.d[a]
13	Real estate	6411-011, 6411-021, 6421-011	710100, 710201	65	1.D
14	Railway passenger transportation	7111-011, 7111-012	650100[a]	—	11.E.a
15	Railway freight transportation	7112-011			11.E.b
16	Road passenger transportation	7121-011, 7121-021, 7131-011	650200, 790100	411, 4141, 412, 4142, 413, 415	11.F.a, 11.F.c[a]
17	Road freight transportation	7122-011, 7122-021, 7132-011	650301	421[a], 422	11.F.b, 11.F.c[a]
18	Water transportation	7141-011, 7142-011, 7142-012, 7143-011	65C	441-444, 448, 449	11.A.a,b,c, 11.B.a,b,c
19	Air transportation	7151-011, 7151-012, 7151-013, 7151-014	65D	451, 452, 458	11.C.a,b,c
20	Storage facility services	7171-011	650301, 650302	421[a], 422	11.H.b

No.	Service				650701	650702		47	417	423		
21	Supporting services for transport	7161-011 7189-011 7189-041 7189-099	7181-011 7189-021 7189-051	7189-031 7189-061	650701 750003	650702 790300		47 752	417	423	1.F.q. 9.C 11.B.e,f 11.E.c,e 11.H.a,c,d	9.B 11.A.e,f 11.C.e 11.F.e 11.I
22	Postal service	7311-011			780100			—			2.A	2.B
23	Telecommunications	7312-011 7319-099	7312-021	7312-031	660100			481	482	489	2.C	
24	Broadcasting	7321-011	7321-021	7321-031	660200	670000		483	484		2.D.c,d	
25	Education	8211-011 8213-011 8213-041	8211-021 8213-021	8213-031	770401[a] 770600[a]	770402[a] 730112[a]	770403[a]	841[a] 823[a] 833[a]	842[a] 824[a] 8731[a]	829[a] 8732[a]	5.A 5.C 10.C.a-n	5.B 5.D
26	Research institutes (natural sciences)	8221-011	8221-031	8221-051							1.C.a	
27	Research institutes (social sciences & humanities)	8221-021	8221-041	8221-061							1.C.b	
28	Research within firms	8222-011									1.C.c	
29	Medical services	8311-011	8311-021	8311-031	770100[a] 770303[a]	770200[a] 770305[a]	770301[a]	80[a]			1.A.h,j 8.B 6.C	8.A 8.C
30	Health and hygiene	8312-011	8312-021	8312-031	770501 770504	770502	770503					
31	Private non-profit organizations' services	8411-011	8411-021					—			12	
32	Advertising	8511-011	8511-012		73D			731			1.F.a	
33	Computer programming and software	8512-011			730104[a]			7371	7372	7373	1.B.b	
34	Information services	8512-012	8512-021		730106[a]			7374-76 7383	7379	7381	1.B.a,c,d,e 10.B	1.F.b
35	Goods and equipment rental and leasing	8513-011 8513-014	8513-012 8513-015	8513-013	730107 760102			735	784		1.E.a,b,d,e	
36	Automobile renting	8514-011			750001			751			1.E.c	
37	Automobile repairing	8515-101			750002			753	754		11.F.d	
38	Machine repairing	8516-011			720204			7378	76		1.F.i,n	
39	Building maintenance services	8519-011			730102			734			1.F.o	
40	Legal and accounting services	8519-021			730301	730303		81	872		1.A.a,b,c	
41	Civil engineering and construction services	8519-031			730302[a]			8712			1.A.d,f	
42	Personnel supply services	8519-041			730103			736			1.F.k	

(*continued*)

Table 13.2 (continued)

Fukao-Ito Industry Code	Definition	1995 Japan I-O Standard Classification	1992 U.S. I-O Standard Classification	1992 BEA Classification for FDI in the U.S. Establishment Data				GATT Secretariat Classification	
43	Other business services	8519-099	730109 730111 730302[a]	733 8711 874	7382 8713	7389 8734		1.A.e,g 1.F,r,s,t 11.D	1.F.c-e,l,m 6.D 11.G.a,b
44	Amusement and recreation services	8611-011 8611-021 8611-031 8611-041 8611-051 8611-061 8611-071 8611-099	760101 760201 760202 760203 760204 760205 760206	781 782 783 792 793 794 7992 7993 7996 7997 7999				2.D.a,b 10.A	10.D
45	Eating and drinking places	8612-011 8612-021 8612-031	74	58				4.D[a]	
46	Hotels and lodging places	8613-011	72A	70 excl. 704				9.A[a]	
47	Individual educational facilities	8619-081	760205	7991				12[a]	
48	Other personal services	8619-011 8619-021 8619-031 8619-041 8619-051 8619-061 8619-071 8619-099	720201 720202 720203 720205 720300 730101 040002 730108	721 726 722 763 764 769 723 724 725 078 729 7384				1.F.p 12[a]	
49	Agricultural services	0131-011 0131-021	770304 040001	07 excl. 078				1.A.i 11.A.d	1.F.f
50	Ship repairing	3611-101	610100 610200	373				11.A.d	11.B.d
51	Aircraft repairing	3622-101	60	372				11.C.d	

Note: Dashes indicate that there is no classification code applicable to the industry.

[a] Indicates an industry that corresponds to more than one industry in the Fukao-Ito classification.

States, sales of services by foreign firms' affiliates in the manufacturing industry accounted for 6 percent of total sales of services by foreign firms' U.S. affiliates in 1996 (U.S. Department of Commerce 1999). The data on the sales of services by JAFFs in the nonservice sector are available from MITI (1998a). We found that such sales were negligible. The data on the sales of services by FAJFs in nonservice sectors are only available for U.S. affiliates. According to the U.S. Department of Commerce (1999), sales of services by affiliates of Japanese firms in manufacturing industry accounted for 4 percent of total service sales of Japanese firms' U.S. affiliates in 1996. Our estimates on service sales by FAJFs are probably smaller than the actual values because of this problem. There are several other industry classification problems in our estimations. For example, since foreign firms supply legal and accounting services to Japan mainly through consulting firms, such activities are classified as "other business services" instead of "legal and accounting services."

13.3.4 Definition of Nationality

As we have already explained, Toyo Keizai adopts multiple criteria in the coverage of Japanese subsidiaries. For listed or unlisted but large subsidiaries, the cutoff capital participation rate is 20 percent. For unlisted and small subsidiaries, the cutoff rate is 49 percent. If we used these data without adjustment, we might obtain biased results. In order to solve this problem, we calculated two sets of estimations for JAFFs, one for JAFFs with a 49 percent and higher foreign capital participation rate, plus all the other establishments directly owned by foreign firms, and the other for JAFFs including all the JAFFs recorded in the Toyo Keizai database plus all the other establishments directly owned by foreign firms.

13.3.5 Cross-Border Transactions of Services by Affiliates

In our estimation, we did not take account of cross-border transactions of services by affiliates. Japanese affiliates of foreign firms provide services not only to Japanese customers but also to foreigners. Foreign affiliates of Japanese firms export their services to Japan. To get consistent statistics, we should subtract these values from sales by JAFFs and sales by FAJFs, respectively. Similarly, Japan's service imports include imports by JAFFs, and Japan's service exports include exports to FAJFs. To avoid double-counting and to make statistics of cross-border transactions of services consistent with our estimates of sales by affiliates, we should subtract these values from Japan's service imports and exports.[10] As table 13.3 shows, JAFFs and FAJFs in service sectors are quite active in international transactions. However, there are no data on what percentage of imports and exports by affili-

---

10. To be more rigorous, we should also take account of transactions among JAFFs and transactions among FAJFs. Kimura and Baldwin (1996) make this point.

Table 13.3   Cross-Border Transactions by Affiliates in Service Sectors, 1997 (%)

Transactions	Business and Personal Services	Transportation and Communication Services, etc.
Exports by JAFFs/Total Sales by JAFFs	3.9	26.8
Imports by JAFFs/Total Procurement by JAFFs	8.2	35.1
Exports to Japan by FAJFs/Total Sales by FAJFs	22.4	11.0
Imports from Japan by FAJFs/Total Procurement by FAJFs	11.0	13.3

Sources: MITI (1999a,b).

ates are service transactions, and there are no data at a more detailed industry classification level. Because of these deficiencies of the statistics, we could not adjust for this factor.

Table 13.4 presents the estimates of sales and employment by JAFFs and FAJFs. In order to compare our estimates on establishment transactions with Japan's cross-border transactions and the size of each industry, we adjusted the data of Japan's 1995 I-O tables to our definitions of sales and industry classifications. Table 13.5 presents data on Japan's cross-border transactions of services and sales and on employment of Japan's service industries. In the I-O tables, the output level of the financial sector is measured by imputed interests and financial transaction fees. We replaced this with the financial sector's total current income, which is reported in MOF's *Annual Report of Financial Institutions* (MOF various years) and the financial report of each firm.

The Japanese government estimates data on sectoral service trade for the I-O tables, using several sources, including balance-of-payments data for internal use, which are confidential and more detailed than publicly available statistics (Kuwabara 1989). In principle, I-O table data on services consist of "special trade (cross-border trade)" and "direct purchases" and do not include factor incomes, such as compensation of employees and construction services provided by nonresidents. For trade in construction services, we used data reported in the balance-of-payments statistics. We did not take account of compensation of employees because detailed industry level data were not available.[11]

In order to compare Japan's purchases of services from foreigners with U.S. purchases, we adjusted corresponding U.S. statistics for the year 1992,

---

11. According to Karsenty (2000), compensation of employees accounts for only 1.4 percent of world total international transactions in services. In several industries, however, such as amusement and recreation, this mode of transaction probably plays a substantial role.

Table 13.4 Japan's International Purchases and Sales of Private Services, 1995: Sales and Employment of Japanese Affiliates of Foreign Firms (JAFFs) and Foreign Affiliates of Japanese Firms (FAJFs) (millions of yen)

| | Japan's Purchases from JAFF and Employment by JAFF | | | | | | | | Sales Abroad and Employment by FAJF | |
| | Sales by: | | | No. of Workers Employed by: | | | | | | |
Industry	Japanese Subsidiaries of Foreign Firms (a)	Branches and Other Establishments of Foreign Firms (b)	JAFFs (a+b)	Japanese Subsidiaries of Foreign Firms (c)	Branches and Other Establishments of Foreign Firms (d)	JAFFs (c+d)	Sales by Majority-Owned Affiliates	No. of Workers Employed by Majority-Owned Affiliates	Sales by FAJFs, Branches and Other Establishments (e)	No. of Workers by FAJFs, Branches and Other Establishments
1 Construction and civil engineering	108,702	12,758	121,460	3,732	438	4,170	77,653	2,666	1,134,973	40,323
2 Electricity	0	0	0	0	0	0	0	0	5,679	210
3 Gas supply	114	0	114	5	0	5	114	5	1,084	40
4 Steam and hot water supply	0	0	0	0	0	0	0	0	0	0
5 Water supply	0	0	0	0	0	0	0	0	0	0
6 Sewerage systems	0	0	0	0	0	0	0	0	0	0
7 Sanitary services	985	0	985	43	0	43	0	0	64	14
8 Wholesale trade	905,849	102,752	1,008,601	73,424	8,309	81,733	856,791	69,428	3,653,874	296,165
9 Retail trade	28,499	3,240	31,739	6,555	732	7,287	26,226	6,019	260,861	60,000
10 Financial intermediary services	172,785	2,226,314	2,399,099	5,100	14,210	19,310	2,359,257	17,921	11,123,551	157,062
11 Life insurance	82,849	138,034	220,883	4,308	4,197	8,505	220,883	8,505	1,201,167	17,748
12 Casualty insurance	36,093	69,213	105,306	1,846	3,540	5,386	105,306	5,386	1,397,069	24,123
13 Real estate	5,204	5,284	10,487	65	66	131	10,087	126	422,193	12,925
14 Railway passenger transportation	0	0	0	0	0	0	0	0	8,287	29
15 Railway freight transportation	253	0	253	3	0	3	253	3	12,287	43
16 Road passenger transportation	0	0	0	0	0	0	0	0	0	0
17 Road freight transportation	44,691	1,181	45,871	530	14	544	45,871	544	93,096	4,165
18 Water transportation	189,465	49,263	238,728	2,111	552	2,663	230,887	2,570	297,059	12,967
19 Air transportation	255,995	681,959	937,954	3,144	8,306	11,450	915,946	11,189	187,522	6,474
20 Storage facility services	8,432	0	8,432	100	0	100	0	0	170,404	7,313
21 Supporting services for transport	40,703	53,800	94,503	1,743	2,018	3,761	94,632	3,501	545,166	26,035

(*continued*)

Table 13.4 (continued)

| Industry | Japan's Purchases from JAFF and Employment by JAFF | | | | | | | | Sales Abroad and Employment by FAJF | |
| | Sales by: | | | No. of Workers Employed by: | | | | | | |
	Japanese Subsidiaries of Foreign Firms (a)	Branches and Other Establishments of Foreign Firms (b)	JAFFs (a+b)	Japanese Subsidiaries of Foreign Firms (c)	Branches and Other Establishments of Foreign Firms (d)	JAFFs (c+d)	Sales by Majority-Owned Affiliates	No. of Workers Employed by Majority-Owned Affiliates	Sales by FAJFs, Branches and Other Establishments (e)	No. of Workers by FAJFs, Branches and Other Establishments
22 Postal service	0	0	0	0	0	0	0	0	0	0
23 Telecommunications	74,654	1,727	76,380	1,643	38	1,681	37,495	838	6,684	421
24 Broadcasting	29,171	0	29,171	642	0	642	4,544	100	6,702	361
25 Education	0	5,656	5,656	0	247	247	5,656	247	0	0
26 Research institute (natural sciences)	0	5,289	5,289	0	231	231	5,289	231	0	0
27 Research institutes (social sciences & humanities)	0	0	0	0	0	0	0	0	0	0
28 Research within firms	2,633	0	2,633	115	0	115	2,633	115	90,774	3,852
29 Medical services	3,934	328	4,262	336	28	364	2,318	104	7,810	322
30 Health and hygiene	0	0	0	0	0	0	0	0	582	98
31 Private non-profit organizations' services	96	0	96	6	0	6	96	6	178	39
32 Advertising	262,323	2,394	264,716	1,863	17	1,880	258,194	1,824	161,203	4,844
33 Computer programming and software	612,381	28,317	640,698	10,647	503	11,150	519,999	8,827	66,986	4,077

34 Information services	406,130	43,936	450,067	9,354	1,012	10,366	286,970	6,158	189,768	105,601
35 Goods and equipment rental and leasing	12,754	916	13,669	557	40	597	13,669	597	185,322	9,857
36 Automobile renting	1,076	0	1,076	47	0	47	1,076	47	7,356	590
37 Automobile repairing	206	572	778	9	25	34	778	34	9,097	1,070
38 Machine repairing	8,678	5,014	13,692	379	219	598	13,692	598	4,167	1,206
39 Building maintenance services	8,220	0	8,220	359	0	359	8,220	359	1,407	1,600
40 Legal and accounting services	0	0	0	0	0	0	0	0	128	28
41 Civil engineering and construction services	687	7,121	7,808	30	311	341	7,808	341	1,401	30
42 Personnel supply services	25,526	35,249	60,775	848	1,171	2,019	51,250	1,603	8,797	286
43 Other business services	126,308	59,528	185,836	5,115	2,467	7,582	162,210	6,528	3,625,729	34,694
44 Amusement and recreation services	47,930	34,398	82,328	673	483	1,156	82,328	1,156	71,646	5,889
45 Eating and drinking places	419,862	6,525	426,387	7,979	124	8,103	233,924	4,620	77,300	22,631
46 Hotels and lodging places	53,441	10,349	63,791	2,334	452	2,786	14,700	642	274,396	40,661
47 Individual educational facilities	29,184	3,084	32,268	1,268	134	1,402	8,662	371	730	76
48 Other personal services	36,149	234	36,382	2,011	13	2,024	36,199	2,016	5,866	877
49 Agricultural services	0	0	0	0	0	0	0	0	111,291	193
50 Ship repairing	0	74	74	0	3	3	0	0	40,370	4,009
51 Aircraft repairing	0	436	436	0	11	11	74	3	0	0
Total	4,041,960	3,594,945	7,636,905	148,923	49,911	198,834	6,702,126	165,238	25,470,031	908,948

Source: See appendix.

Notes: "Majority-owned foreign affiliates" refers to those affiliates in which foreign investors' ownership share is 49 percent or more.

Table 13.5   Japan's International Purchases and Sales of Private Services, 1995: Cross-Border Trade, Size of Industry, and "Revealed Comparative Advantage" (millions of yen)

Industry	Cross-Border Trade		Size of Industry		Japan's International Purchases and Sales		"Revealed Comparative Advantage"	
	Imports (f)	Exports (g)	Total Domestic Output (h)	No. of Employees	Purchases from Foreigners (a+b+f)[a]	Japan's Sales to Foreigners (e+g)[b]	(Sales by FAJFs – Sales by JAFFs)/ Total Domestic Output (%)	(Exports – Imports)/ Total Domestic Output (%)
1 Construction and civil engineering	301,900	620,000	88,149,287	7,046,117	423,360	1,754,973	1.150	0.361
2 Electricity	274	24,593	16,737,515	13,472	274	30,272	0.034	0.145
3 Gas supply	904	131	1,968,145	49,184	1,018	1,215	0.049	−0.039
4 Steam and hot water supply	0	0	104,384	1,778	0	0	0.000	0.000
5 Water supply	572	3,130	2,900,361	91,045	572	3,130	0.000	0.088
6 Sewerage systems	69	483	1,658,461	34,126	69	483	0.000	0.025
7 Sanitary services	0	415	3,094,654	256,638	985	479	−0.030	0.013
8 Wholesale trade	2,099,751	3,078,626	63,201,010	5,110,711	3,108,352	6,732,500	4.185	1.549
9 Retail trade	10,759	20,952	39,120,545	8,838,477	42,498	281,813	0.586	0.026
10 Financial intermediary services	1,676,742	999,376	56,272,142	1,375,573	4,075,841	12,122,927	15.504	−1.204
11 Life insurance	137,151	4,663	5,275,873	529,579	358,034	1,205,830	18.581	−2.511
12 Casualty insurance	60,894	78,437	3,250,105	191,173	166,200	1,475,506	39.745	0.540
13 Real estate	4,491	5,151	64,185,198	683,186	14,978	427,344	0.641	0.001
14 Railway passenger transportation	81,477	19,061	6,100,164	267,391	81,477	27,348	0.136	−1.023
15 Railway freight transportation	0	26	185,463	9,695	253	12,313	6.489	0.014
16 Road passenger transportation	127,869	21,092	10,184,846	667,492	127,869	21,092	0.000	−1.048
17 Road freight transportation	0	5,901	17,409,419	1,521,601	45,871	98,997	0.271	0.034
18 Water transportation	956,300	890,900	4,562,409	192,703	1,195,028	1,187,959	1.279	−1.433
19 Air transportation	1,119,200	343,500	2,414,322	57,735	2,057,154	531,022	−31.083	−32.129
20 Storage facility services	0	125	1,604,686	122,026	8,432	170,529	10.094	0.008
21 Supporting services for transport	1,437,067	1,279,547	7,652,467	467,136	1,531,570	1,824,713	5.889	−2.058
22 Postal service	7,413	9,201	2,142,138	194,657	7,413	9,201	0.000	0.083
23 Telecommunications	67,630	38,668	9,941,337	366,386	144,010	45,352	−0.701	−0.291
24 Broadcasting	0	16	2,679,336	69,143	29,171	6,718	−0.839	0.001
25 Education	156	36	22,229,403	2,441,916	5,812	36	−0.025	−0.001

26 Research institutes (natural sciences)	29,316	19,602	1,718,560	196,646	34,605	19,602	-0.308	-0.565
27 Research institutes (social sciences & humanities)	3,309	1,932	153,952	18,744	3,309	1,932	0.000	-0.894
28 Research within firms	0	0	9,145,081	578,465	2,633	90,774	0.964	0.000
29 Medical services	748	59	29,814,230	2,553,400	5,010	7,869	0.012	-0.002
30 Health and hygiene	0	0	692,307	73,680	0	582	0.084	0.000
31 Private non-profit organizations' services	39,342	47,139	4,658,723	522,564	39,438	47,317	0.002	0.167
32 Advertising	337,106	102,314	6,952,700	193,050	601,822	263,517	-1.489	-3.377
33 Computer programming and software	59,623	27,653	4,208,484	373,312	700,321	94,639	-13.632	-0.760
34 Information services	227,355	111,803	3,356,042	269,379	677,422	301,571	-7.756	-3.443
35 Goods and equipment rental and leasing	226,823	102,787	9,720,931	198,576	240,492	288,109	1.766	-1.276
36 Automobile renting	16	1	942,393	29,499	1,092	7,357	0.666	-0.002
37 Automobile repairing	236	120	6,845,341	668,227	1,014	9,217	0.122	-0.002
38 Machine repairing	6	1	5,960,245	229,443	13,698	4,168	-0.160	0.000
39 Building maintenance services	0	0	2,458,526	371,067	8,220	1,407	-0.277	0.000
40 Legal and accounting services	127,224	47,240	2,168,840	274,714	127,224	47,368	0.006	-3.688
41 Civil engineering and construction services	153,051	120,264	4,917,179	547,427	160,859	121,665	-0.130	-0.667
42 Personnel supply services	0	55	995,809	232,861	60,775	8,852	-5.220	0.006
43 Other business services	428,333	296,833	14,164,779	1,595,626	614,169	3,922,562	24.285	-0.928
44 Amusement and recreation services	218,910	26,493	13,517,060	846,133	301,238	98,139	-0.079	-1.424
45 Eating and drinking places	954,507	129,314	22,894,947	3,548,471	1,380,894	206,614	-1.525	-3.604
46 Hotels and lodging places	1,633,060	278,316	7,004,908	592,493	1,696,851	552,712	3.007	-19.340
47 Individual educational facilities	502	127	1,972,389	568,397	32,770	857	-1.599	-0.019
48 Other personal services	3,528	985	8,783,951	1,740,629	39,910	6,851	-0.347	-0.029
49 Agricultural services	0	0	676,113	88,664	0	111,291	16.460	0.000
50 Ship repairing	12,892	38,451	305,995	12,487	12,966	78,821	13.169	8.353
51 Aircraft repairing	10	8,408	160,514	4,046	446	8,408	-0.272	5.232
Total	12,546,516	8,803,927	597,213,669	46,926,940	20,183,421	34,273,958	2.986	-0.627

Source: See appendix.

[a] Purchases from foreigners are the sum of figures in columns (a) and (b) in table 13.4 and figures in column (f) in this table.

[b] Japan's sales to foreigners are the sum of figures in column (e) in table 13.4 and figures in column (g) in this table.

which are reported in U.S. Department of Commerce (1995a,c) to our definition of sales and industry classifications. The results are reported in table 13.6. We should note that U.S. data on inward direct investment cover all the subsidiaries that are more than 10 percent foreign-owned: that is, the coverage of U.S. data is broader than Japan's data in the case of purchases from affiliates. For United States-Japan comparison, we also prepared table 13.7 in which we compared sales and number of employees of majority-owned foreign affiliates in the U.S. and Japan. The U.S. data are taken from the U.S. Department of Commerce (1995b). Since the U.S. data are not available at the three-digit industry level, the United States-Japan comparison in table 13.7 is done at the more aggregated industry level.

13.4 An Overview of Japan's International Sales and Purchases of Services

According to our new statistics (tables 13.4 and 13.5), JAFFs in the service sector employed 199,000 workers in 1995, which is about three times greater than the number reported in MITI (1999b).

Imbalances between the activities of JAFFs and those of FAJFs are also smaller than those reported in the MOF FDI statistics. In terms of employment, the JAFF-FAJF ratio is 0.22 (199,000/909,000). In terms of sales, the ratio is 0.30 (7.6 trillion yen/25.5 trillion yen). The MOF statistics exaggerate the gap, probably for the following reasons.

First, during the second half of the 1980s, Japanese firms engaged in a large amount of FDI in the tertiary sector, especially in the United States. Stock and real estate bubbles in Japan at this period enabled real estate companies, general construction companies, institutional investors, and other small investors to borrow large funds to invest in foreign real estate (Wilkins 1990; Kenneth Leventhal and Company 1994). During this period, Japanese firms in the tertiary sector, especially banks and general construction companies, also expanded their business in purely domestic markets in foreign countries, such as retail banking in California or Britain or the development of shopping malls in the United States (Wilkins 1990; Graham and Krugman 1991). Because a substantial part of FDI in the real estate sector was conducted as portfolio investment, activities by affiliates measured by sales or employment are relatively small compared with capital flows. Moreover, although many of Japan's FDI projects in the tertiary sector resulted in failure afterward, withdrawals of equity investment or repayments of loans or bonds are not subtracted from the MOF statistics, which are gross data. These factors exaggerate Japan's outward FDI in the MOF statistics.

Second, as we have already pointed out, because of Japanese authorities' regulations, many foreign banks and insurance companies entered Japan through the setting up of branches instead of the founding of subsidiary companies. This fact makes their investment flows relatively small

Table 13.6 Purchases from Foreigners: U.S. (1992)–Japan (1995) Comparison

Industry	Ratio of Imports to Total Domestic Output		Ratio of No. of Workers Employed by Affiliates of Foreign Firms to Total No. of Workers (Inward FDI Penetration)			Ratio of Sales by Affiliates of Foreign Firms to Total Domestic Output		
	Japan (a)	United States (b)	Japan	Japan, Majority-owned	United States (more than 10% foreign owned)	Japan (c)	Japan, Majority-owned	United States (more than 10% foreign owned) (d)
1 Construction and civil engineering	0.003	0.000	0.001	0.000	0.020	0.001	0.001	0.029
2 Electricity	0.000	0.004	0	0	0.002	0	0	0.002
3 Gas supply	0.000	0	0.000	0.000	0.007	0.000	0.000	0.035
4 Steam and hot water supply	0	0	0	0	0.070	0	0	0.041
5 Water supply	0.000	0	0	0	0.087	0	0	0.015
6 Sewerage systems	0	0	0	0	0.087	0	0	0.015
7 Sanitary services	0	0	0.000	0.000	0.070	0.000	0.000	0.041
8 Wholesale trade	0.033	0.095	0.016	0.014	0.084	0.016	0.014	0.084
9 Retail trade	0.000	0	0.001	0.001	0.038	0.001	0.001	0.038
10 Financial intermediary services	0.030	0.003	0.014	0.013	0.066	0.043	0.042	0.066
11 Life insurance	0.026	0.005	0.016	0.016	0.143	0.042	0.042	0.072
12 Casualty insurance	0.019	0.005	0.028	0.028	0.143	0.032	0.032	0.072
13 Real estate	0.000	0	0.000	0.000	0.020	0.000	0.000	0.006
14 Railway passenger transportation	0.013	0.036	0	0	0	0	0	0
15 Railway freight transportation	0	0.036	0.000	0.000	0	0.001	0.001	0
16 Road passenger transportation	0.013	0.041	0	0	0.067	0	0	0.026
17 Road freight transportation	0	0.008	0.000	0.000	0.019	0.003	0.003	0.021
18 Water transportation	0.210	0.488	0.014	0.013	0.083	0.052	0.051	0.085
19 Air transportation	0.464	0.082	0.198	0.194	0.120	0.388	0.379	0.022
20 Storage facility services	0	0.008	0.001	0.000	0.019	0.005	0.000	0.021
21 Supporting services for transport	0.188	0.187	0.091	0.091	0.087	0.012	0.012	0.116
22 Postal service	0	0	0	0	0	0	0	0
23 Telecommunications	0.007	0.034	0.005	0.002	0.004	0.008	0.004	0.005
24 Broadcasting	0	0	0.009	0.001	0.013	0.011	0.002	0.061
25 Education	0.000	0.008	0.000	0.000	0.064	0.000	0.000	0.003
26 Research institutes (natural sciences)	0.017	0.008	0.001	0.001	0.064	0.003	0.003	0.003

(continued)

Table 13.6 (continued)

Industry	Ratio of Imports to Total Domestic Output		Ratio of No. of Workers Employed by Affiliates of Foreign Firms to Total No. of Workers (Inward FDI Penetration)			Ratio of Sales by Affiliates of Foreign Firms to Total Domestic Output		
	Japan (a)	United States (b)	Japan	Japan, Majority-owned	United States (more than 10% foreign owned)	Japan (c)	Japan, Majority-owned	United States (more than 10% foreign owned) (d)
27 Research institutes (social sciences & humanities)	0.021	0.008	0	0	0.064	0	0	0.003
28 Research within firms	0	0	0.000	0.000	0.048	0.000	0.000	0.038
29 Medical services	0.000	0.000	0.000	0.000	0.027	0.000	0.000	0.006
30 Health and hygiene	0	0.000	0	0	0.027	0	0	0.006
31 Private non-profit organizations' services	0.008	0	0.000	0.000	0	0.000	0.000	0
32 Advertising	0.048	0.004	0.010	0.009	0.075	0.038	0.037	0.011
33 Computer programming and software	0.014	0.002	0.030	0.024	0.041	0.152	0.124	0.042
34 Information services	0.068	0.002	0.038	0.038	0.041	0.134	0.086	0.042
35 Goods and equipment rental and leasing	0.023	0	0.003	0.003	0.054	0.001	0.001	0.074
36 Automobile renting	0.000	0	0.002	0.002	0.057	0.001	0.001	0
37 Automobile repairing	0.000	0.000	0.000	0.000	0.006	0.000	0.000	0.003
38 Machine repairing	0.000	0	0.003	0.003	0.029	0.002	0.002	0.081
39 Building maintenance services	0	0	0.001	0.000	0.078	0.003	0.003	0.049
40 Legal and accounting services	0.059	0.003	0.001	0.001	0.001	0.002	0.002	0.001
41 Civil engineering and construction services	0.031	0.005	0.001	0.001	0.014	0.002	0.002	0.004
42 Personnel supply services	0	0.017	0.009	0.007	0.068	0.061	0.051	0.054
43 Other business services	0.030	0.004	0.005	0.004	0.041	0.013	0.011	0.052
44 Amusement and recreation services	0.016	0.002	0.001	0.001	0.043	0.006	0.006	0.040
45 Eating and drinking places	0.042	0.021	0.002	0.001	0.027	0.019	0.010	0.019
46 Hotels and lodging places	0.233	0.196	0.005	0.001	0.100	0.009	0.002	0.120
47 Individual educational facilities	0.000	0	0.002	0.001	0.009	0.016	0.004	0.003
48 Other personal services	0.000	0.000	0.001	0.001	0.013	0.004	0.004	0.012
49 Agricultural services	0	0.001	0	0	0.008	0	0	n.a.
50 Ship repairing	0.042	0.015	0.000	0.000	0.024	0.000	0.000	0.028
51 Aircraft repairing	0.000	0.119	0.003	0.003	0.041	0.003	0.003	0.027
Total (weighted average)	0.018	0.021	0.005	0.005	0.044	0.010	0.008	0.040

Industry	Ratio of Total Purchases from Foreigners to Total Domestic Output (Foreign Sales Penetration)		Share of Imports in Total Purchases from Foreigners		FDI Restrictiveness Index	
	Japan $(a+c)$	United States $(b+d)$	Japan $(a/[a+c])$	United States $(b/[b+d])$	Japan	United States
1 Construction and civil engineering	0.005	0.030	0.713	0.013	0.000	0.050
2 Electricity	0.000	0.006	1	0.631	1.000	0.300
3 Gas supply	0.001	0.035	0.888	0	1.000	0.217
4 Steam and hot water supply	0	0.041	n.a.	0	0.625	0.300
5 Water supply	0.000	0.015	1	0	0.250	0.300
6 Sewerage systems	0.000	0.015	1	0	0.000	0.050
7 Sanitary services	0.000	0.041	0	0	0.250	0.050
8 Wholesale trade	0.049	0.178	0.676	0.530	0.250	0.098
9 Retail trade	0.001	0.038	0.253	0.037	0.250	0.098
10 Financial intermediary services	0.072	0.069	0.411	0.037	0.500	0.525
11 Life insurance	0.068	0.077	0.383	0.064	0.500	0.264
12 Casualty insurance	0.051	0.077	0.366	0.064	0.500	0.264
13 Real estate	0.000	0.006	0.300	0	0.000	0.050
14 Railway passenger transportation	0.013	0.036	1	1	1.000	0.050
15 Railway freight transportation	0.001	0.036	0	1	1.000	0.050
16 Road passenger transportation	0.013	0.067	1	0.608	1.000	1.000
17 Road freight transportation	0.003	0.028	0	0.269	0.625	0.775
18 Water transportation	0.262	0.574	0.800	0.852	1.000	1.000
19 Air transportation	0.852	0.103	0.544	0.791	1.000	1.000
20 Storage facility services	0.005	0.028	0	0.269	0.250	1.000
21 Supporting services for transport	0.200	0.303	0.938	0.618	0.533	0.797
22 Postal service	0.003	0	1	n.a.	1.000	0.763
23 Telecommunications	0.014	0.039	0.470	0.862	0.750	0.525
24 Broadcasting	0.011	0.061	0	0	1.000	0.406
25 Education	0.000	0.012	0.027	0.723	0.150	0.680
26 Research institutes (natural sciences)	0.020	0.012	0.847	0.723	1.000	1.000
27 Research institutes (social sciences & humanities)	0.021	0.012	1	0.723	0.000	1.000
28 Research within firms	0.000	0.038	0	0	1.000	1.000

Table 13.6 (continued)

Industry	Ratio of Total Purchases from Foreigners to Total Domestic Output (Foreign Sales Penetration)		Share of Imports in Total Purchases from Foreigners		FDI Restrictiveness Index	
	Japan (a+c)	United States (b+d)	Japan (a/[a+c])	United States (b/[b+d])	Japan	United States
29 Medical services	0.000	0.006	0.149	0.004	1.000	0.860
30 Health and hygiene	0	0.006	n.a.	0.004	0.000	0.050
31 Private non-profit organizations' services	0.008	0	0.998	n.a.	1.000	1.000
32 Advertising	0.087	0.016	0.560	0.282	0.000	0.050
33 Computer programming and software	0.166	0.044	0.085	0.041	0.250	0.288
34 Information services	0.202	0.044	0.336	0.041	0.167	0.208
35 Goods and equipment rental and leasing	0.025	0.074	0.943	0	0.500	0.549
36 Automobile renting	0.001	0.025	0.015	0	0.000	0.050
37 Automobile repairing	0.000	0.003	0.233	0.017	0.250	0.050
38 Machine repairing	0.002	0.081	0.000	0	0.500	0.525
39 Building maintenance services	0.003	0.049	0	0	0.000	0.050
40 Legal and accounting services	0.059	0.003	1	0.829	0.250	0.217
41 Civil engineering and construction services	0.033	0.009	0.951	0.561	0.125	0.050
42 Personnel supply services	0.061	0.071	0	0.236	0.625	0.050
43 Other business services	0.043	0.057	0.697	0.079	0.345	0.401
44 Amusement and recreation services	0.022	0.043	0.727	0.056	0.063	0.169
45 Eating and drinking places	0.060	0.040	0.691	0.513	0.125	0.050
46 Hotels and lodging places	0.242	0.316	0.962	0.621	0.000	0.050
47 Individual educational facilities	0.017	0.003	0.015	0	1.000	1.000
48 Other personal services	0.005	0.013	0.088	0.029	0.500	0.525
49 Agricultural services	0	n.a.	n.a.	n.a.	1.000	0.525
50 Ship repairing	0.042	0.043	0.994	0.353	1.000	0.525
51 Aircraft repairing	0.003	0.146	0.022	0.816	0.000	0.050
Total (weighted average)	0.028	0.061	0.445	0.206	0.493	0.419

Source: See appendix.

Notes: "Majority-owned foreign affiliates" refers to those affiliates in which foreign investors' ownership share is 49 percent or more. The correlation coefficient between foreign sales penetration ratio in Japan and the United States is 0.4205. n.a. = not available.

Table 13.7 Sales and Number of Employees of Majority-Owned Foreign Affiliates: U.S. (1992)–Japan (1995) Comparison

Fukao-Ito Industry Code	Sectors	Ratio of No. of Workers Employed by Majority-Owned Foreign Affiliates to Total No. of Workers		Ratio of Sales by Majority-Owned Foreign Affiliates to Total Domestic Output	
		Japan	United States	Japan	United States
1	Construction	0.000	0.010	0.001	0.016
8	Wholesale trade	0.014	0.067	0.014	0.067
9	Retail trade	0.001	0.033	0.001	0.033
10	Finance, except depository institutions[a]	0.013	0.012	0.042	n.a.
13	Real estate	0.000	0.028	0.000	0.013
14, 15, 16, 17, 18, 19, 20, 21	Transportation	0.005	0.022	0.026	0.027
	Services				
46	Hotels and other lodging places	0.002	0.021	0.011	0.019
33, 34	Computer and data processing services	0.001	0.073	0.002	0.094
44	Motion pictures, including television tape and film	0.023	0.014	0.107	0.020
29, 30	Health services	0.001	0.038	0.006	0.098
28, 32, 35, 36, 37, 38, 39, 40, 41, 42, 43	Business services	0.000	0.007	0.000	n.a.
45, 47, 48	Other services	0.002	0.032	0.008	0.019
	Total (Weighted-average)	0.001	0.005	0.008	n.a.
		0.004	0.028	0.012	0.027

Sources: U.S. Department of Commerce (1995b); see also tables 13.4 and 13.5.

Note: "Majority-owned foreign affiliates" refers to those affiliates in which foreign investors' ownership share is 49 percent or more for Japan, and 50 percent or more for the United States. n.a. = not available.

[a]The ratio of sales for the U.S. financial sector was not calculated because definitions of output in U.S. I-O tables differ from those of sales in U.S. establishment data. For details, see appendix.

compared with the actual sizes of their affiliates' activities measured by sales or employment.

Using table 13.6, we can compare Japan's and the U.S.'s purchases of services from foreigners. For the service sector as a whole, Japan's ratio of imports to total domestic output is 1.8 percent, which is almost at the same level as the corresponding U.S. ratio, 2.1 percent. However, in the case of purchases from majority-owned foreign affiliates (table 13.7), Japan's ratio of purchases from affiliates to total domestic output is 1.2 percent, which is less than half of the corresponding U.S. ratio of 2.7 percent. In terms of employment, Japan's ratio of the number of workers employed by majority-owned foreign affiliates to the total number of workers is 0.4 percent, which is one-seventh of the corresponding U.S. ratio of 2.8 percent. It seems that Japan's market for services is more closed for establishment transactions than for cross-border transactions.

In order to test whether Japan's market for services is more closed for establishment transactions than for cross-border transactions, we estimated gravity models for both the direction of U.S. service exports and the regional distribution of sales of services by U.S. firms' foreign affiliates.[12,13] The results are summarized in table 13.8. The dependent variables are the logarithm of U.S. exports and sales by affiliates. As explanatory variables, we use the logarithm of each country's gross domestic product (GDP), the logarithm of per capita GDP, the logarithm of distance from the United States, and a dummy for Japan. The equations are estimated for 1992 and 1997. The Japan dummies are not significant both in the U.S. export equations and in sales-by-affiliates equations. In other words, we cannot conclude that Japan's market for services is significantly more closed to sales by U.S. firms than other countries' markets. However, it seems that the signs of the estimated coefficients of Japan dummies are consistent with our findings from the United States-Japan comparison based on table 13.6 and table 13.7. The coefficients of the Japan dummies take a positive value in the case of the export equations and a negative value in the case of equations for sales by affiliates. The results imply that Japan's purchases of services through establishment transactions from U.S. firms in 1997 were about 50 percent less than the predicted value.

Next, we study Japan's purchases of services from foreigners by industry. Figure 13.1 shows the industry composition of Japan's purchases. Purchases are concentrated in a limited number of industries. Four indus-

---

12. There are several empirical studies that estimated an econometric model explaining the regional distribution of U.S. direct investment abroad and found that a Japan dummy is negative and significant. However, these studies are based either on data on FDI in manufacturing industries (Grubert and Mutti 1991) or on data on FDI in all the industries (Eaton and Tamura 1994). On this issue, also see Lawrence (1993) and Development Bank of Japan (1997).

13. Francois (1999) estimates gravity models for the direction of U.S. exports of business and financial services and construction services.

Table 13.8   **Determinants of U.S. Cross-Border Sales of Services and Sales of Services by Foreign Affiliates of U.S. Firms: Cross-Country Estimation Based on Gravity Models**

	1992		1997	
	ln(EX92)	ln(OFDI92)	ln(EX97)	ln(OFDI97)
ln(GDP92)	0.5577	0.6543		
	(5.279)***	(2.701)**		
ln(GDPPC92)	0.1783	0.7330		
	(2.180)**	(3.394)***		
ln(GDP97)			0.6054	0.6441
			(6.187)***	(3.742)***
ln(GDPPC97)			0.1897	0.6973
			(2.523)**	(5.432)**
ln(DIST)	−0.4460	0.3503	−0.3305	−0.0184
	(−1.747)*	(0.480)	(−1.532)	(−0.036)
DJPN	0.7112	−0.6982	0.4637	−0.6018
	(1.093)	(−0.567)	(0.810)	(−0.666)
_cons	8.3935	−0.8909	7.3418	2.9577
	(3.217)***	(−0.117)	(3.284)***	(0.558)
N	32	25	32	25
F	21.23***	11.36***	22.59***	17.05***
Adj. R^2	0.723	0.6333	0.7358	0.7279

Sources: U.S. Department of Commerce (1999); IMF (various issues).

Note: t-statistics are in parentheses. Definition of variables: EX92 = U.S. cross-border sales of services in 1992; OFDI92 = Sales of services by foreign affiliates of U.S. firms in 1992; EX97 = U.S. Cross-border sales of services in 1997; OFDI97 = Sales of services by foreign affiliates of U.S. firms in 1997; GDP92 = 1992 nominal GDP in U.S. dollars; GDPPC92 = 1992 nominal GDP per capita in U.S. dollars; GDP97 = 1997 nominal GDP in U.S. dollars; GDPPC97 = 1997 nominal GDP per capita in U.S. dollars; DIST = Distance between each country's capital city and Washington D.C.; DJPN = Japan dummy.

***p = .01 (two-tailed test)
**p = .05 (two-tailed test)

tries—financial intermediary services, wholesale trade, air transportation, and hotels and lodging places—account for 54 percent of Japan's total purchases of services from foreigners. In the case of financial services, most foreign banks and insurance companies entered Japan through setting up branches (see table 13.4). In 1995, Citibank employed 1,100 workers and earned an annual current income of 326 billion yen. Goldman Sachs Ltd. And Salomon Brothers Asia Ltd. employed 510 and 450 workers, respectively, at their Tokyo branches. Almost all the air passenger transportation services by foreign firms are conducted through their Japanese branches. However, in the case of airfreight transportation and water transportation there are several large affiliates. In 1995, Federal Express Japan and United Parcel Service Yamato employed 852 and 650 workers, respectively. A European water transportation company, Maersk, employed 360 workers.

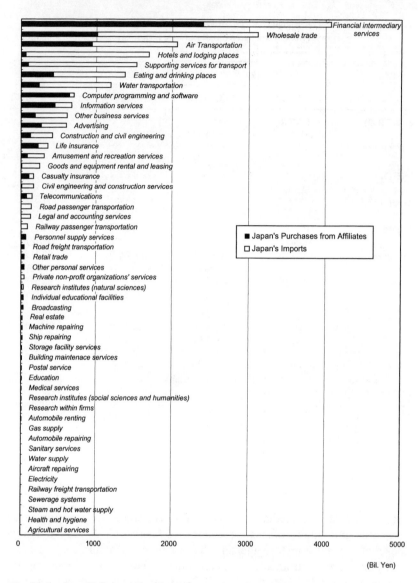

Fig. 13.1   Japan's international purchases of services, 1995
Source: Tables 13.4 and 13.5.

Foreign manufacturing firms set up large wholesale affiliates in order to promote their sales in Japan. For example, Caterpillar Mitsubishi Construction Machinery employed 2,235 workers at its wholesale affiliates. Kodak Japan Ltd. employed 1,078 workers.

In figure 13.2, we compare Japan's and the United States' sectoral importance of purchases from foreigners, which we measure by a ratio of to-

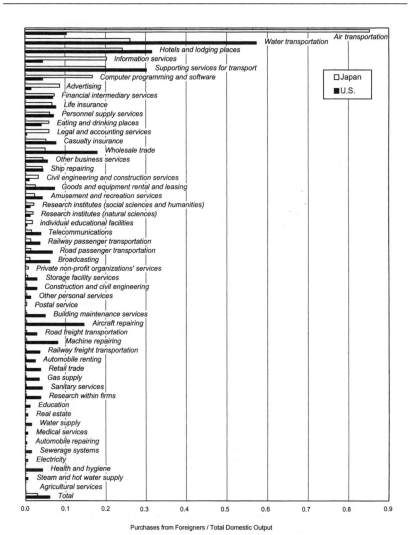

Fig. 13.2   **Purchases from foreigners: Japan (1995) and United States (1992) comparison**
Source: Table 13.6.

tal purchases from foreigners to total domestic output. In Japan, differences in this ratio among industries are more remarkable than in the United States. Japan's variation coefficient of this ratio among industries is 2.42, compared to a variation coefficient of only 1.59 for the United States.

Figure 13.3 shows Japan's "revealed comparative advantage" measured as the ratio of net exports to total domestic output and the ratio of net purchases from affiliates (sales by FAJFs minus sales by JAFFs) to total domestic output. According to figure 13.3, Japan is most competitive in in-

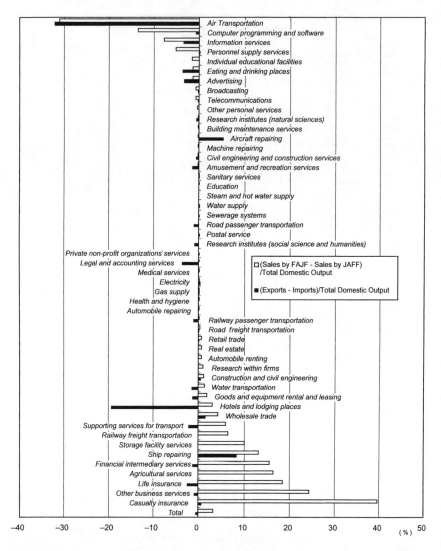

Fig. 13.3   Japan's "revealed comparative advantage"
Source: Table 13.5.

dustries that support Japan's international activities, such as casualty and
life insurance, other business services, agricultural services,[14] and financial
intermediary services. Among all of Japan's FDI, investment in these kinds
of supporting industries for Japan's international activities has the longest
history. Japan's large trading companies (*sogo shosha*), banks, insurance
companies, and transportation companies started their FDI before the sec-

14. Japan's large trading companies (*sogo shosha*) own several warehouse companies in the
United States for imports of agricultural products.

ond World War. The Japanese government sometimes backed up this type of investment. Figure 13.3 also shows that Japan is least competitive in air transportation, computer programming and software, and information services, both in international trade and in establishment transactions.

As we have already seen, for the service sector as a whole Japan's ratio of the number of workers employed by majority-owned foreign affiliates to the total number of workers is one-seventh of the corresponding U.S. ratio. Among our fifty-one service-sector categories, in which categories is the Japanese market more closed to international establishment transactions than the U.S. market? Figure 13.4 shows the differences in Japan's inward FDI penetration and the corresponding U.S. penetration by industry. In order to minimize the bias in our cross-industry comparisons, we use the data for majority-owned affiliates for Japan's penetration. We should note that the U.S. data cover all affiliates whose foreign ownership ratio is 10 percent or higher. There are some similarities between figure 13.3 and figure 13.4. Japan has a higher penetration ratio than the United States in air transportation, computer programming and software, and information services. Japan has a lower penetration ratio than the United States in casualty and life insurance, financial intermediary services, hotels and lodging places, and supporting services for transport.

So far, our analysis was static and mainly based on Japan's 1995 data. We should note, however, that FDI into Japan is growing at amazing speed. Table 13.9 shows MOF statistics on FDI flows into Japan. According to the statistics, the inward direct investment stock in Japan's nonmanufacturing sector has grown eightfold in the last ten years. The total of FDI flows in the last three years is greater than the FDI stock at the end of the 1997 fiscal year. In recent years, the number of cases of cross-border M&A has been increasing especially.[15] In 1999, AT&T and British Telecom jointly bought a combined 30 percent share of Nippon Telecom. A British company, Cable and Wireless, acquired International Digital Communications (IDC) by a takeover bid.

Probably the following two factors have contributed to the recent increase of inward FDI. First, in recent years, the Japanese government promoted important deregulatory and related measures in order to transform Japan's socioeconomic system into a new system that is more open to the international community and based on the rules of self-responsibility and market principles. As a part of this deregulation program, the Japanese government alleviated or abolished several regulations on inward FDI. For example, all restrictions on foreign ownership and on foreign board members in Type 1 telecommunications carriers (except for Nippon Telegraph and Telephone and Kokusai Denshin Denwa) including their radio station licenses, removed in 1998. In 1999, all restrictions on foreign capital and the appointment of foreign directors in all cable television businesses were re-

---

15. According to MITI (2000), there were 129 investments into Japan through cross-border mergers and acquisitions in 1999.

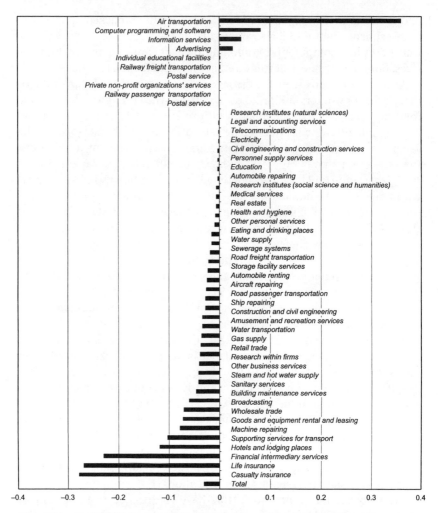

Fig. 13.4   Share of workers employed by affiliates: Japan (1995) and United States (1992) comparison ([Japan's no. of workers employed by majority-owned affiliates/Japan's total no. of workers] minus [no. of U.S. workers employed by affiliates/total no. of U.S. workers])

Source: Table 13.6.

moved. Second, the recent stagnation of Japan's land and stock prices created a kind of "fire-sale" situation, from which foreign investors benefited.[16]

As we have seen in section 13.2, MOF FDI statistics are not appropriate measures for JAFFs' activities. Therefore, using Toyo Keizai data, we compared JAFFs' employment in 1997 with that in 1990. Table 13.10 and figure

16. For more detail on Japan's recent deregulation measures, see Japan Investment Council (various years).

Table 13.9   FDI Flows into Japan (billions of yen)

Fiscal Year	1950–90	1991	1992	1993	1994	1995	1996	1997	1998	1999	2000	Total
Construction	12.9	3.1	0.0	0.1	0.4	0.1	0.0	0.3	1.4	2.2	0.0	20.5
Real estate	115.8	9.4	30.7	10.7	3.2	1.6	26.5	48.2	41.6	16.8	34.6	339.0
Commerce	416.6	107.3	155.4	100.5	113.5	67.9	166.4	99.6	175.9	348.5	276.1	2,027.8
Business and personal services	150.3	73.7	106.7	24.0	37.4	49.1	236.0	88.8	318.1	205.8	236.5	1,526.4
Transportation services	19.8	3.5	2.5	5.1	0.8	1.2	1.0	0.4	6.1	2.2	5.7	48.3
Communication services	20.8	13.6	6.3	3.2	3.0	5.3	2.1	3.3	16.8	330.0	750.8	1,155.1
Finance and insurance	96.4	120.3	19.0	4.0	68.7	100.1	27.3	161.6	456.9	511.5	1,029.3	2,595.2
Others	110.4	1.1	1.8	27.4	0.3	3.2	0.2	8.7	11.1	2.5	1.3	168.0
Nonmanufacturing total	942.7	331.9	322.5	175.0	227.3	228.4	459.5	410.8	1,027.8	1,419.6	2,334.4	7,880.0
Manufacturing	1,666.5	257.7	208.1	183.6	205.4	141.2	311.1	267.4	312.6	979.7	790.7	5,324.0
Total amount	2,608.5	589.6	530.6	358.6	432.7	369.7	770.7	678.2	1,340.4	2,399.3	3,125.1	13,203.3

Sources: MOF (1999); [http://www.mof.go.jp].

Note: FDI flows approved or notified from 1950 onward.

Table 13.10   **Recent Trends in JAFF's Employment and Japan's Imports, 1990–97 (millions of yen)**

Industry	No. of JAFF 1990	No. of JAFF 1997	No. of Workers Employed by JAFF 1990	No. of Workers Employed by JAFF 1997	Imports 1990	Imports 1997
Agriculture	2	1	154	198	2,825,836	2,863,929
Mining	0	2	0	70	7,735,520	8,185,535
Manufacturing	965	828	314,299	286,933	23,265,941	32,849,284
Services and others	2,181	2,456	150,206	203,940	9,253,169	7,984,945
Construction	13	18	2,070	2,026	n.a.	660,100
Wholesale trade	1,321	1,380	75,575	78,900	327,447	350,615
Retail trade	23	46	2,065	10,910	n.a.	n.a.
Finance	215	248	19,949	25,356	700,947	1,090,322
Insurance	22	37	11,970	14,298	54,476	246,100
Real estate	13	12	85	115	7,726	n.a.
Eating and drinking places	13	11	5,281	8,388	813,644	127,748
Advertising	23	25	1,864	4,912	289,852	295,448
Electricity	0	0	0	0	2,249	n.a.
Gas and steam supply	1	2	4	15	697	n.a.
Water supply	0	0	0	0	955	n.a.
Sanitary services	1	4	0	44	0	n.a.
Transportation	62	58	4,851	8,088	2,306,259	1,612,056
Supporting services for transport	55	17	1,884	1,097	167,769	72,807
Telecommunications	13	36	815	2,926	47,036	180,270
Broadcasting	0	1	0	6	153	n.a.
Research institutes	5	1	283	n.a.	17,597	17,980
Medical and health services	8	10	170	677	930	n.a.
Private nonprofit organizations' services	0	1	0	6	28,108	28,463
Information services[a]	172	326	11,378	25,676	218,713	n.a.
Goods and equipment rental and leasing	2	3	371	489	151,981	241,400
Other business services	169	173	6,025	13,455	385,959	815,999
Amusement and recreation services	14	12	622	1,807	266,458	205,003
Hotels and lodging places	11	14	1,603	1,655	1,478,421	341,682
Other personal services	18	21	3,166	3,094	7,823	1,722
Not classified	7		175		1,977,969	1,697,230
Total	3,148	3,287	464,659	491,141	43,080,466	51,883,693

Sources: Toyo Keizai Shinpo-sha (various years); Economic Planning Agency (1999); Japanese Government (1994).

Notes: The correlation coefficient between the percentage change in the number of employees and the percentage change in imports (1990–97) is 0.3534. The data on JAFFs partially cover Japanese branches and other establishments directly owned by foreign firms. n.a. = not available.

[a]Information services imports for 1997 are included in other business services.

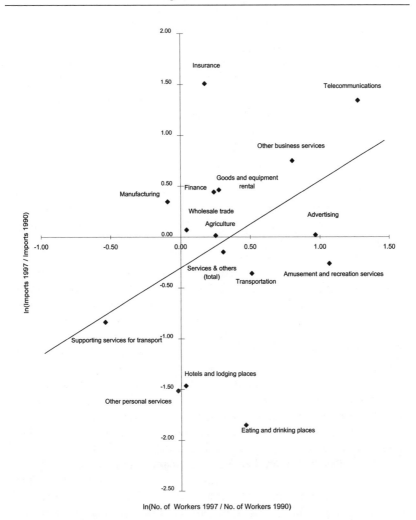

Fig. 13.5   Recent trends in JAFF's employment and Japan's imports: 1990–97
Source: Table 13.10.

13.5 show changes in the number of workers employed by JAFFs and changes in Japan's imports of services. According to table 13.10, the number of workers employed by JAFFs in nonmanufacturing sectors excluding primary industries increased by 36 percent, which is substantially smaller than MOF FDI statistics indicate.[17] According to MOF statistics, inward FDI stocks tripled from the end of 1990 to the end of 1997. Probably, MOF statistics exaggerate the increase of JAFFs' activities in recent years.

17. On the other hand, U.S. firms, for example, increased their sales of services through their affiliates in Japan by 122 percent in this period (U.S. Department of Commerce 1999).

According to table 13.10 and figure 13.5, increases of JAFFs' employment in service sectors are quite uneven among industries. Japanese affiliates of foreign firms' employment in retail trade, advertising, telecommunications, information services, and other business services has doubled, whereas that in wholesale trade, hotels and lodging places, and insurance industries was relatively stagnant.

13.5   Econometric Analysis of Determinants of Inward Foreign Direct Investment Penetration

As we have seen in the previous section, there are significant differences in inward FDI penetration in the various service industries. What industry characteristics affect the inward FDI penetration of each industry? In this section, we conduct an empirical study on this issue.

This type of cross-industry analysis on FDI into Japan has been conducted by Lawrence (1993); Weinstein (1996); Nakamura, Fukao, and Shibuya (1997); and Horaguchi (1995).[18] One of the most hotly debated issues in these studies was whether Japan's *keiretsu* relationships impede inward FDI. It has been argued that *keiretsu* relationships reduce inward FDI through cross-shareholdings and long-term supplier relationships. Using MITI (1991) data on only ten industries, Lawrence (1993) did a cross-industry regression and found that *keiretsu* relationships significantly impeded inward FDI. By constructing a panel data based on MOF data, Weinstein (1996) conducted a similar kind of regression and found that the coefficient on the shares of financial group member sales in each sector is negative but not significant in many cases. By using their newly compiled statistics on Japan's inward FDI penetration (the share of sales by JAFFs in total sales) in fifty-eight manufacturing industries from microdata of MITI's *Kigyo Katsudo Kihon Chosa* (*Basic Survey on Business Activities by Enterprises*), Nakamura, Fukao, and Shibuya (1997) conducted a cross-industry regression. They found that sales concentration as measured by the Herfindahl index has significant negative effects on Japan's inward FDI penetration, whereas capital intensity and skilled-worker intensity have significant positive effects on the FDI penetration. They also found that *keiretsu* variables and a government barrier dummy variable based on the OECD (various issues) do not have a significant effect on FDI penetration. Horaguchi (1995) also found that a coefficient on the *keiretsu* share was not significant.

These previous empirical studies mainly focused on the manufacturing sectors. No empirical analysis on inward FDI penetration in the service sectors has been conducted. The lack of analysis on the service sectors is probably due to the deficiency of data, as we have already suggested in section 13.2.

18. In the case of FDI into the United States, Ray (1989), Kogut and Chang (1991), and Pugel, Kragas, and Kimura (1996) conducted similar types of cross-industry analyses.

In this section we estimate an empirical model that explains the determinants of Japan's inward FDI penetration. The variables of this estimation are defined in table 13.11, and the estimation results are shown in table 13.12. Further details on the definitions and sources of the variables are provided in the appendix. We use Japan's FDI penetration ratio in the ser-

Table 13.11   Definition of Variables for Analysis on Inward FDI Penetration

Dependent Variable

Japan's inward FDI penetration FDIJA	Share of workers employed by majority-owned JAFF in Japan's total workers: 1995	

Independent Variables		Expected Sign of Coefficients
U.S. inward FDI penetration FDIUS	Share of workers employed by foreign firms' U.S. affiliates in U.S. total workers: 1992	+
FDI Restrictiveness RINVJAUS	Japan's FDI restrictiveness minus U.S. FDI restrictiveness: 1994	–
Public Services PUBEMP	Share of workers employed by local or central governments in Japan's total workers: 1996	–
Productivity DPROD	Japan's productivity level (United States = 1): 1990	–/+
Locational advantage LAND	Land intensity (land input [book value] per employee: industry average: 1995)	–
UNIV	Skilled-labor intensity (share of university graduates in total workers: 1992)	+
Labor market structure JOBSEP	Job separation rate: 1995	+
Advantages in the managerial resources ADINT	Advertisement intensity (ratio of advertising expenses to the gross value-added: 1995)	+
RDINT	R&D intensity (ratio of R&D expenses to the gross value-added: 1995)	+
Keiretsu KRETS	Share of workers employed by horizontal or vertical *keiretsu* firms in total workers: 1998	–
HORIZ	Share of workers employed by horizontal *keiretsu* firms in total workers: 1998	–
VERT	Share of workers employed by vertical *keiretsu* firms in total workers: 1998	–

Source: See appendix.

Note: "Majority-owned foreign affiliates" refers to those affiliates in which foreign investors' ownership share is 49 percent or more.

Table 13.12   **Determinants of Japan's Inward FDI Penetration: Tobit Estimation**

	Japan's Inward FDI Penetration (Dependent variable: FDIJA)					
	Eq. (1)	Eq. (2)	Eq. (3)	Eq. (4)	Eq. (5)	Eq. (6)
FDIUS	0.280	0.301	0.326	0.160	0.162	0.171
	(2.454)**	(2.512)**	(2.405)**	(3.367)***	(2.968)***	(2.685)***
RINVJAUS	−0.026	−0.026	−0.027	−0.016	−0.016	−0.017
	(−2.306)**	(−2.354)**	(−2.427)**	(−1.696)*	(−1.668)*	(−1.666)*
PUBEMP	−0.084	−0.090	−0.091	−0.046	−0.047	−0.048
	(−3.079)***	(−2.951)***	(−3.060)***	(−2.218)**	(−2.055)**	(−2.029)**
DPROD	0.041	0.043	0.042	0.020	0.020	0.020
	(2.821)***	(2.781)***	(2.937)***	(1.894)*	(1.791)*	(1.800)*
LAND	−0.058	−0.072	−0.047	−0.088	−0.089	−0.082
	(−0.720)	(−0.849)	(−0.577)	(−1.497)	(−1.441)	(−1.384)
UNIV	−0.057	−0.058	−0.057	−0.011	−0.012	−0.012
	(−1.621)	(−1.638)	(−1.691)*	(−0.637)	(−0.630)	(−0.639)
JOBSEP	−0.657	−0.641	−0.898	−0.403	−0.402	−0.473
	(−1.565)	(−1.496)	(−1.792)*	(−1.535)	(−1.540)	(−1.486)
ADINT	1.527	1.521	1.550	0.730	0.731	0.748
	(2.751)***	(2.818)***	(2.978)***	(2.557)**	(2.544)**	(2.524)**
RDINT	−0.161	0.053	−0.234	−0.104	−0.086	−0.170
	(−0.603)	(0.150)	(−0.739)	(−0.569)	(−0.468)	(−0.822)
KRETS		−0.016			−0.001	
		(−1.169)			(−0.178)	
HORIZ			−0.076			−0.017
			(−1.612)			(−0.756)
VERT			0.023			0.009
			(1.072)			(1.050)
_cons	−0.031	−0.031	−0.026	−0.014	−0.014	−0.013
	(−1.957)*	(−1.935)*	(−1.665)*	(−1.922)*	(−1.880)*	(−1.750)*
N	45	45	45	44	44	44
Wald χ^2	18.40**	20.65**	23.53**	26.33***	30.09***	30.91***
Log likelihood	80.450	80.753	81.853	97.495	97.501	97.728

Notes: The numbers in parentheses are z-statistics based on the Huber-White-sandwich robust standard errors. The following six industries are excluded from the estimations due to the availability of some variables: postal service, education, research institutes (natural sciences), research institutes (social sciences and humanities), health and hygiene, and private nonprofit organizations' services. The air transportation industry is excluded from the estimations for equations (4), (5), and (6).

***$p = .01$ (two-tailed test)
**$p = .05$ (two-tailed test)
*$p = .10$ (two-tailed test)

vice industries as the dependent variable.[19] Japan's FDI penetration is defined by Japan's ratio of the number of workers employed by majority-owned foreign affiliates to the total number of workers.

19. On the theoretical foundation of cross-industry estimation, see Kogut and Chang (1991), Petri (1991), and Lawrence (1993). On *keiretsu,* also see Saxonhouse (1993).

In order to control for differences in the tradability of different services, we used FDIUS (U.S. inward FDI penetration). We expect a positive coefficient for this variable.

To know the effects of government regulations on inward FDI, we prepared the variable RINVJAUS (Japan's FDI restrictiveness minus U.S. FDI restrictiveness). Following Hoekman (1996), we compiled a frequency measure for FDI restrictiveness at the three-digit industry level, using data from General Agreement on Trade in Services (GATS) schedules for Japan and the United States (World Trade Organization [WTO] 1997), Asia-Pacific Economic Cooperation (APEC; 1996), OECD (various issues), Japan Investment Council (various years), and the Japanese Government (various years). The two countries' FDI restrictiveness indexes are reported in table 13.6. RINVJAUS is defined as the difference between Japan's and the United States' FDI restrictiveness. We expect a negative coefficient for this variable. Inward FDI in an industry will be limited, if establishments owned by government dominate the industry. To study this effect, we used the variable PUBEMP (share of workers employed by local or central government). We expect a negative coefficient for PUBEMP.

In cases in which cross-border transactions of services are not difficult, multinational corporations will choose the location where the production costs are lowest.[20] Therefore, the inward FDI penetration ratio will be affected by Japan's locational advantage for each industry. Because Japan's land prices and wages of unskilled workers are relatively high, Japan probably has a locational disadvantage for land-intensive or unskilled-worker-intensive industries. Consequently, we would expect a positive coefficient for UNIV (skilled-labor intensity) and a negative coefficient for LAND (land intensity). It has been argued that firm-specific skills play a more important role in Japanese firms and that this feature has hindered the development of the secondary labor market in Japan. This fact might impede the new entry of foreign firms (Weinstein 1996). In order to take this factor into account, we prepared JOBSEP (job separation rate). We expect a positive coefficient for this variable.

In order to take into account the effects of *keiretsu,* we used three *keiretsu* variables, HORIZ (the share of workers employed by horizontal *keiretsu* firms), VERT (the share of workers employed by vertical *keiretsu* firms), and KRETS (the share of workers employed by horizontal or vertical *keiretsu* firms). If *keiretsu* impede inward FDI, we will have negative coefficients.

The standard FDI theory (see, e.g., Caves 1982; Dunning 1988) emphasizes intangible assets, such as the stock of technological knowledge accumulated by research and development (R&D) or the accumulation of

---

20. Brainard (1993, 1997) discusses this issue for the case of manufacturing products. For the issue of locational advantage, also see Dunning (1988).

marketing know-how from past advertising as the source of multinational enterprises' advantages. When a firm moves production overseas, it is in a disadvantageous position in relation to local firms because of differences in terms of language, customs, and institutions. Multinational enterprises will exist only if the foreign establishments they control and operate attain lower costs or higher revenue productivity than the same establishments functioning under local management. According to this theory, we will observe more active FDI in R&D-intensive or advertisement-intensive industries. We would expect positive coefficients for ADINT (advertisement intensity) and RDINT (R&D intensity). If Japanese firms' productivity level is higher than that of foreign firms, Japanese firms will have a higher sales share in the world market, and inward FDI will be limited. To take this factor into account, we used DPROD (an index comparing Japan's productivity in each industry with the U.S. equivalent), which was taken from Kawai (1996). It is problematic to use this variable for the following reasons. First, since Japanese firms compete not only with U.S. firms but also with other countries' firms, DPROD is not an appropriate variable. Second, in Kawai's methodology, if Japan's absolute producer price level in one industry is higher than the corresponding U.S. price level and if this gap cannot be explained by Japan-United States differences in factor prices and prices of intermediate inputs, then Japan's productivity in that industry is inferred to be lower compared to the United States. However, there is a possibility that Japan's high absolute price level (relatively low DPROD) might reveal either Japan's higher industry rent or Japan's higher fixed costs. Third, there might exist a reverse causality. High inward FDI penetration might increase DPROD by either reducing the industry rent or improving that industry's productivity.

Since there exists a lower bound, zero, for our dependent variable, we conduct a tobit estimation. The results are summarized in table 13.12. Among our fifty-one industries, we were unable to obtain data for six industries, that is, postal services, education, research institutes on natural sciences, research institutes on social sciences and humanities, health and hygiene, and private nonprofit organizations' services. Therefore, the maximum sample size is forty-five. As we have seen in figure 13.4, inward FDI in Japan's air transportation industry stands out and seems to be an outlier. We checked the robustness of our results by excluding the air transportation industry from our sample.

In the case of policy variables, we achieved significant results. The estimated coefficients of RINVJAUS (Japan's FDI restrictiveness minus U.S. FDI restrictiveness) and PUBEMP (the share of workers employed by local or central government) are negative and significant. These results imply that by eliminating its restrictions on inward FDI and reducing government activities, Japan can increase inward FDI.

In the case of locational advantage variables, the estimated coefficient of LAND is negative, as we expected, but is not significant. Contrary to our expectations, the coefficients of UNIV (skilled-labor intensity) and JOBSEP (job separation rate) are negative but insignificant in many cases. The coefficient of DPROD is positive and significant.

In the case of the variables that stand for the importance of intangible assets, the estimated coefficient of ADINT (advertisement intensity) is positive and significant. Consistent with the standard theory of FDI, Japan's inward FDI penetration is relatively high in industries that have higher advertisement intensity. The coefficient of RDINT (R&D intensity) is not significant. In the case of *keiretsu* variables, we did not get significant results, suggesting that *keiretsu* do not work as an impediment to inward FDI in Japan's service sector.

13.6 Conclusions

In this paper, we estimated the sales and employment of JAFFs and FAJFs in the service sector at the three-digit industry level for the year 1995.

We found that imbalances between activities of JAFFs and FAJFs are smaller than those reported in the MOF FDI statistics. In terms of employment, the JAFF-FAJF ratio is 0.22. We compared Japan's purchases of services from foreigners with U.S. purchases. For the service sector as a whole, Japan's ratio of imports to total domestic output is 1.8 percent, which is almost at the same level as the corresponding U.S. ratio, 2.1 percent. However, in the case of purchases through establishment transactions, Japan's ratio of the number of workers employed by majority-owned foreign affiliates to the total number of workers is 0.4 percent, which is one-seventh of the corresponding U.S. ratio, 2.8 percent. It seems that Japan's market for services is more closed for establishment transactions than for cross-border transactions

We also found that, compared with the United States, Japan's purchases from foreigners are concentrated in a limited number of industries. Four industries—financial intermediary services, wholesale trade, air transportation, and hotels and lodging places—account for about 54 percent of Japan's total purchases of services from foreigners. From the viewpoint of "revealed comparative advantage," Japan is most competitive in industries that support Japan's international activities, such as casualty and life insurance, other business services, and financial intermediary services. Japan is least competitive in air transportation, computer programming and software, and information services, both in international trade and in establishment transactions.

Using our cross-industry data, we estimated an empirical model explaining the determinants of Japan's inward FDI penetration. We found that inward FDI penetration is closely related to several characteristics of industries. Japan's inward FDI penetration is relatively high in industries that have higher advertisement intensity, a lower presence of government activities, and a lower presence of official restrictions on inward FDI. We found that the presence of *keiretsu* does not have significant negative effects on FDI penetration.

We should note that our new estimates may contain large estimation errors due to statistical deficiencies, as we pointed out in section 13.3. We hope that the Japanese government will make greater efforts to improve its statistics on Japan's international sales and purchases of services. Some fundamental improvements can be achieved without great cost. For example, as we have already discussed in section 13.3, the Japanese government could easily compile reliable statistics on the number of workers employed by majority-owned JAFFs for all the industries at the four-digit industry level by making use of the microdata of *Jigyosho Kigyo Tokei Chosa* (*Establishment and Enterprise Census of Japan*) conducted by the Japan Management and Coordination Agency.

Appendix
Description of Variables and Data Sources

Size of Industry

Our data on total domestic output, total domestic demand, and number of workers for each industry were taken from *1995 Input-Output Tables* (Japanese Government 1999). In I-O tables, the output level of the financial sector is measured by imputed income from interest and transaction fees. We replaced this with the financial sector's total current income. We calculated the domestic total current income of the financial intermediary services industry by summing up all banks' current incomes, all securities companies' operating revenues, and all other financial institutions' operating revenues (MOF *Annual Report of Financial Institutions* [various years]; MOF *Annual Report of Securities Companies* [various years]).

Sales and Employment by Japanese Affiliates of Foreign Firms

Our data on the number of workers employed by foreign firms' Japanese subsidiaries were taken from the Toyo Keizai's *Directory of Japanese Subsidiaries of Foreign Firms* (Toyo Keizai Shinpo-sha various years). Our data on the number of workers employed in Japanese branches and other establishments directly owned by foreign firms were taken from the Statistics Bureau of the Japan Management and Coordination Agency (1998). We estimated the sales of those Japanese subsidiaries for which such data were not available in the Toyo Keizai database as well as the sales of Japanese branches and establishments directly owned by foreign firms.

For details of estimation procedures, see section 13.3.

Sales and Employment by Foreign Affiliates of Japanese Firms

Our data on the number of workers employed by Japanese firms' foreign subsidiaries were taken from Toyo Keizai's *Directory of Japanese Subsidiaries Abroad.* Using the Toyo Keizai database, we estimated foreign subsidiaries' sales in the same way as JAFFs' sales. Moreover, we referred to MITI's (1999a) microdata in our estimate of FAJFs' sales when data from Toyo Keizai were not available. For details of the estimation procedures, see section 13.3.

Cross-Border Trade

Our data on Japan's services imports and exports were primarily taken from statistics on Japan's special trade and direct purchases that are included in the *1995 Input-Output Tables* (Japanese Government 1999).

In the context of our analysis, cross-border service trade statistics in Japan's I-O tables have the following shortcomings:

1. Imports and exports in I-O tables do not include payments and receipts for construction services, which, if provided by nonresidents, should be considered as service imports.

2. As merchandise imports are on a cost plus insurance and freight (c.i.f.) basis, I-O output tables omit those services—transportation and insurance—that are associated with the import of goods and already included in the value of goods imports.

3. The value of overseas wholesalers' activities is included in the value of goods imports either on a free on board (f.o.b.) basis or on a c.i.f. basis, while the value of domestic wholesalers' activities for exported goods is properly summed up in the output of wholesale trade sector.

In order to solve these problems, we used Bank of Japan (various issues) data on trade of construction and civil engineering, water transportation, and air transportation services. For imports of wholesale trade services that are included in the value of goods imports, we estimated distribution margins in the following way. We calculated the ratio of distribution margins for exported goods to total exports on an f.o.b. basis and estimated margins on imported goods by multiplying imports on an f.o.b. basis by the commercial margin ratio. We obtained the value of goods imports on an f.o.b. basis from Bank of Japan (various issues).

In the case of financial intermediary services, we calculated a measure of import quantities that is comparable to our measure of activities for this sector, that is, current income. We derived it by multiplying this industry's import-output ratio of the I-O tables with this industry's total current income.

U.S. Imports and Total Domestic Output

Our data on U.S. imports and total domestic output were taken from the 1992 U.S. input-output tables (U.S. Department of Commerce 1995a). Due to the same shortcomings as in the case of Japan's input-output tables, we revised the data of the I-O tables, using data on cross-border transactions from *U.S. International Services* (U.S. Department of Commerce 1999) for construction and civil engineering, railway passenger and freight transportation, road passenger and freight transportation, water and air transportation, and supporting services for transport. Data on imports of financial intermediary services, telecommunications, eating and drinking places, and hotels and lodging places were also taken from U.S. Department of Commerce (1999). For imports of wholesale trade services, we estimated distribution margins that are included in the value of goods imports in the same way as with Japan's imports. We should note that import data in U.S. Department of Commerce (1999) exclude imports from U.S. firms' foreign affiliates.

Sales by Foreign Firms' U.S. Affiliates

The data on sales by foreign firms' U.S. affiliates were taken from U.S. Department of Commerce (1995c). We derived sales data for industries in which these are confidential by multiplying the number of workers employed by foreign-owned establishments by the sales-employee ratio of all establishments. As with the estimation of Japan's purchases from JAFFs, sales of the wholesale and retail trade were adjusted to be based on margins, using total U.S. output and the number of workers employed by all establishments in the United States.

U.S. Ratio of Total Purchases from Foreigners to Total Domestic Output

This ratio is defined by "(Sales by foreign firms' U.S. affiliates + imports)/ total domestic output." For financial intermediary services and insurance industries, definitions of output in U.S. I-O tables differ from those of sales in U.S. establishment data in the same way as in Japanese I-O tables. Hence, we used the number of workers as a measure of activities in these industries, as follows: U.S. ratio of total purchases from foreigners to total domestic output = (the number of workers employed by foreign firms' U.S. affiliates / total number of workers) + (the value of imports / total domestic output).

Japan's Inward Foreign Direct Investment Penetration (FDIJA)

The variable FDIJA is defined as the share of the number of workers employed by majority-owned JAFFs in Japan's total number of workers in 1995. Our data on Japan's total number of workers were taken from the *1995 Input-Output Tables* (Japanese Government 1999).

U.S. Inward Foreign Direct Investment Penetration (FDIUS)

The variable FDIUS is defined as the share of the number of workers employed by foreign firms' U.S. affiliates in the U.S. total number of workers in 1992. The data were taken from the U.S. Department of Commerce (1995c).

Skilled Labor Intensity (UNIV)

UNIV is defined as the ratio of the number of university graduate employees to the total number of employees in that particular industry. The data were taken from Prime Minister's Office (1998) and Ministry of Labor (1996).

Land Intensity (LAND)

Our data on LAND are taken from the Development Bank of Japan (2000) and Nikkei QUICK Information Technology (2000). We first calculated the ratio of the book value (unit: billions yen) of owned land to the number of employees for each firm. LAND is a weighted average of the land-employee ratio in each industry. We used the number of employees of each firm as a weight. For water supply and sewerage systems industries, we calculated the land-employee ratio using MOF (1996). We first regressed the ratio calculated by the Development Bank of Japan data on the ratio calculated by MOF data for the industries for which the ratios calculated by both data were available. We then took the adjusted ratios for water supply and sewerage systems industries by using the estimated regression equation.

Differences between Japanese and U.S. Foreign Direct Investment Restrictiveness (RINVJAUS)

Following Hoekman (1996), we compiled a frequency measure for FDI restrictiveness at the three-digit industry level, using data from GATS schedules for Japan and the United States (WTO 1997). The GATS schedule of each country shows to which service sectors and under what conditions the basic principles of the GATS—market access and national treatment—are applied in that country. The GATS schedule covers 155 service sectors. The commitments and limitations are in every case entered with respect to each of the four modes of supply, cross-border supply, consumption abroad, commercial presence, and presence of natural persons. It seems that commitments on the commercial presence mode of supply have the most significant impact on inward FDI, so we used only information on this mode of supply. For sectors uncovered by the GATS schedule, we acquired information on each country's FDI restrictiveness from APEC (1996), OECD (various issues), Japan Investment Council (various years), and the Japanese Government (various years). RINVJAUS is defined as the difference between Japan's and the U.S. FDI restrictiveness.

Share of Public Services (PUBEMP)

PUBEMP is defined as the ratio of the number of workers employed by the establishments owned by the central or local governments to the total number of employees in that particular industry in Japan. The data were taken from the Statistics Bureau of Japan Management and Coordination Agency (1998).

Productivity (DPROD)

DPROD is defined as the productivity of a particular industry in Japan relative to that in the United States. The data are based on Kawai (1996). For these data, also see Kawai and Urata (1997).

Advertisement Intensity (ADINT)

ADINT is defined as the ratio of advertising expenses to the gross value added in each industry. The data were taken from the *1995 Input-Output Tables* (Japanese Government 1999). The advertising expenses are defined as the amount of input from the advertising industry to each industry.

Research and Development Intensity (RDINT)

RDINT is defined as the ratio of R&D expenses to the gross value added in each industry. The data were taken from the *1995 Input-Output Tables* (Japanese Government 1999). The R&D expenses are defined as the amount of input from the research industry to each industry.

Keiretsu (KRETS)

KRETS is defined as the share of workers employed by *keiretsu* firms in the total work force. The data on *keiretsu* were taken from Toyo Keizai Shinpo-sha (1992, 2000). We treated all the firms that belong to horizontal or vertical *keiretsu* groups and all the subsidiaries of such firms as *keiretsu* firms.

Horizontal *Keiretsu* (HORIZ)

HORIZ is defined as the share of workers employed by horizontal *keiretsu* firms in the total work force. The data on *keiretsu* were taken from Toyo Keizai Shinpo-sha (1992, 2000). We treated all the firms that belong to the *Shacho-kai* (President Clubs) of seven corporate groups (Mitsui, Mitsubishi, Sumitomo, Fuyo, Sanwa, Ichikan, and Tokai) and all the subsidiaries of such firms as horizontal *keiretsu* firms.

Vertical *Keiretsu* (VERT)

VERT is defined as the share of workers employed by vertical *keiretsu* firms in the total work force. The data on *keiretsu* were taken from Toyo

Keizai Shinpo-sha (1992, 2000). We treated all the firms that belong to forty-three independent corporate groups (Toyota, Nissan, Hitachi, Toshiba, Matsushita, Taisei, etc.) and all the subsidiaries of such firms as vertical *keiretsu* firms.

Job Separation Rate (JOBSEP)

The data on job separation rates were taken from Ministry of Labor (1995).

References

Asia-Pacific Economic Cooperation (APEC). 1996. *Guide to the investment regimes of the APEC member economies.* Singapore: APEC Committee on Trade and Investment.
Ascher, Bernard, and G. Obie Whichard. 1991. Developing a data system for international sales of services: Progress, problems, and prospects. In *International economic transactions: Issues in measurement and empirical research,* ed. Peter Hopper and J. David Richardson, 203–36. Chicago: University of Chicago Press.
Bank of Japan. Various issues. *Balance of payment statistics monthly.* Tokyo: Bank of Japan.
Brainard, S. Lael. 1993. A simple theory of multinational corporations and trade with a trade-off between proximity and concentration. NBER Working Paper no. 4269. Cambridge, Mass.: National Bureau of Economic Research, February.
———. 1997. An empirical assessment of the proximity-concentration trade-off between multinational sales and trade. *American Economic Review* 87 (4): 520–44.
Caves, Richard E. 1982. *Multinational enterprise and economic analysis.* Cambridge: Cambridge University Press.
Development Bank of Japan. 1997. Tainichi Chokusetsu Toshi to Gaishi-kei Kigyo no Bunseki (The analysis of inward foreign direct investment and foreign-owned affiliates in Japan). *Chosa* no. 225. Tokyo: Development Bank of Japan.
———. 2000. *Kigyo Zaimu data bank* (Financial data of enterprises). Tokyo: Development Bank of Japan.
Dunning, John H. 1988. *Explaining international production.* London: Unwin Hyman.
Eaton, Jonathan, and Akiko Tamura. 1994. Bilateralism and regionalism in Japanese and U.S. trade and direct foreign investment patterns. *Journal of the Japanese and International Economies* 8 (4): 478–510.
Economic Planning Agency. 1999. *Heisei 8-nen-9-nen SNA Sangyo Renkan-hyo* (1996–97 SNA input-output tables). Tokyo: Economic Research Institute, Japan Economic Planning Agency.
Francois, Joseph. 1999. Estimates of barriers to trade in services. Erasmus University, Rotterdam, Department of Economics. Mimeograph.
General Agreement on Tariffs and Trade (GATT). 1991. *Services sectional classification list.* GATT Secretariat's Document no. MTN.GNS/W/120 (10 July). Geneva: GATT Secretariat.
———. 1995. *Trade policy review: Japan 1994.* Geneva: GATT Secretariat.
Graham, Edward M., and Paul R. Krugman. 1991. *Foreign direct investment in the United States,* 2nd ed. Washington, D.C.: Institute for International Economics.

Grubert, Harry, and John Mutti. 1991. Taxes, tariffs, and transfer pricing in multinational corporate decision making. *Review of Economics and Statistics* 73: 285–93.

Hoekman, Bernard. 1996. Assessing the General Agreement on Trade in Services. In *The Uruguay Round and the developing countries,* ed. Will Martin and L. Alan Winters, 88–124. Cambridge: Cambridge University Press.

Horaguchi, Haruo. 1995. Tainichi Chokusetsu Toshi: Keiretsu ha Sogai Yoin ka (Inward FDI to Japan: Is *keiretsu* an impediment?). In *Nippon no Sangyo-Soshiki,* ed. Masu Uekusa, 265–86. Tokyo: Yuhikaku.

International Monetary Fund (IMF). Various issues. *International financial statistics.* Washington, D.C.: IMF.

Japanese Government. Various years. *Japan's APEC individual action plan.* Tokyo: Japanese Government.

———. 1994. *1990 input-output tables.* Tokyo: Management and Coordination Agency, Japanese Government.

———. 1999. *1995 input-output tables.* Tokyo: Management and Coordination Agency, Japanese Government.

Japan Investment Council. Various years. *Yearbook of the Japan Investment Council.* Tokyo: Japan Investment Council.

Karsenty, Guy. 2000. Assessing trade in services by mode of supply. In *GATS 2000: New directions in services trade liberalization,* ed. Pierre Sauve and Robert M. Stern, 33–56. Washington, D.C.: Brookings Institution.

Kawai, Hiroki. 1996. Shijo Kaiho no Ippan-Kinko Bunseki (A general equilibrium analysis of market liberalization in Japan). *Nippon Keizai Kenkyu* 31:133–65.

Kawai, Hiroki, and Shujiro Urata. 1997. The cost of regulation in the Japanese service industry. IDE APEC Study Center Working Paper Series 96/97 no. 17. Tokyo: IDE APEC Study Center.

Kenneth Leventhal and Company. 1994. *Japanese investment in U.S. real estate 1993.* Los Angeles: Kenneth Leventhal and Company.

Kimura, Fukunari, and Robert E. Baldwin. 1996. Application of nationality-adjusted net sales and value added framework: The case of Japan. NBER Working Paper no. 5670. Cambridge, Mass.: National Bureau of Economic Research, July.

Kogut, Bruce, and Sea Jin Chang. 1991. Technological capabilities and Japanese foreign direct investment in the United States. *Review of Economics and Statistics* 73 (3): 401–13.

Kravis, Irving B., and Robert E. Lipsey. 1988. Production and trade in services by U.S. multinational firms. NBER Working Paper no. 2615. Cambridge, Mass.: National Bureau of Economic Research, June.

Kuwabara, Hiromi. 1989. Showa 60-nen Sangyo Renkan Hyo niokeru Tokushu Boeki Bunrui no Suikei ni Tsuite (On estimation of special trade for 1985 I-O tables). In *Kokusai Sangyo Renkan Hyo no Sakusei to Riyo,* ed. Takao Sano and Chiharu Tamamura, 13–24. Tokyo: Institute of Developing Economies.

Lawrence, Robert Z. 1993. Japan's low levels of inward investment: The role of inhibitions on acquisitions. In *Foreign direct investment,* ed. Kenneth A. Froot, 85–111. Chicago: University of Chicago Press.

Ministry of Finance (MOF). 1996. *Zaisei Kinyu Tokei Geppo* (Ministry of Finance statistics monthly) no. 532 (August). Tokyo: Printing Office, Japan Ministry of Finance.

———. 1999. *Zaisei Kinyu Tokei Geppo* (Ministry of Finance statistics monthly) no. 572 (December). Tokyo: Printing Office, Japan Ministry of Finance.

———. Kinyu Nenpo Henshu lin-kai (Editorial Committee on Annual Report of

Financial Institutions). Various years. *Kinyu Nenpo* (Annual Report of Financial Institutions). Tokyo: Kinzai Institute for Financial Affairs, Inc.

———. Shoken Nenpo Henshu lin-kai (Editorial Committee on Annual Report of Securities Companies). Various years. *Shoken Nenpo* (Annual Report of Securities Companies). Tokyo: Kinzai Institute for Financial Affairs, Inc.

Ministry of International Trade and Industry (MITI). 1991. *Dai 24-kai Gaishi-kei Kigyo no Doko* (The twenty-fourth report on trends of business activities by Japanese subsidiaries of foreign firms). Tokyo: Printing Office, Japan Ministry of Finance.

———. 1998a. *Kigyo Katsudo Kihon Chosa* (Basic survey on business activities by enterprises). Tokyo: Printing Office, Japan Ministry of Finance.

———. 1998b. *Tsusho Hakusho Heisei 10-nen ban* (White paper on international trade: 1998). Tokyo: Printing Office, Japan Ministry of Finance.

———. 1999a. *Dai 27-kai Wagakuni Kigyo-no Kaigai Jigyo Katsudo* (The twenty-seventh report on trends of Japan's business activities abroad). Tokyo: Printing Office, Japan Ministry of Finance.

———. 1999b. *Dai 31-kai Gaisi-kei Kigyo no Doko* (The thirty-first report on trends of business activities by Japanese subsidiaries of foreign firms). Tokyo: Printing Office, Japan Ministry of Finance.

———. 1999c. *The Japan-U.S. input-output table, 1995.* Tokyo: Tsusan Tokei Kyokai.

———. 2000. *Tsusho Hakusho Heisei 12-nen ban* (White paper on international trade: 2000). Tokyo: Printing Office, Japan Ministry of Finance.

Ministry of Labor. 1995. *Maigetsu Kinro Tokei Sokuho/Zenkoku Chosa* (Monthly labor survey/national survey). Tokyo: Policy Planning and Research Department, Minister's Secretariat, Japan Ministry of Labor.

———. 1996. *Heisei 7-nen Chingin Kozo Kihon Chosa* (Basic survey on wage structure: 1995). Tokyo: Policy Planning and Research Department, Minister's Secretariat, Japan Ministry of Labor.

Nakamura, Yoshiaki, Kyoji Fukao, and Minoru Shibuya. 1997. Tainichi Toshi ha Naze Sukunai ka? Keiretsu, Kisei ga Mondai ka? (Why is the FDI into Japan so small? Is it because of *keiretsu* or restrictions?). *Tsusho Sangyo Kenkyusho Kenkyu Series* no. 31. Tokyo: MITI.

Nijhowne, Shaila, and David Usher. 1999. Classification, the measurement of production and international trade in services, and GATS. Paper presented at World Services Congress. 1–3 November, Atlanta, Georgia.

Nikkei QUICK Information Technology. 2000. *Nikkei Kigyo Data* (Nikkei data of enterprises). Tokyo: Nikkei QUICK Information Technology.

Organization for Economic Cooperation and Development (OECD). 1999a. *International direct investment statistics yearbook 1999.* Paris: OECD.

———. 1999b. *Measuring globalization: The role of multinationals in OECD economies* (1999 ed.). Paris: OECD.

———. Various issues. *Code of liberalization of capital movements.* Paris: OECD.

Petri, Peter A. 1991. Market structure, comparative advantage, and Japanese trade under the strong yen. In *Trade with Japan: Has the door opened wider?* ed. Paul R. Krugman, 51–84. Chicago: University of Chicago Press.

Prime Minister's Office, Government of Japan. 1998. *Heisei 9-nendo Shugyo Kozo Kihon Chosa* (1997 employment status survey). Tokyo: Statistics Bureau, Japan Prime Minister's Office.

Pugel, Thomas A., Erik S. Kragas, and Yui Kimura. 1996. Further evidence on Japanese direct investment in U.S. manufacturing. *Review of Economics and Statistics* 78:208–13.

Ray, Edward J. 1989. The determinants of foreign direct investment in the United States, 1979–85. In *Trade policies for international competitiveness*, ed. Robert C. Feenstra, 53–83. Chicago: University of Chicago Press.

Saxonhouse, Gary R. 1993. What does Japanese trade structure tell us about Japanese trade policy? *Journal of Economic Perspectives* 7:21–44.

Statistics Bureau, Japan Management and Coordination Agency. 1993. *Standard industrial classification for Japan: List of classification categories and explanatory notes*. Tokyo: Statistics Bureau, Japan Management and Coordination Agency.

———. 1998. *Heisei 8-nen Jigyosho Kigyo Tokei Chosa* (1996 establishment and enterprise census of Japan). Tokyo: Statistics Bureau, Japan Management and Coordination Agency.

Stern, Robert M. 2000. U.S.-Japan trade policy and FDI issues. Paper presented at preconference meeting, *Analytic and Negotiating Issues in U.S.-Japan International Economic Relations*. 19–20 May, Keio University, Tokyo.

Toyo Keizai Shinpo-sha. 1992. *Kigyo Keiretsu Soran 1990* (Directory of corporate groups 1990). Tokyo: Toyo Keizai Shinpo-sha.

———. 2000. *Nippon-no Kigyo Group 1990 and 2000* [CD-ROM-ban] (Japanese corporate groups, 1990 and 2000 [CD-ROM]). Tokyo: Keizai Shinpo-sha.

———. Various years. *Kaigai Shinshutsu Kigyo Soran* [CD-ROM-ban] (Directory of Japanese subsidiaries abroad [CD-ROM]). Tokyo: Toyo Keizai Shinpo sha.

———. Various years. *Gaishi-kei Kigyo Soran* [CD-ROM-ban] (Directory of Japanese subsidiaries of foreign firms [CD-ROM]). Tokyo: Toyo Keizai Shinpo-sha.

United Nations (UN) Department of International Economic and Social Affairs, Statistical Office. 1991. Provisional central product classification. Statistical Papers, Series M, no. 77. New York: UN.

U.S. Congress. Office of Technology Assessment. 1986. *Trade in services: Exports and foreign revenues*. Special report. OTA-ITE-316. Washington, D.C.: GPO.

U.S. Department of Commerce. 1995a. *Benchmark input-output accounts of the United States, 1992*. Retrieved from [http://www.bea.doc.gov/bea/uguide. htm#_1_15]. 19 April 2000.

———. 1995b. *Foreign direct investment in the United States: 1992 Benchmark survey results*. Retrieved from [http://www.bea.doc.gov/bea/ai/0794iid/maintext/ html]. 3 June 2000.

———. 1995c. *Foreign direct investment in the United States: Establishment data for 1992*. Retrieved from [http://www.bea.doc.gov/bea/uguide.htm#_1_23]. 19 April 2000.

———. 1999. *U.S. international services: Cross-border trade and sales through affiliates, 1986–98*. Retrieved from [http://www.bea.doc.gov/bea/di/1001serv/intlserv. html]. 26 February 2000.

Weinstein, David. 1996. Structural impediments to investment in Japan: What have we learned over the last 450 years? In *Foreign direct investment in Japan*, ed. Masaru Yoshitomi and Edward M. Graham, 136–92. Cheltenham, U.K.: Elgar.

Wilkins, Mira. 1990. Japanese multinationals in the United States: Continuity and change, 1879–1990. *Business History Review* 64:585–629.

World Trade Organization (WTO). 1997. *Guide to reading the GATS schedules of specific commitments and the list of Article II (MFN) exemptions*. Geneva: WTO.

Comment   Mario B. Lamberte

Doing analysis on trade in services, which in recent years has expanded tremendously, is not going to be as easy as doing analysis on trade in goods. First, the conceptual issues need to be addressed so that trade in services can be accurately represented in the economy. Second, the measurement issues need to be tackled. Third, detailed data on trade in services must be collected. These are the tasks that the paper by Fukao and Ito tries to accomplish before they proceeded with their descriptive and econometric analysis.

The authors' aim was to improve the statistics on trade in services in Japan by including not only cross-border transactions, which are captured in existing balance-of-payments tables, but also establishment transactions, by using a much more comprehensive data set than the ones currently used by official bodies, such as the Ministry of Finance. They were able to estimate the sales and employment of Japanese affiliates of foreign firms (JAFFs) and foreign affiliates of Japanese firms (FAJFs) in the service sector at the three-digit industry level for the years when data were available to them. The authors are frank in pointing out that their estimates of establishment transactions have several drawbacks owing to the statistical deficiencies. Nonetheless, they deem their estimates much more reliable than those that are currently available for Japan. More specifically, their estimates of establishment transactions show that imbalances between the activities of JAFFs and FAJFs are much smaller than the estimates provided by earlier studies. This suggests that the Japanese economy is much more open to inward direct investment in the service industries than previous studies have portrayed it to be. This finding is important because it points out that inaccurate statistics can lead to faulty analysis and erroneous conclusions.

I am sure their approach can provide a useful guide to countries that want to improve their statistics on trade in services. For developing economies, the task of gathering information on establishment transactions will likely be focused on those of the affiliates of foreign firms. Because of regulations, many foreign firms have entered the service sector of developing economies by setting up branches rather than organizing subsidiaries. Although most developing countries have begun liberalizing their service sector, nonetheless establishment transactions through affiliates will likely remain substantial in these countries. Thus, efforts of capital-exporting countries to improve their statistics on both cross-border and establishment transactions can benefit developing economies if such information is shared with them.

The authors have mainly relied on Toyo Keiza's data as basic statistics for

Mario B. Lamberte is president of the Philippine Institute for Development Studies.

their estimation of sales and employment data for both JAFFs and FAJFs in the service sector. It might help the readers appreciate the reliability of the Toyo Keizai data if the authors could add a paragraph in the appendix describing the sample size, the collection method, and the response rate. Whenever possible, the authors made adjustments in the Toyo Keizai data set to complete the information required for their analysis. For instance, information on sales was not available for many subsidiaries covered by the Toyo Keizai data. To remedy this problem, the authors "calculated each industry's average value of sales per worker from data on subsidiaries, for which both the number of workers and the sales were available" and used these values "to estimate the sales of subsidiaries for which data on sales were not available in the Toyo Keizai data set and sales of Japanese branches and other establishments directly owned by foreign firms." This procedure will likely produce biased results. My own experience with this kind of survey in the Philippines is that large firms often do not voluntarily provide vital information about their companies, such as sales or profits. However, such information can be obtained from the Philippine Securities and Exchange Commission, which requires all corporations to submit financial statements on a regular basis. I am sure Japan has similar requirement for all corporations, including subsidiaries of foreign firms. If the Toyo Keizai data set includes identification of the individual firms in the survey, then the authors can use the data available at the securities and exchange commission to fill in the information on sales not reported by many subsidiaries. However, this is going to be a long and painful process. The alternative is to construct an econometric model with sales on the left-hand side and some characteristics of the firms on the right-hand side and apply this model to firms that have complete information. The estimated model can be used in estimating sales of those firms that did not provide such information.

The nationality requirement of Japanese subsidiaries in the Toyo Keizai data set, 20 percent capital participation rate in listed and unlisted but large subsidiaries and 49 percent for unlisted and small subsidiaries, appears to be restrictive. Some foreign firms may be interested only in gaining a small foothold in domestic firms to secure much larger collateral business with them. This is prevalent in banking, insurance, retail, and transport service sectors, especially because most countries restrict equity participation of foreign firms in these sectors. As the authors pointed out, the U.S. data on inward direct investment cover all subsidiaries that are more than 10 percent foreign-owned, which obviously is a much broader coverage than the data set used by the authors. If it is not possible to adjust the Toyo Keizai data set to conform to the U.S. definition of subsidiaries, future data collection efforts could perhaps take this into account.

One of the interesting results observed by the authors is that Japan's purchases of services from foreigners are concentrated in the four industries, namely, finance, wholesale trade, water transport, and air transportation. It might be worthwhile to dig deeper into this result. Was entry by foreigners into the other industries (e.g., life insurance, advertising, telecommunications) more restrictive than the four mentioned above?

The econometric analysis of the determinants of inward foreign direct investment (FDI) penetration is instructive. The authors found that *keiretsu* impede inward FDI. I expected from the authors some discussion about the policy implications of this finding. Should *keiretsu* relationships be prohibited in order to encourage more inward FDI? Or should *keiretsu* be subjected to market discipline to give foreign firms a fair chance in competing with Japanese firms?

The authors should tighten a bit the role of the three-firm concentration ratio, CR3, in their model. The coefficient of CR3 could take either a positive value, if it indicates the existence of economies of scale, or a negative value, if incumbent firms block the entry of potential new competitors, including foreign firms. The results seem to support the first interpretation. However, one should take a closer look at the interpretation of CR3 suggested by the authors. The existence of economies of scale in the industry can actually deter any entrants, in which case the coefficient of CR3 should have been negative. The other thing that needs to be examined closely is why every time CR3 is dropped from the equation the *keiretsu* variable, GRP, loses its explanatory power. There must be something going on between these two variables that needs to be sorted out.

Comment   Chong-Hyun Nam

This is a very interesting paper, and not only interesting but also very informative. One learns a lot from this kind of empirical work.

This paper tries to be very ambitious. A lot of effort was put into establishing basic data, such as sales of services by Japanese affiliates of foreign firms and by foreign affiliates of Japanese firms. The sales by these subsidiary firms are called establishment transactions, and they are compared with cross-border transactions.

Based on this data set, a great deal of interesting empirical work was conducted. A revealed comparative advantage profile was estimated for the Japanese service industries, and, using a gravity model, the authors investigated determinants of U.S. exports of services through cross-border sales or through establishment sales. Determinants of inward foreign direct investment (FDI) were also examined.

I have only a few brief comments to make. My first comment goes to data

Chong-Hyun Nam is professor of economics at Korea University.

preparation, which is a major part of the work in the study. In the paper, I find that nationality of subsidiary firms is being determined on a somewhat arbitrary basis, and that sales by these subsidiary firms may have been subject to a large measurement error. For instance, in order to be listed as foreign subsidiaries in Japan, the cutoff foreign capital participation rate is 20 percent for large subsidiaries and 49 percent for small subsidiaries. The capital participation cutoff rate is even lower (at 10 percent) to be listed as overseas subsidiaries of the U.S. firms. However, sales by these subsidiaries are all 100 percent counted as establishment sales, although their mother company's capital participation rate may vary between 10 percent and 100 percent. I wonder how one could justify this.

My second comment falls on some empirical findings as shown in table 13.6 and table 13.10. The estimation results in table 13.6 suggest that Japan has relatively fewer barriers to cross-border trade among sample countries but has relatively greater impediments to the establishment of foreign subsidiaries in Japan. The estimation results in table 13.10 suggest that such market structure variables as entry rate, concentration ratio, and the presence of *keiretsu* have played an important role in determining the level of inward FDI in Japan. Contrary to our expectation, however, other economic variables, such as unskilled-labor intensity, land intensity, or FDI restrictiveness, all turned out to be rather unimportant.

I think, despite all the econometric problems associated with estimation, these findings are the highlight of the paper, and they are telling, indeed. It is so much so because the current market structure is really the historical outcome of past economic policy environments in Japan, and, therefore, domestic policy reforms must be instituted to fix the unusually low level of inward FDI in Japan.

In fact, I noticed from table 13.7 that inward FDI has been skyrocketing lately in such industries as telecommunication services and finance and insurance services. An explanation for this would be interesting. I suspect that domestic policy changes, including regulatory reforms, must have played an important role for such phenomena.

My final comment is concerned with the measurement of revealed comparative advantage, as shown in figure 13.3. According to the results, Japan seems highly competitive in such traditional service industries as casualty insurance, business services, and financial intermediary services, but least competitive in such more modern service industries as air transportation, computer programming and software, and information services.

Again, I think, the paper can get some good mileage out of exploring further the determinants of revealed comparative advantage. It seems to me very interesting to find out whether such a comparative advantage profile has been affected by domestic policy factors in any significant way.

Contributors

Kun-Ming Chen
Department of International Trade
National Chengchi University
Taipei 11623, Taiwan

Ji Chou
Center for Economic Forecasting
Chung-Hua Institution for Economic
    Research
75 Chang Hsing Street
Taipei, Taiwan

Philippa Dee
Trade and Economics Studies
Productivity Commission
Level 3 Nature Conservation
    House
P.O.B 80
Belconnen Act 2616
Canberra, Australia

Kyoji Fukao
The Institute of Economic
    Research
Hitotsubashi University
Naka 2-1, Kunitachi-shi
Tokyo 186-8603, Japan

Kevin Hanslow
Trade and Economics Studies
Productivity Commission
Level 3 Nature Conservation House
Corner Emu Bank and Benjamin Way
Belconnen Act 2616
Canberra, Australia

Li-min Hsueh
Chung-Hua Institution for Economic
    Research
75 Chang-Hsing Street
Taipei 106, Taiwan

Sang In Hwang
Kangnung National University
123 Jibyon-Dong, Kangnung-Si
Kangwon-Do, South Korea 210-702

Ponciano S. Intal, Jr.
The DLSU Angelo King Institute for
    Economic and Business Studies
De La Salle University
2401 Taft Avenue
Manila, Philippines

Keiko Ito
International Centre for the Study of
    East Asian Development
11-4 Otemachi, Kokurakita,
    Kitakyushu
803-0814 Japan

Takatoshi Ito
Institute of International Economics
Hitotsubashi University
Naka 2-1, Kuntachi
186-8603 Tokyo, Japan

Kazumasa Iwata
Department of Advanced Social
    and International Studies
University of Tokyo
3-8-1, Komaba
Meguro-ku, Tokyo, Japan

Jong-Il Kim
Department of Economics
Dongguk University
3-26 Phildong, Joonggu
Seoul, 100-715 Korea

June-Dong Kim
Korea Institute for International
    Economic Policy
300-4 Yumgok-Dong
Seocho-Gu, Seoul, 137-747 Korea

Fukunari Kimura
Keio University
Faculty of Economics
Mita 2-15-45
Minato-ku, Tokyo 108-8345,
    Japan

Kozo Kiyota
Faculty of Business
    Administration
Yokohama National University
79-4 Tokiwadai Hodogaya-ku
Yokohama 240-8501, Japan

Anne O. Krueger
International Monetary Fund
700 19th Street, NW
Suite 12-300F
Washington, DC 20431

Nai-Fong Kuo
Jin Wen Institute of Technology
99, An Chung Road, Hsintien
Taipei County 231, Taiwan 900

Edwin Lai
Department of Economics and
    Finance
City University of Hong Kong
83 Tat Chee Avenue
Yau Yat Tsuen, Kowloon, Hong Kong

Mario B. Lamberte
Philippine Institute for Development
    Studies
Room 403, NEDA sa Makati Bldg.
106 Amorsolo Street
Legaspi Village
Makati City, Philippines

Nae-Chan Lee
Korea Information Society
    Development Institute
1-1 Juam-Dong, Kwachun-City
Kyunggi-Do, 427-710, Korea

Han-Young Lie
Korea Information Society
    Development Institute
1-1 Juam-Dong, Kwachun-City
Kyunggi-Do, 427-710, Korea

An-loh Lin
Chung-Hua Institution for Economic
    Research
No. 75 Chang-Hsing Street
Taipei, Taiwan, 106

Aaditya Mattoo
The World Bank
MC3-327
1818 H Street NW
Washington, DC 20433

Chong-Hyun Nam
Department of Economics
Korea University
5-1, Anam-dong, Sungbuk-ku
Seoul, 136-701, Korea

Tien Phamduc
School of International Business
    and Asian Studies
Griffith University
Nathan, Queensland 4111 Australia

Ramonette B. Serafica
Department of Economics
LS207
De La Salle University
2401 Taft Avenue
1004 Manila, Philippines

Inseok Shin
Korea Development Institute
207-41 Chongnyang, Dongdaemun-ku
Seoul, Korea

Richard H. Snape
Deputy Chairman
Productivity Commission
Locked Bag 2
Collins Street East
Melbourne Vic, 8003 Australia

Chayun Tantivasadakarn
Faculty of Economics
Thammasat University
Prachan Road
Bangkok 10200 Thailand

Shujiro Urata
School of Social Sciences
Waseda University
1-13-19 Nishikamakura
Kamakura, Kanagawa
Tokyo, Japan 248-0035

Kuo-Liang Wang
Department of Economics
National Cheng-Chi University
Taipei, Taiwan, 116

Shiu-Tung Wang
Department of Shipping and
    Transportation Management
National Taiwan Ocean University
2 Pei-Ning Road
Keelung, Taiwan

Su-wan Wang
Chung-Hua Institution for Economic
    Research
No. 75 Chang Hsing Street
Taipei, Taiwan, 106

Clement Yuk Pang Wong
City University of Hong Kong
Department of Economics and
    Finance
83 Tat Chee Road
Kowloon Tong, Hong Kong

Chung-Shu Wu
The Institute of Economics
Academia Sinica
Nankang, Taipei, Taiwan 11529

Jungho Yoo
Korea Development Institute
207-41 Cheongnyang,
    Dongdaemun-ku
Seoul, Korea

Mahani Zainal-Abidin
Faculty of Economics and
    Administration
University of Malaya
50603 Kuala Lumpur, Malaysia

Anming Zhang
City University of Hong Kong and
    Faculty of Commerce and Business
    Administration
University of British Columbia
Vancouver, BC V6T 1Z2 Canada

Author Index

485

Subject Index